Discrete Calculus

Leo J. Grady · Jonathan R. Polimeni

Discrete Calculus

Applied Analysis on Graphs
for Computational Science

Dr. Leo J. Grady
Siemens Corporate Research
755 College Road East
Princeton, NJ 08540-6632
USA
leo.grady@siemens.com

Dr. Jonathan R. Polimeni
Athinoula A. Martinos Center
 for Biomedical Imaging
Department of Radiology
Massachusetts General Hospital
Harvard Medical School
Charlestown, MA 02129
USA
jonp@nmr.mgh.harvard.edu

ISBN 978-1-84996-289-6   e-ISBN 978-1-84996-290-2
DOI 10.1007/978-1-84996-290-2
Springer London Dordrecht Heidelberg New York

British Library Cataloguing in Publication Data
A catalogue record for this book is available from the British Library

Library of Congress Control Number: 2010930985

© Springer-Verlag London Limited 2010
Apart from any fair dealing for the purposes of research or private study, or criticism or review, as permitted under the Copyright, Designs and Patents Act 1988, this publication may only be reproduced, stored or transmitted, in any form or by any means, with the prior permission in writing of the publishers, or in the case of reprographic reproduction in accordance with the terms of licenses issued by the Copyright Licensing Agency. Enquiries concerning reproduction outside those terms should be sent to the publishers.
The use of registered names, trademarks, etc., in this publication does not imply, even in the absence of a specific statement, that such names are exempt from the relevant laws and regulations and therefore free for general use.
The publisher makes no representation, express or implied, with regard to the accuracy of the information contained in this book and cannot accept any legal responsibility or liability for any errors or omissions that may be made.

Cover design: KünkelLopka GmbH

Printed on acid-free paper

Springer is part of Springer Science+Business Media (www.springer.com)

Leo (John) Grady dedicates the book to his very patient wife, Amy Grady, daughter Alexandra Grady, son Leo James Grady (who arrived during the course of writing) and the memory of his late father, Leo Joseph Grady.

Preface

The goal of this book is to present the topic of discrete calculus to scientists and engineers and to show how the theory can be applied to solving a wide variety of real-world problems. We feel that discrete calculus allows us to unify many approaches to data analysis and content extraction while being accessible enough to be widely applied in many fields and disciplines. This project initially began as a tutorial on discrete calculus and its applications, and we hope that this work can provide an introduction to discrete calculus and demonstrate its effectiveness in many problem domains.

This book benefited enormously from the friends and colleagues who provided software, data, and their time in reading the manuscript. In particular, for software we would like to thank Sébastien Bougleux for help with total variation code for an arbitrary graph and Camille Couprie for power watershed code. For use of their data, we wish to thank Jason Bohland, James Fowler, David Gleich, and Robert Sumner. For comments on the manuscript we would like to thank Christopher Alvino, Mukund Balasubramanian, Jason Bohland, Sébastien Bougleux, Gaëlle Desbordes, Mathieu Desbrun, Noha El-Zehiry, Gareth Funka-Lea, David Gleich, Oliver Hinds, Anil Hirani, Hiroshi Ishikawa, Robert Kotiuga, Elliot Saltzman, Dheeraj Singaraju, Ganesh Sundaramoorthi and Enzo Tonti. We thank Eric Schwartz for all his help and support over the years. We would like to thank Wayne Wheeler, Simon Rees, and Catherine Brett at Springer for their enthusiasm and for initially proposing this project. Leo Grady would also like to thank Chenyang Xu for convincing him to take on this project and to Gareth Funka-Lea for his patience and support throughout the writing of this book.

Princeton, New Jersey	Leo Grady
Charlestown, Massachusetts	Jonathan Polimeni

Contents

1 Discrete Calculus: History and Future 1
    1.1 Discrete Calculus 1
        1.1.1 Origins of Vector Calculus 2
        1.1.2 Origins of Discrete Calculus 3
        1.1.3 *Discrete* vs. *Discretized* 4
    1.2 Complex Networks 6
    1.3 Content Extraction 7
    1.4 Organization of the Book 8
    1.5 Intended Audience 8

Part I A Brief Review of Discrete Calculus

2 Introduction to Discrete Calculus 13
    2.1 Topology and the Fundamental Theorem of Calculus 14
    2.2 Differential Forms 16
        2.2.1 Exterior Algebra and Antisymmetric Tensors 17
        2.2.2 Differentiation and Integration of Forms 26
        2.2.3 The Hodge Star Operator 32
        2.2.4 Differential Forms and Linear Pairings 37
    2.3 Discrete Calculus 38
        2.3.1 Discrete Domains 39
        2.3.2 Discrete Forms and the Coboundary Operator 48
        2.3.3 Primal and Dual Complexes 50
        2.3.4 The Role of a Metric: the Metric Tensor, the Discrete
              Hodge Star Operator, and Weighted Complexes 54
        2.3.5 The Dual Coboundary Operator 59
        2.3.6 The Discrete Laplace–de Rham Operator 60
    2.4 Structure of Discrete Physical Laws 63
    2.5 Examples of Discrete Calculus 63
        2.5.1 Fundamental Theorem of Calculus and the Generalized
              Stokes' Theorem 65
        2.5.2 The Helmholtz Decomposition 73

		2.5.3 Matrix Representation of Discrete Calculus Identities . . .	77

        2.5.3 Matrix Representation of Discrete Calculus Identities . . . 77
        2.5.4 Elliptic Equations 79
        2.5.5 Diffusion 82
        2.5.6 Advection 86
  2.6 Concluding Remarks 88

3 Circuit Theory and Other Discrete Physical Models 91
  3.1 Circuit Laws 93
  3.2 Steady-State Solutions 94
        3.2.1 Dependent Sources 97
        3.2.2 Energy Minimization 99
  3.3 AC Circuits 101
  3.4 Connections Between Circuit Theory and Other Discrete Domains 104
        3.4.1 Spring Networks 104
        3.4.2 Random Walks 106
        3.4.3 Gaussian Markov Random Fields 110
        3.4.4 Tree Counting 112
        3.4.5 Linear Algebra Applied to Circuit Analysis 117
  3.5 Conclusion 121

Part II Applications of Discrete Calculus

4 Building a Weighted Complex from Data 125
  4.1 Determining Edges and Cycles 126
        4.1.1 Defining an Edge Set 126
        4.1.2 Defining a Cycle Set 129
  4.2 Deriving Edge Weights 135
        4.2.1 Edge Weights to Reflect Geometry 135
        4.2.2 Edge Weights to Penalize Data Outliers 136
        4.2.3 Edge Weights to Cause Repulsion 143
        4.2.4 Edge Weights to Represent Joint Statistics 144
        4.2.5 Deducing Edge Weights from Observations 144
  4.3 Obtaining Higher-Order Weights to Penalize Outliers 147
        4.3.1 Weights Beyond Flows 149
  4.4 Metrics Defined on a Complex 150
  4.5 Conclusion 153

5 Filtering on Graphs 155
  5.1 Fourier and Spectral Filtering on a Graph 156
        5.1.1 Graphs that Are Not Shift-Invariant 159
        5.1.2 The Origins of High Frequency Noise 162
  5.2 Energy Minimization Methods for Filtering 163
        5.2.1 The Basic Energy Minimization Model 163
        5.2.2 Extended Basic Energy Model 167
        5.2.3 The Total Variation Model 168

	5.3	Filtering with Implicit Discontinuities 170
	5.4	Filtering with Explicit, but Unknown, Discontinuities 172
	5.5	Filtering by Gradient Manipulation 174
	5.6	Nonlocal Filtering 174
	5.7	Filtering Vectors and Flows 175
		5.7.1 Translating Scalar Filtering to Flow Filtering 176
	5.8	Filtering Higher-Order Cochains 179
	5.9	Applications 180
		5.9.1 Image Processing 180
		5.9.2 Three-Dimensional Mesh Filtering 185
		5.9.3 Filtering Data on a Surface 187
		5.9.4 Geospatial Data 192
		5.9.5 Filtering Flow Data—Brain Connectivity 193
	5.10	Conclusion 197
6	**Clustering and Segmentation** 199	
	6.1	Targeted Clustering 200
		6.1.1 Primal Targeted Clustering 201
		6.1.2 Dual Targeted Clustering 210
	6.2	Untargeted Clustering 215
		6.2.1 Primal Untargeted Clustering 216
		6.2.2 Dual Untargeted Clustering 220
	6.3	Semi-targeted Clustering 221
		6.3.1 The k-Means Model 221
	6.4	Clustering Higher-Order Cells 227
		6.4.1 Clustering Edges 227
	6.5	Applications 229
		6.5.1 Image Segmentation 229
		6.5.2 Social Networks 235
		6.5.3 Machine Learning and Classification 236
		6.5.4 Gene Expression 240
	6.6	Conclusion 242
7	**Manifold Learning and Ranking** 243	
	7.1	Manifold Learning 244
		7.1.1 Multidimensional Scaling and *Isomap* 245
		7.1.2 Laplacian Eigenmaps and Spectral Coordinates 247
		7.1.3 Locality Preserving Projections 249
		7.1.4 Relationship to Clustering 251
		7.1.5 Manifold Learning on Edge Data 251
	7.2	Ranking 253
		7.2.1 PageRank 253
		7.2.2 HITS 256
	7.3	Applications 257

	7.3.1	Shape Characterization 257
	7.3.2	Point Correspondence 260
	7.3.3	Web Search 262
	7.3.4	Judicial Citation 264
7.4	Conclusion	 266

8 Measuring Networks 267

8.1	Measures of Graph Connectedness 268
	8.1.1	Graph Distance 268
	8.1.2	Node Centrality 269
	8.1.3	Distance-Based Properties of a Graph 270
8.2	Measures of Graph Separability 274
	8.2.1	Clustering Measures 274
	8.2.2	Small-World Graphs 277
8.3	Topological Measures 279
8.4	Geometric Measures 281
	8.4.1	Discrete Gaussian Curvature 282
	8.4.2	Discrete Mean Curvature 283
8.5	Applications 285
	8.5.1	Social Networks 285
	8.5.2	Chemical Graph Theory 286
8.6	Conclusion 288

Appendix A Representation and Storage of a Graph and Complex 291

A.1	General Representations for Complexes 291
	A.1.1	Cells List Representation 291
	A.1.2	Operator Representation 292
A.2	Representation of 1-Complexes 293
	A.2.1	Neighbor List Representation 293

Appendix B Optimization 295

B.1	Real-Valued Optimization 296
	B.1.1	Unconstrained Direct Solutions 297
	B.1.2	Constrained Direct Solutions 298
	B.1.3	Descent Methods 308
	B.1.4	Nonconvex Energy Optimization over Real Variables . . . 312
B.2	Integer-Valued Optimization 312
	B.2.1	Linear Objective Functions 312
	B.2.2	Quadratic Objective Functions 314
	B.2.3	General Integer Programming Problems 318

Appendix C The Hodge Theorem: A Generalization of the Helmholtz Decomposition 319

C.1	The Helmholtz Theorem 319
C.2	The Hodge Decomposition 326

Summary of Notation 331

References 335

Index 353

Color Plates 361

Acronyms

AC	Alternating Current
BEM	Basic Energy Model
CDP	Constitutive Determination Problem
CFL	Courant–Friedrichs–Lewy
CRT	Chromatographic Retention Time
CT	Computed Tomography
DC	Direct Current
DFT	Discrete Fourier Transform
DNA	Deoxyribonucleic Acid
fMRI	functional Magnetic Resonance Imaging
GMRF	Gaussian Markov Random Field
GPU	Graphics Processing Unit
HITS	Hyperlink-Induced Topic Search
IRLS	Iteratively Reweighted Least Squares
KCL	Kirchhoff's Current Law
KM	K-Means
KVL	Kirchhoff's Voltage Law
LE	Laplacian Eigenmaps
LII	Legal Information Institute
LP	Linear Programming
LPP	Locality Preserving Projections
LTI	Linear Time Invariant
MAP	Maximum *A Posteriori*
MDS	Multidimensional Scaling
ML	Maximum Likelihood
MRF	Markov Random Field
MRI	Magnetic Resonance Imaging
MPCV	MultiPhase Chan–Vese
mRNA	messenger Ribonucleic Acid
MS	Mumford–Shah
MSF	Maximum Spanning Forest

NP-Hard	Non-deterministic Polynomial-Time Hard
PCA	Principle Components Analysis
PDE	Partial Differential Equation
QP	Quadratic Programming
QSAR	Quantitative Structure–Activity Relationships
QSPR	Quantitative Structure–Property Relationships
RGB	Red Green Blue
RLC	Resistor–Inductor–Capacitor
SDP	Semidefinite Programming
SNF	Smith Normal Form
SPD	Symmetric Positive Definite
SPECT	Single Photon Emission Computed Tomography
TV	Total Variation

Chapter 1
Discrete Calculus: History and Future

Abstract The goal of this book is to bring together three active areas of current research into a single framework and show how each area benefits from more exposure to the other two. The areas are: discrete calculus, complex networks, and algorithmic content extraction. Although there have been a few intersections in the literature between these areas, they have largely developed independently of one another. However, we believe that researchers working in any one of these three areas can strongly benefit from the tools and techniques being developed in the others. We begin this book by outlining each of these three areas, their history and their relationship to one another. Subsequently, we outline the structure of this work and help the reader navigate its contents.

1.1 Discrete Calculus

The term "*discrete calculus*" is one of many expressions, along with "discrete exterior calculus" and "mimetic discretization", that describes the body of literature that has focused on finding a proper set of definitions and differential operators that makes it possible to operate the machinery of multivariate calculus on a finite, discrete space. In contrast to traditional goals of finding an accurate *discretization* of conventional multivariate calculus, discrete calculus establishes a separate, equivalent calculus that operates purely in the discrete space without any reference to an underlying continuous process. Therefore, the purpose of this field has been to establish a fully *discrete* calculus rather than a *discretized* calculus. The standard setting for this discrete calculus is a cell complex, of which a graph or network is a special case.

Although the tools of discrete calculus have risen to prominence more recently, the concepts in discrete calculus were historically developed in parallel with the conventional calculus. In fact, the origins of both conventional vector calculus and discrete calculus have their origins in studies of *spatial* representations and relationships as well as the description of physical systems associated with space. In order to understand the relationship of conventional calculus to discrete calculus, we believe that it is useful to briefly examine the history of development in both areas.

The term "discrete calculus" appears to be a recent invention which we have reluctantly adopted. Our reluctance is due to possible confusion with *discretized* methods, which have a different goal. The term *combinatorial calculus* might be more appropriate, but "discrete calculus" seems established at this point and the term "combinatorial calculus" has been used previously in a different context.

1.1.1 Origins of Vector Calculus

Modern, conventional calculus consists of several components. One component of calculus employs the notion of the *infinitesimal*, using limits and infinite series to develop the theory. Naturally, these concepts are associated with an underlying continuous space. In contrast, other components of calculus do not depend on the concept of the infinitesimal. For example, the Fundamental Theorem of Calculus describes essentially a topological relationship, given by the operation of integration, between the integrand and the domain of integration. In the context of describing a discrete calculus, in which the domain is discrete and finite, the aspects of conventional calculus that focus attention on the infinitesimal will play a smaller role in this exposition. Instead, our focus in reviewing conventional calculus will be in those aspects that can be used to describe *space* and the behavior of functions defined in space.

Historically, univariate or one-dimensional calculus that was developed by Newton and Leibniz was extended to describe two-dimensional space by using the real and imaginary parts of a complex number to represent the two dimensions of the plane (known as the **complex plane**). This development has variously been attributed to Caspar Wessel or Jean-Robert Argand, both of whom worked in the late 18th century (and as a result the complex plane is sometimes called the **Argand plane**). Other sources attribute to Gauss the introduction of the complex plane to represent two-dimensional space, although Gauss' work was mainly in the early 19th century.

Unfortunately, physical space is three-dimensional and therefore the two-dimensional representation by the complex plane was insufficient to describe all physical processes. Furthermore, it was unclear how to extend the concept of complex numbers to three dimensions. This problem was finally solved in 1844 by William Hamilton who defined *four*-dimensional complex numbers called **quaternions** [185]. Quaternions formed the basis for modern vector calculus by defining a *scalar* quantity as the real part of a quaternion and a *vector* quantity as the imaginary part of a quaternion. Hamilton's student Peter Tait continued to develop and promote quaternions after Hamilton's death in 1865, but later researchers Josiah Gibbs and Oliver Heaviside (independently) stripped out the quaternion focus of the work and presented a simplified form of vector calculus. This simplified form, without any explicit reference to complex numbers or quaternions, is today the conventional vector calculus taught in school. However, it is important to note that, because of its origin, conventional vector calculus was derived explicitly to describe space in *three dimensions*. This point will be emphasized again in Chap. 2 as we proceed to develop the tools of discrete calculus.

During the mathematical development of quaternions (and later vector calculus), James Maxwell was developing his theory of electromagnetism. Maxwell immediately recognized the value of quaternions in his work and seized upon this new mathematics to help him describe the physical behavior of electric and magnetic fields. Therefore, the description of space provided by vector calculus was immediately used by Maxwell to describe the behavior of *functions* associated with that space. In fact, the use of vector calculus in physics became so successful that connections were made between various physical theories, showing that quantities in one area of physics behaved analogously to quantities in a different area of physics. These analogies were later explained in 1976 by Enzo Tonti who suggested that the reason for these analogies was that each analogous quantity was associated with the same unit of *space* [378]. Consequently, we see again the close connection between the mechanics of vector calculus and the mathematical description of space. Ultimately, it is this connection that will allow for the development of a discrete calculus on a discrete domain.

Vector calculus was further generalized to describe surfaces and also extended beyond three dimensions. The development of calculus on surfaces belongs to the classical discipline of *differential geometry*. The abstraction of calculus and extension to higher dimensions is sometimes called *exterior calculus* or the *theory of differential forms*, which was first developed by Élie Cartan in the early 20th century. This more abstract and generalized form of calculus is where we begin our exposition in Chap. 2 to develop the discrete calculus.

From an early stage in the development of vector calculus, there was interest in *discretizing* the equations of vector calculus so that they could be solved in pieces. A major motivation for this approach is that many of the differential operators are *linear*. In this case, linearity implies that the act of applying an operator to a function may be subdivided into small, local operators and then reassembled to produce the result. In 1928, Courant, Friedrichs and Lewy published the **finite differences** approach to discretizing differential equations [92], which became the standard method for discretization and was heavily developed during the middle of the 20th century. Courant later planted the seed for what later became known as the **finite element method** in 1943 [91] and it was later formalized [361].

The rise of ubiquitous computing has propelled a sustained interest in the discretization of differential equations to model everything from airplanes to medical implants. More recently, development of discretization has moved toward formulation of differential calculus on a more general cell complex rather than at a series of point locations, which is sometimes known as *mimetic discretization* [40, 47, 178, 249, 275].

For more details on the history of the development of vector calculus, see [95].

1.1.2 Origins of Discrete Calculus

The origins of discrete calculus also began with a study of *space* in the context of graph theory [37]. Specifically, the study by Euler of the Königsberg Bridge Problem modeled the two banks and islands of Königsberg as *nodes* in a graph and the bridges connecting them as *edges* [126]. Therefore, from its earliest beginnings, graph theory was also modeling space and the neighborhood connections between different areas.

The first application of graph theory to the modeling of physical systems came from Kirchhoff, who both developed the basic laws of circuit theory and also made fundamental contributions to graph theory [232]. Kirchhoff's work on applying graph theory to model circuits in 1841 predated the development of quaternions, vector calculus and Maxwell's Laws. At the end of the 19th century Poincaré published his work on *analysis situs* [307] in which he analyzed simplicial complexes, simplicial homology and laid the foundation for the subject of algebraic topology. Poincaré also was concerned with representing space by discrete elements and, in fact, the term *analysis situs* is Latin for "analysis of position" or "analysis of location". Algebraic topology was further developed in the early 20th century by many contributors, including Whitney, de Rham, Cartan and Lefschetz (see [107, 217] for more history on this development).

Circuit theory continued to develop using graph theory and the concepts of algebraic topology [118, 400]. In 1955, Roth directly connected algebraic topology to electrical circuits and used the theory to establish conditions under which a circuit will have a solution (i.e., be realizable) [323]. This achievement, coupled with the unconventional work of Kron [250], caused electrical engineers to begin viewing electrical circuits as a alternative to conventional vector calculus in which all of the laws of vector calculus were discrete. This viewpoint came together in the review article by Branin [59] who explicitly posited that electrical circuits (and "higher-dimensional" circuits) had the same structure as conventional vector calculus.

As technology has allowed increased representation and computational power, these tools from discrete calculus have received recent attention. In particular, the area of computer graphics has seen a strong interest in the concepts of discrete calculus [103, 105, 200, 258, 374, 421], although the interest has been by no means limited to that field [81, 123, 161, 219, 420]. The rise of linear algebra packages, such as MATLAB, make the use of discrete calculus operators and algorithms quite convenient, since the primary operators take the form of large, sparse matrices. Additionally, the demonstrated ability of parallel computational devices such as GPUs to efficiently solve problems in linear algebra with sparse matrices holds strong promise for discrete calculus operations in the future as these parallel computing devices become increasingly common.

1.1.3 Discrete *vs.* Discretized

It is remarkable that both conventional vector calculus and discrete calculus developed around the representation of *space* and the manipulation of functions defined on that space. As we see in Chap. 2, the definition of the underlying space actually *defines* the structure of the differential operators in both conventional vector

1.1 Discrete Calculus

calculus and discrete calculus. Additionally, it is remarkable that both continuous vector calculus and discrete calculus were initially adopted by researchers seeking to understand the behavior of electricity, with Maxwell initially adopting the early conventional vector calculus to describe electromagnetism and Kirchhoff (and later researchers) adopting the early discrete calculus to describe circuit theory.

The focus of historical study in vector calculus and partial differential equations has been on producing *analytical*, closed-form solutions to problems. In contrast, the finite nature of discrete calculus and rise of computational power has driven that area to be less focused on analytical solutions and to place more emphasis on algorithms for finding solutions. In Chap. 2, we give the discrete calculus expressions for such classic topics as integration by parts and Green's Theorems. The truth is that these fundamental tools in conventional, analytical calculus are simply not as important in discrete calculus because there is less need to find analytical solutions to equations in the discrete calculus setting. However, these classical techniques can be useful in the sense that the intuition behind these concepts in conventional vector calculus can be re-used in the discrete calculus setting, and also these classical analytic tools can sometimes be used in proving behavioral properties about certain algorithms (see, e.g., [161]).

Before moving on, we want to stress again the importance of distinguishing *discretization* of conventional calculus from the *discrete* calculus treated in this work. In the first case, the goal is to compute a solution to some problem on a continuous space. However, an analytic solution is too difficult to find and so a discretization strategy is employed that allows for a computer to produce an approximate solution. Therefore, the main goal in discretization methods is the fidelity of the discretized, computed approximation to the desired analytical solution. Consequently, an important technique for proving the correctness of a discretization strategy is to show that as the discretization becomes finer and finer (i.e., closer to the continuum) that the solution obtained by discretization in the limit approaches the known analytical solution. This discretization approach is commonly used in mimetic discretizations [103, 105, 200, 258, 374, 421] and in modern finite element methods [40, 47, 178, 249, 275].

In contrast, *discrete* calculus treats a discrete domain (e.g., a graph) as entirely its own entity with no reference to an underlying continuum. For example, a social network (such as a citation network) is not associated with any continuous space in the sense that the network is not viewed as a discretization or sampling of an underlying continuum. However, the tools of discrete calculus can still be used to analyze the structure of the network and the behavior of functions associated with the network. Consequently, traditional discretization concerns about approaching a continuous solution in the limit are meaningless in the context of discrete calculus.

Neither conventional calculus nor discrete calculus are subordinate to each other. Both frameworks can be used to describe physical systems, e.g., with conventional vector calculus describing the behavior of electromagnetic fields and discrete calculus describing the behavior of electrical circuits. Chapter 3 goes into greater detail about the connection between discrete calculus, circuit theory, and other discrete systems. Additionally, the history of 20th century physics has shown that there are legitimate philosophical questions about the appropriateness of treating space, and quantities associated with that space, as continuous or discrete entities. We go no further in addressing these issues except to state that the focus of this work will be on discrete calculus, its relationship to conventional calculus, and the occasional intersections with discretization methods.

1.2 Complex Networks

The term "complex network" is used to describe any non-trivial network.[1] Examples of "trivial" networks are **regular graphs** (where every node has the same number of incident edges), lattices, or random graphs. Traditionally, trivial networks were the focus of study because they are easier to study analytically. However, networks obtained from the real world are often nontrivial, and the availability of modern computers has allowed us to represent and analyze huge networks.

The current level of interest in complex networks began in the late 1990s. During this period the Internet (along with the World Wide Web) was on the rise and there were many groups looking at network structure for purposes of designing a more secure, efficient network as well as techniques for analyzing the structure of the network for tasks like Internet search. During this same period, a series of influential papers by Watts, Strogatz, Albert and Barabási [20, 362, 396] spurred interest in the description of complex networks.

One major effect of the interest in complex networks has been the recognition that complex networks may be used to model a huge array of phenomena across all scientific and social disciplines. Examples include the World Wide Web, citation networks, social networks (e.g., Facebook), recommendation networks (e.g., Netflix), gene regulatory networks, neural connectivity networks, oscillator networks, sports playoff networks, road and traffic networks, chemical networks, economic networks, epidemiological networks, game theory, geospatial networks, metabolic networks, protein networks and food webs, to name a few. The ubiquity of complex networks and the importance of understanding their structure has been the focus of several books in popular science [21, 64, 77, 398].

This book is not about complex networks directly. However, there are ideas which have been developed in the field of complex networks throughout the book (particularly in Chap. 8). Instead, the examples used in the applications chapters borrow heavily from the problems which have been studied in this field. In effect, our goal is to show how the tools of discrete calculus and the algorithms developed here can be applied to the vast array of problems which have been uncovered in the literature on complex networks, as well as show how some of the concepts developed in the complex network literature relate to discrete calculus. Furthermore, in contrast to image processing or computer graphics applications, we have been careful to develop all

---

[1] Throughout this work we use the terms *network* and *graph* interchangeably to mean exactly the same thing—a 1-complex comprised of nodes and edges. A complex will be defined in Chap. 2.

of our tools *without assuming a network embedding* so that the tools developed here may by applied to an arbitrary complex network.

1.3 Content Extraction

The third area addressed in this book is content extraction. The term **content extraction** has a broad meaning that can encompass many different problems and disciplines. In our case, we use the term *content extraction* to indicate any algorithm in which the goal is to extract information from a dataset and/or network.[2] Examples of content extraction algorithms covered in the book include filtering (denoising), clustering, manifold learning, ranking, and network characterization.

Content extraction can be used to analyze the structure of data associated with a network (sometimes called *attributed graphs*) or the structure of the network itself. An important methodology described in this book for analyzing data associated with a network is to use the data to define *weights* on the network and then use an algorithm that analyzes the structure of the network to draw conclusions about the data. Chapter 4 describes how weights may be generated from the data. For example, to perform clustering of data associated with nodes, in Chap. 6 we show how that data may be used to establish edge weights, after which any algorithm that clusters a weighted network may be applied to produce a clustering of the data. In particular, image content may be clustered using this approach.

Many of the algorithms developed for content extraction were developed in the context of image processing or computer graphics. In both of these cases, traditional one-dimensional signal processing must be substantially modified to operate in multiple spatial dimensions (generally in two or three dimensions). Consequently, the algorithms developed in these fields explicitly account for spatial interactions. In many ways, the work in this book may be viewed as continuing the development of the variational algorithms based on *active contours* [222] and *level sets* [339] which dominated image processing (among other fields) for many years. These methods cast content extraction problems as energy minimization problems in which the optimum solution to the energy minimization problem provided a solution to the content extraction problem. Level sets provided a mechanism for optimizing these energies, using tools from the study of partial differential equations. The book by Sethian [339] demonstrated the remarkable number of applications that could be treated by the energy minimization methodology of level set techniques. Similar to this body

---

[2]In many disciplines the network structure is itself considered the dataset (since this information must be collected). Our distinction here between *data* and *network* is simply due to the fact that in several of the disciplines we consider there is data associated with each node or edge in the network (as in the case of image processing), in which case the network defines the spatial domain upon which the data is processed. However, we address algorithms that can be used to process data associated with a network and algorithms that can be used to analyze the structure of the network itself. Consequently, algorithms that process data associated with networks or algorithms that analyze the network structure itself are both considered to be content extraction algorithms.

of work, we also equate energy optimization with the solution to content extraction problems. However, the use of discrete calculus to formulate the energy minimization problems affords us the major advantage of generalizing the utility of these content extraction methods to arbitrary discrete domains (e.g., graphs). This generalization allows us to apply the energy minimization methodology to tackle the problems of the future being defined in the field of complex networks. Additionally, this formulation in terms of discrete calculus may also be applied in the same areas that were conventionally treated by level sets by viewing a Cartesian domain as a special case of the more general framework (i.e., a lattice). In fact, recent work has demonstrated that energies which were conventionally formulated using vector calculus and optimized with level sets could be dramatically outperformed by formulating the same energies using discrete calculus and performing the optimization using techniques in combinatorial optimization [163].

1.4 Organization of the Book

In the first part of the book, we present a brief review of discrete calculus with a focus on those key concepts that are required for the successful application of discrete calculus. This is by no means an exhaustive treatment of this topic, but is included to establish the notation and terminology used throughout the subsequent chapters, and to make our treatment reasonably self-contained. We provide reference throughout to the literature for readers who would like to delve deeper into the vast topics of differential forms and discrete calculus.

In the second part of the book, we redevelop many of the standard tools in image processing on a generalized, unembedded network. In these chapters, the generalized Laplacian operator plays a central, consistent role. Specifically, we show how the discrete calculus provides a natural definition of "low-frequency" on a discrete space, which then yields filtering and denoising algorithms. These algorithms are also developed from the standpoint of local interaction models between neighboring nodes. We then show how filtering algorithms can give rise to clustering algorithms. Clustering algorithms are then used to develop manifold learning and data discovery methods. Finally, ranking algorithms and algorithms for analyzing the structure of a network are also addressed. In addition to generalizing this set of tools to arbitrary networks, we believe that the context of discrete calculus has allowed us to unify very many standard image processing algorithms into a common framework. Therefore, the reader who is interested purely in image processing will find a unified framework for viewing a wide variety of standard algorithms in filtering, clustering, and manifold learning.

1.5 Intended Audience

This book is intended for graduate students, researchers, and engineers who are familiar with the basics of vector calculus, graph theory, and linear algebra. For

1.5 Intended Audience

researchers interested in discrete calculus, we intend for this book to tie algorithms and applications to the theory. For researchers in the domain of complex networks, we intend this book to provide an introduction to the foundations of discrete calculus on a network and a set of theoretical and algorithmic tools for analyzing networks. For researchers interested in image processing and computer graphics, we intend to introduce the foundations of discrete calculus, argue why algorithms should be developed on a more general graph, and demonstrate how to reformulate traditional algorithms defined in the continuum onto a discrete structure.

In each of the applications chapters in the second part of the book, we present several worked examples of how to use discrete calculus to analyze real data, in which multiple algorithms are applied. Naturally, the best algorithm for any given data set will depend on the application and on the nature of the data. Therefore, our intention is not so much to determine which algorithm is the best for a particular application, but rather to demonstrate the wide applicability of the framework and to present multiple processing strategies to give the reader a sense for the performance and behavior of the algorithms.

The primary content of this book is a review of work which has occurred in several fields and an attempt to bring them all into the same framework with a standardized notation. However, there are also aspects of a research monograph in the sense that some of the material has not previously appeared in the literature to the knowledge of the authors. Significant new material includes our generalization of algorithms and concepts used to analyze nodes and node data to novel analyses of edges and edge data. Additionally, we view the running unification of several ideas (and algorithms) into a single framework as a useful contribution. Finally, our goal is to provide the reader with the ability to understand the concepts being described and also an idea of how to implement them. Where algorithms are not fully described, citations are provided such that the interested reader may find more details.

Chapter 2 forms the basis of our exposition for discrete calculus and therefore every subsequent chapter depends to some degree on this chapter. Chapter 3 extends the exposition of discrete calculus to the description of physical systems, with a focus on circuit theory. Although some concepts from circuit theory will reappear in later chapters (e.g., effective resistance), this chapter primarily stands on its own. Chapter 4 marks the beginning of the application sections, in the sense that it details how a weighted edge set or cycle set may be derived for a particular application. The usefulness of this chapter ultimately depends on the particular application that a reader may be pursuing. Chapter 5 introduces the concept of filtering on an arbitrary graph, which then forms the basis for Chap. 6 on clustering. Chapter 7 continues to build directly on the clustering and filtering concepts to introduce manifold learning and ranking techniques. Chapter 8 breaks from the stream of the previous three chapters to provide various methods for measuring connectivity, separability, and topological and geometric properties of a network. Appendix A contains useful notes for the implementation of the algorithms described in the text and Appendix B provides an introduction to the set of optimization techniques used throughout the book. Finally Appendix C ties most closely back to Chap. 2 by going into further details on the Hodge Decomposition.

Part I
A Brief Review of Discrete Calculus

Chapter 2
Introduction to Discrete Calculus

Abstract In this chapter we review conventional vector calculus from the standpoint of a generalized exposition in terms of exterior calculus and the theory of forms. This generalization allows us to distill the important elements necessary to operate the basic machinery of conventional vector calculus. This basic machinery is then redefined in a discrete setting to produce appropriate definitions of the domain, boundary, functions, integrals, metric and derivative. These definitions are then employed to demonstrate how the structure of the discrete calculus behaves analogously to the conventional vector calculus in many different ways.

Calculus is often introduced along with the concept of the *infinitesimal*, and thus calculus is typically associated with the continuum. Similarly, the role of the differential, dx, can be understood in terms of the Riemann Integral, which provides an intuitive explanation for how an integral can, in the limit, express the area under the curve. However, the expression of the Fundamental Theorem of Calculus does not require any concept of limit or infinitesimal but essentially states a topological property of the integral that can be phrased for either continuous or discrete spaces. The goal of this chapter is to demonstrate that many aspects of differential and integral calculus, including the Fundamental Theorem, can be phrased either in a continuous or discrete setting. That is, these central concepts are not tethered to either the continuous or discrete representation, but rather can be instantiated in either framework. With this perspective, the discrete formulation is not subordinate to the more familiar continuous calculus, but rather the two formulations are equally legitimate and both capture the essential character of the calculus.

It is assumed that the reader is generally more familiar with vector calculus in the continuous setting. Therefore, we begin by introducing the central concepts of the discrete calculus that will be used throughout the remaining chapters. While this chapter draws upon several fields of mathematics to provide an adequate overview of the discrete calculus, unfortunately it is beyond the scope of this text to give a full treatment of all relevant topics, including differential forms, exterior calculus, analysis on manifolds, and algebraic topology. However, for the interested reader, there are several insightful and thorough texts on these topics, including, for example, Refs. [99, 139, 142, 290, 335].

We begin the treatment of discrete calculus by considering the formalism of differential forms, which provides the structure for generalizing standard vector calculus—which is defined only for up to three spatial dimensions—to arbitrary dimensions. This formalism will be shown to be equally applicable to continuous or discrete spaces, and therefore provides the basic framework for discrete calculus.

2.1 Topology and the Fundamental Theorem of Calculus

The underlying topological nature of calculus can be seen by considering integration as a *pairing* between two objects: the domain of integration and the function or differential to be integrated on that domain. For example, the definite volume integral

$$\int_V \rho \, dv \tag{2.1}$$

consists of a pairing of the domain V with the integrand $\rho \, dv$ which is some function defined over this domain. The pairing evaluates to a scalar quantity and the evaluation of the integral depends equally on both components: the domain and the function.

The equal importance of the domain and the function is made clear by the **Fundamental Theorem of Calculus**, which, in effect, defines the relationship between integration and differentiation. The Fundamental Theorem states that, if $f(t)$ is a continuous function defined on the closed interval of $[a, b]$, and the indefinite integral

$$\int_a^x f(t) \, dt = F(x) \tag{2.2}$$

for any $x \in [a, b]$, then the function F is differentiable and is the **antiderivative** of f, i.e., $F'(x) = f(x)$ in $[a, b]$. This is sometimes known as the *First Fundamental Theorem of Calculus* [10]. Extending the theorem to the boundary of the closed interval, we arrive at the *Second Fundamental Theorem of Calculus* [10],

$$\int_a^b f(t) \, dt = F \Big|_a^b = F(b) - F(a), \tag{2.3}$$

and therefore the definite integral of $f(t)$ over this region is simply the antiderivative $F(t)$ evaluated at the *boundary* of the domain of integration, which in this case consists of the two distinct points, a and b. In other words, if the integrand is the derivative of some function, then the integral over a closed region only depends on the behavior of the antiderivative on the boundary of the region. This

2.1 Topology and the Fundamental Theorem of Calculus

implies that if we subdivide the domain $[p_0, p_N]$ into N subintervals with points $\{p_0, p_1, \ldots, p_{N-1}, p_N\}$ such that p_0 and p_N bound the domain then, by the linearity of integration,

$$\int_{p_0}^{p_N} f(t)\,dt = \sum_{i=0}^{N-1} \int_{p_i}^{p_{i+1}} f(t)\,dt$$

$$= [F(p_1) - F(p_0)] + [F(p_2) - F(p_1)] + \cdots + [F(p_N) - F(p_{N-1})]$$
$$= F(p_N) - F(p_0), \tag{2.4}$$

demonstrating that all evaluations of antiderivative F on the interior points cancel in the final sum, leaving only the boundary term $F(p_N) - F(p_0)$.

Several well-known theorems[1] from vector calculus are higher-dimensional instantiations of the Fundamental Theorem, including the *Gradient Theorem*

$$\int_{\mathcal{C}} (\nabla \phi) \cdot d\vec{r} = \phi(b) - \phi(a) \tag{2.5}$$

where curve $\mathcal{C}$ begins at point a and ends at point b, *Green's Theorem*

$$\iint_{\mathcal{S}} \left(\frac{\partial v}{\partial x} - \frac{\partial u}{\partial y} \right) d\vec{S} = \int_{\partial \mathcal{S}} (u\,dx + v\,dy), \tag{2.6}$$

Stokes' Theorem

$$\iint_{\mathcal{S}} (\nabla \times \vec{H}) \cdot d\vec{S} = \int_{\partial \mathcal{S}} \vec{H} \cdot d\vec{r}, \tag{2.7}$$

and the Divergence Theorem or *Gauss's Theorem*,

$$\iiint_{\mathcal{V}} \nabla \cdot \vec{H}\,dV = \iint_{\partial \mathcal{V}} \vec{H} \cdot d\vec{S}. \tag{2.8}$$

Each of these expressions of the Fundamental Theorem demonstrates that the pairing between the domain of integration and the quantity to be integrated requires that the dimension of the integrand and the domain must correspond.

These expressions phrased in the language of vector calculus all share a common structure that relates the vector fields to the topology of the underlying space in

---

[1] The familiar **integration by parts** formula,

$$\int_a^b u\,dv = uv \Big|_a^b - \int_a^b v\,du$$

is simply a corollary to the Fundamental Theorem that uses the product rule of differentiation.

a way that is *independent of the dimension* of the space. Indeed, the Fundamental Theorem holds as well for integration over curves, surfaces, or volumes. A common framework exists for expressing all of these relations that highlights their essential topological character.

The theory of *differential forms* is a natural generalization of standard vector calculus to arbitrary dimensions. In this formalism, the dimensionality of the differential forms as well as the dimensionality of the domain of integration are explicit, and integration always pairs a differential form and a domain of the same dimension. In this framework, the Fundamental Theorem of Calculus can be stated concisely in what is known as the *Generalized Stokes' Theorem* in terms of differential forms as

$$\int_S \mathrm{d}\tilde{\omega} = \int_{\partial S} \tilde{\omega} \quad (2.9)$$

where $\tilde{\omega}$ is a differential form and d represents the derivative operator for differential forms, to be defined later. In this expression the derivative d on the left side is exchanged for the boundary operator ∂ on the right side, suggesting that these two operators are strongly related. Furthermore, this exchange suggests a topological character of the derivative. Indeed, the theory of differential forms is important in *both* the mathematical fields of algebraic topology and analysis on manifolds!

Both the definitions of the differential form and the corresponding derivative operator are invariant (in the sense of tensor analysis) to changes of coordinates and do not require the specification of a metric. In the next section, we shall demonstrate that the derivative of differential forms is a generalization of the common differential dx defined along one coordinate (e.g., x) to a differential measuring the change of a function along all coordinates (e.g., x, y, and z) simultaneously. Once this formalism is established, the instantiation of the Fundamental Theorem in the discrete setting is direct and transparent.

2.2 Differential Forms

The generalization of the derivative provided by the theory of differential forms in arbitrary dimensions is motivated by the requirement that it must measure how a function changes in all directions simultaneously, just as $\mathrm{d}f/\mathrm{d}x$ measures how a function f changes in the x coordinate direction. This requirement leads directly to the antisymmetry property of differential forms and the exterior algebra that is based on the measurement of volume enclosed by a set of vectors, which we demonstrate below.

In this section we review the basic properties of differential forms. We begin by recalling the basic definitions of p-vectors and p-forms in n dimensions as antisymmetric tensors of contravariant and covariant type, respectively, and the metric tensor that provides a mapping between p-vectors and p-forms. The exterior algebra of antisymmetric tensors motivates the exterior derivative for differential forms, which leads to Generalized Stokes' Theorem and the integrability conditions encapsulated by the Poincaré lemma. Finally, the Hodge star operator provides a mechanism for computing the inner product of two p-forms.

2.2.1 Exterior Algebra and Antisymmetric Tensors

In this section we survey the theory of higher-order vectors, their associated dual quantities, known as forms, and the corresponding algebra. These so-called p-vectors and p-forms are special cases of antisymmetric tensors, so this theory could be viewed simply as a subset of tensor analysis, yet their elegant geometric interpretation and their suitability for faithfully representing the physical character of many of the quantities studied in several branches of physics makes them a powerful addition to the tools available for modeling physical systems (see, e.g., [197, 218, 395, 399]).

2.2.1.1 The Vector Spaces of p-Vectors and p-Forms

The *inner product* of a pair of vectors can be used to measure the angle between the vectors, and similarly the *exterior product* provides a means to measure the area of a parallelogram defined by a pair of vectors or the volume of a parallelepiped produced by a group of vectors in higher dimensions. Let V be a vector space over the real numbers $\mathbb{R}$. For two elements[2] $\bar{x}$ and $\bar{y}$ of V the **exterior product** (or wedge product)

$$\bar{x} \wedge \bar{y} \tag{2.10}$$

is an *anticommutative* product on the elements of V and linear in each of its arguments. Therefore for $\bar{x}_1, \bar{x}_2, \bar{x}_3 \in V$ and $a, b \in \mathbb{R}$,

$$(a\bar{x}_1 + b\bar{x}_2) \wedge \bar{x}_3 = a(\bar{x}_1 \wedge \bar{x}_3) + b(\bar{x}_2 \wedge \bar{x}_3), \tag{2.11}$$

$$\bar{x}_1 \wedge \bar{x}_2 = -\bar{x}_2 \wedge \bar{x}_1 \tag{2.12}$$

from which it follows that $\bar{x} \wedge \bar{x} = 0$. It is thus straightforward to demonstrate that the exterior product is associative. If $\bar{e}_1, \bar{e}_2, \ldots, \bar{e}_n$ constitute a basis for V, then $\bar{x} = \sum x^i \bar{e}_i$ and $\bar{y} = \sum y^i \bar{e}_i$ and from the rules of the exterior product we can see that

$$\bar{x} \wedge \bar{y} = \sum_{i<j} (x^i y^j - x^j y^i) \bar{e}_i \wedge \bar{e}_j. \tag{2.13}$$

---

[2]In this section we will denote elements of general vector spaces with the common "overbar" notation, as in $\bar{x}$. This is to distinguish them from vectors in $\mathbb{R}^3$ encountered in standard vector calculus, which will be denoted as $\vec{F}$. We will adopt similar convention for forms, using the "tilde" notation, as in $\tilde{\omega}$.

0-vector   1-vector   2-vector

Fig. 2.1 Graphical representations of low-order p-vectors. A 0-vector is a zero-dimensional point, and a 1-vector is identical to the vectors encountered in standard vector calculus which represents both a magnitude (depicted by the length of the vector) and a direction. A 2-vector, however, can be viewed as the plane spanned by the pair of constituent 1-vectors and its magnitude is the area of the parallelogram whose sides are the pair of 1-vectors

In two dimensions (i.e., $n = 2$) and assuming a standard Euclidean metric, $(x^1 y^2 - x^2 y^1)$ is the area of the parallelogram with sides $\bar{x}$ and $\bar{y}$.

The exterior product can be used to generate the product of larger numbers of elements of V. For $p = 0, 1, \ldots, n$ we define a new vector space denoted as $\bigwedge^p V$ whose elements are the space of p-**vectors** over V, where p is called the **degree**. The vector space $\bigwedge^0 V = \mathbb{R}$, $\bigwedge^1 V = V$, and $\bigwedge^n V = \mathbb{R}$, and for arbitrary $2 \leq p \leq (n-1)$ the vector space is spanned by elements represented as

$$\bar{x}_1 \wedge \bar{x}_2 \wedge \cdots \wedge \bar{x}_p \tag{2.14}$$

where each $\bar{x}_i \in V$. Thus vector spaces of higher-order vectors can be assembled from lower-order vectors. Note that the dimension of the vector space $\bigwedge^p V$ is simply $\binom{n}{p}$, since this is the number of unique basis elements that can be defined through the rules of the exterior algebra.

By extending the rules of the exterior product outlined above, the product $\bar{x}_1 \wedge \bar{x}_1 \wedge \cdots \wedge \bar{x}_p = 0$ if, for any $i \neq j$, $\bar{x}_i = \bar{x}_j$, and the product $\bar{x}_1 \wedge \bar{x}_2 \wedge \cdots \wedge \bar{x}_p$ changes sign if any two $\bar{x}_i$ are interchanged. The exterior product can be applied to arbitrary orders of p-vectors. If $\bar{v} \in \bigwedge^a V$ is an a-vector and $\bar{w} \in \bigwedge^b V$ is a b-vector then

$$\bar{v} \wedge \bar{w} = (-1)^{ab}(\bar{w} \wedge \bar{v}) \tag{2.15}$$

and $\bar{v} \wedge \bar{w}$ is a vector of order $a + b$. Graphical representations of low-order p-vectors are provided in Fig. 2.1.

A remarkable property of the exterior product is its relationship to the determinant of a matrix. Consider a linear transformation $A : V \to V$, which induces a unique mapping g_A from the Cartesian product V^n to $\bigwedge^n V$,

$$g_A(\bar{x}_1, \bar{x}_2, \ldots, \bar{x}_n) = A\bar{x}_1 \wedge A\bar{x}_2 \wedge \cdots \wedge A\bar{x}_n$$
$$= f(\bar{x}_1 \wedge \bar{x}_2 \wedge \cdots \wedge \bar{x}_n). \tag{2.16}$$

Since g_A is multilinear, and $\bigwedge^n V$ is $\mathbb{R}$ and therefore one-dimensional, the mapping f consists of multiplication by a scalar, denoted by $|A|$,

$$A\bar{x}_1 \wedge A\bar{x}_2 \wedge \cdots \wedge A\bar{x}_n = |A|(\bar{x}_1 \wedge \bar{x}_2 \wedge \cdots \wedge \bar{x}_n). \tag{2.17}$$

Based on the properties of the exterior product, it can be shown [99, 139] that $|A|$ is the *determinant* of the linear mapping A, and is a generalization of the relation

2.2 Differential Forms

given in (2.13). Thus the wedge product provides a convenient means for expressing linear independence, since the p 1-vectors $\bar{x}_1, \bar{x}_2, \ldots, \bar{x}_p$ are linearly dependent if and only if $\bar{x}_1 \wedge \bar{x}_2 \wedge \cdots \wedge \bar{x}_p = 0$. In cases where a Euclidean metric is employed, a p-vector $\bar{x}_1 \wedge \bar{x}_2 \wedge \cdots \wedge \bar{x}_p$ represents the p-dimensional subspace of $\mathbb{R}^n$ spanned by the vectors $\bar{x}_i$ weighted by the volume of the parallelepiped whose edges are the $\bar{x}_i$.

The exterior product produces an associative algebra known as the **exterior algebra** or **Grassmann algebra**. This algebra can be similarly applied to the elements of the space that are *dual* to the p-vectors. We first define a real-valued **linear functional** $\tilde{a}$ on V to be a linear transformation of elements of V to the real numbers $\mathbb{R}$, i.e., $\tilde{\alpha} : V \to \mathbb{R}$. Thus for $\bar{v}, \bar{w} \in V$ and $a, b \in \mathbb{R}$,

$$\tilde{\alpha}(a\bar{v} + b\bar{w}) = a\tilde{\alpha}(\bar{v}) + b\tilde{\alpha}(\bar{w}). \tag{2.18}$$

The collection of all such linear functionals on V constitutes a vector space termed the **dual space** to V, denoted as V^*. It is a vector space under the operations

$$(\tilde{\alpha} + \tilde{\beta})(\bar{v}) = \tilde{\alpha}(\bar{v}) + \tilde{\beta}(\bar{v}) \tag{2.19a}$$

and

$$(c\tilde{\alpha})(\bar{v}) = c\tilde{\alpha}(\bar{v}) \tag{2.19b}$$

for $\tilde{\alpha}, \tilde{\beta} \in V^*$, $\bar{v} \in V$, and $c \in \mathbb{R}$. If $\bar{e}_1, \bar{e}_2, \ldots, \bar{e}_n$ form a basis of V, then we may define a **dual basis** $\tilde{\sigma}^1, \tilde{\sigma}^2, \ldots, \tilde{\sigma}^n$ of V^* by setting

$$\tilde{\sigma}^i(\bar{e}_j) = \begin{cases} 1 & \text{if } i = j, \\ 0 & \text{otherwise} \end{cases} \tag{2.20}$$

and, thus, by linearity,

$$\tilde{\sigma}^i(\bar{v}) = \tilde{\sigma}^i \left(\sum_j \bar{e}_j v^j \right) = \sum_j \tilde{\sigma}^i(\bar{e}_j) v^j = v^i \tag{2.21}$$

therefore $\tilde{\sigma}^i$ extracts the ith component of the vector $\bar{v}$. When viewed as elements of the dual vector space V^*, these linear functionals are often called **forms**.

The Grassmann algebra may be extended to forms through the dual basis defined above, which allows the construction of higher-order forms. Thus the product

$$\tilde{\alpha} \wedge \tilde{\beta} \tag{2.22}$$

is well-defined and is also an anticommutative product but on the elements of V^*. As with vectors, for $p = 0, 1, \ldots, n$ we may define a new vector space denoted as $\bigwedge^p V^*$ whose elements are the space of p-**forms** over V^*. Here $\bigwedge^0 V^*$ is the space of scalar-valued functions, $\bigwedge^1 V^*$ is the space of linear functional on vectors, and

$\bigwedge^n V^*$ is the space of scalar-valued functions. For other values of p the vector space $\bigwedge^p V^*$ is spanned by elements of the form

$$\tilde{\omega}_1 \wedge \tilde{\omega}_2 \wedge \cdots \wedge \tilde{\omega}_p. \tag{2.23}$$

A p-form thus maps a p-*tuple* of vectors into a scalar. The dimension of each space $\bigwedge^p V^*$ is also $\binom{n}{p}$.

It is important to note that although forms are commonly thought of as supplying a "measure" for vectors, they do not require a metric for their evaluation.

Example 2.1 (Evaluating the operation of a differential form on a vector) Consider a 2-vector $\bar{v} \in \bigwedge^2 V$ in $\mathbb{R}^3$ defined in terms of the standard basis set $\bar{v} = 2\bar{e}_1 \wedge \bar{e}_2 + 3\bar{e}_2 \wedge \bar{e}_3 - \bar{e}_3 \wedge \bar{e}_1$, and a 2-form $\tilde{\omega} \in \bigwedge^2 V^*$ in $\mathbb{R}^3$ defined in terms of the standard *dual* basis set $\tilde{\omega} = 7\tilde{\sigma}^1 \wedge \tilde{\sigma}^2 - 6\tilde{\sigma}^2 \wedge \tilde{\sigma}^3 + 4\tilde{\sigma}^3 \wedge \tilde{\sigma}^1$. Then $\tilde{\omega}(\bar{v}) = -8$.

While any expression of a p-form or a p-vector used for calculation as in the example above does require a choice of basis elements or a coordinate system, the operations of a p-forms are independent of any particular coordinate system such that, e.g., the evaluation of $\tilde{\omega}(\bar{v})$ will be the same regardless of the particular coordinate system used for the calculation. That is, they exhibit *coordinate invariance* in the sense of tensor analysis: the physical meaning of a form or vector must be invariant under a change of coordinates. However, the behavior of p-forms and p-vectors under changes of coordinate system are quite different—yet they manage to balance each other out. We will revisit this topic of coordinate independence below.

The above definition of the exterior product and p-vectors holds for a given vector space V, but to extend these concepts to more general spaces we will require the notion of a manifold and its associated tangent spaces.

2.2.1.2 Manifolds, Tangent Spaces, and Cotangent Spaces

A general **manifold** is a topological space that is "locally Euclidean". For instance, the plane and the sphere are both common examples of two-dimensional manifolds. A manifold consists of a collection or "atlas" of homeomorphisms to Euclidean space called *charts*. These charts must be compatible such that the composition of a chart with the inverse of an overlapping chart must also be a homeomorphism, and this homeomorphism is termed the *transition map*. In order for the manifold to be suitable for calculus it must be a **differentiable manifold** which requires that the transition maps be differentiable. Typically we will be considering submanifolds of some Euclidean space. A subset $\mathcal{M} = \mathcal{M}^n \subset \mathbb{R}^{n+r}$ is an n-dimensional (differentiable) **submanifold** of $\mathbb{R}^{n+r}$ if locally $\mathcal{M}$ can be described by r coordinates differentiably in terms of the remaining n coordinates. Note that while these definitions are phrased in terms of coordinates, they merely require that *some* such coordinates exists, and thus a manifold need not be equipped with a global coordinate system (for a both rigorous and intuitive definition of a manifold, see Ref. [142]).

2.2 Differential Forms

Because a manifold is locally Euclidean, we can define at each point $q \in \mathcal{M}^n$ a **tangent space** to the manifold $\mathcal{M}^n$ at q, denoted as $T\mathcal{M}_q^n$, as the real vector space consisting of all tangent vectors (i.e., 1-vectors) to $\mathcal{M}^n$ at q. The union of tangent spaces at every point in the manifold is termed the *tangent bundle* $T\mathcal{M}^n$.

The tangent space can be considered as a generalization of the familiar concept of a tangent vector to a curve in the differential geometry of curves, which is often used to define the instantaneous velocity of a point traveling along the curve at a given point q. For this reason, the tangent space of a manifold at a point q can be viewed as representing the set of differential operators that measure the instantaneous velocity of curves contained within the manifold at the point q. A vector $\bar{v}$ of the tangent space $T\mathcal{M}_q^n$ can be used to *differentiate* a real-valued function f on $\mathcal{M}^n$, $f : \mathcal{M}^n \to \mathbb{R}$, via the **directional derivative**

$$(D_{\bar{v}} f)(q) \equiv \frac{\mathrm{d}}{\mathrm{d}t}[f(q + t\bar{v})]_{t=0} \qquad (2.24)$$

at point q. If x is any local Cartesian coordinate system specified by the n-tuple of coordinates $(x^1, \ldots, x^n)$, then the directional derivative can be expressed as

$$(D_{\bar{v}} f)(q) = \sum_j \frac{\partial f}{\partial x^j}(q) \, v^j. \qquad (2.25)$$

Note that the components v^j of the vector $\bar{v}$ appear in the above expression, and there is a one-to-one correspondence between tangent vectors $\bar{v}$ in $T\mathcal{M}_q^n$ and differential operators on differentiable functions near q. The tangent vector can therefore be expressed as

$$\bar{v}|_q = \sum_j v^j \frac{\partial}{\partial x^j}\bigg|_q$$

and each of the n operators $\partial/\partial x^j$ can be thought of as a vector, thus we will now denote each vector as $\boldsymbol{\partial}/\boldsymbol{\partial} x^j$, at each point q. The tangent vector can then be expressed as

$$\bar{v} = \sum_j v^j \frac{\boldsymbol{\partial}}{\boldsymbol{\partial} x^j}. \qquad (2.26)$$

The vectors $\boldsymbol{\partial}/\boldsymbol{\partial} x^j$ can be used to define a basis for the tangent space $T\mathcal{M}_q^n$ (and below will be used interchangeably with the standard basis set $\bar{e}_j \in \bigwedge^1(T\mathcal{M}_q^n)$).

Given this generalization of vectors to higher-dimensional manifolds, we will now consider the generalization of *functions* defined on these vectors. A **scalar differential form** is a linear functional that maps an element of the tangent space to a scalar.[3]

---

[3] The distinction between a "differential form" and a "form" (a.k.a. an "algebraic" form or a "linear" form) stems from whether the vector space V is viewed as the tangent space to a manifold, i.e.,

Thus for each tangent space on a manifold we may define a *dual vector space* of differential forms. The **cotangent space** $T^*\mathcal{M}_q^n$ at a point $q \in \mathcal{M}^n$ is the dual space to the tangent space $T\mathcal{M}_q^n$ at q such that $\tilde{\omega} \in T^*\mathcal{M}^n : T\mathcal{M}^n \to \mathbb{R}$. (The union of cotangent spaces at every point in the manifold is termed the *cotangent bundle* $T^*\mathcal{M}^n$.) This space is spanned by the standard dual basis $\tilde{\sigma}^i$ defined relative to the standard basis of tangent vectors $\bar{e}_i$ via

$$\tilde{\sigma}^i(\bar{e}_j) = \begin{cases} 1 & \text{if } i = j, \\ 0 & \text{otherwise.} \end{cases} \tag{2.27}$$

Note that it is also possible to construct vector spaces of scalar differential forms to define "vector-valued differential forms" (see, e.g., Ref. [260]), however we shall restrict our discussion to differential forms that are scalar-valued and therefore will use the term "differential form" to refer only to scalar differential forms.

We will now explore why forms defined in this way are termed differential forms. Recall that on $\mathcal{M}^n$ a vector $\bar{v}$ at point q defines a differential operator, the directional derivative, on functions defined near q. Then if f is such a function, $f : \mathcal{M}^n \to \mathbb{R}$, we define the **differential** of f at the point $q \in \mathcal{M}^n$ as the linear function $\mathrm{d}f : T\mathcal{M}_q^n \to \mathbb{R}$ defined as

$$\mathrm{d}f(\bar{v})|_q \equiv \bar{v}_q(f) = (D_{\bar{v}}f)(q). \tag{2.28}$$

Here the differential $\mathrm{d}f$ is defined independent of any basis but requires only the vector $\bar{v}$ also defined at q. Note that the differential $\mathrm{d}f$ of f evaluated on the vector $\bar{v}$ is equivalent to the directional derivative $D_{\bar{v}}$ of $\bar{v}$ evaluated on the function f.

Above in (2.26) we noted that the differential operation defined by a tangent vector was equivalent to the tangent vector itself. The differential can therefore be expressed in terms of local coordinates x as

$$\mathrm{d}f\left(\sum v^i \frac{\partial}{\partial x^j}\right)\bigg|_q = \sum v^i \frac{\partial f}{\partial x^j}(q). \tag{2.29}$$

If we consider the special case of the differential $\mathrm{d}x^i$ of the coordinate function x^i, then

$$\mathrm{d}x^i\left(\frac{\partial}{\partial x^j}\right) = \frac{\partial x^i}{\partial x^j} = \begin{cases} 1 & \text{if } i = j, \\ 0 & \text{otherwise} \end{cases} \tag{2.30}$$

and thus

$$\mathrm{d}x^i(\bar{v}) = \mathrm{d}x^i\left(\sum_j v^j \frac{\partial}{\partial x^j}\right) = \sum_j v^j \mathrm{d}x^i\left(\frac{\partial}{\partial x^j}\right) = v^i \tag{2.31}$$

---

if $V = T\mathcal{M}$, in which case the elements of $T\mathcal{M}$ are the differential operators for curves along $\mathcal{M}$. Indeed the exterior algebra of forms and many of their geometrical interpretations hold even if V is not considered as a tangent space.

which indicates that, for each i, the linear functional $\mathrm{d}x^i$ operating on $\bar{v}$ extracts the ith component of $\bar{v}$. Thus the differentials of the coordinate functions are dual to the tangent vectors and are therefore exactly the elements of the cotangent space, i.e.,

$$\mathrm{d}x^i = \tilde{\sigma}^i, \qquad (2.32)$$

and any expression in terms of these basis elements, i.e., $\tilde{\omega} = \sum \omega_i \, \mathrm{d}x^i$, is a differential form. Naturally the above definitions can be extended to higher order vectors and forms using the exterior product, i.e., we may define the vector space of tangent vectors of degree p, $\bigwedge^p(T\mathcal{M}_q^n)$, in the tangent space $T\mathcal{M}_q^n$, and the dual vector space of differential forms of degree p, $\bigwedge^p(T^*\mathcal{M}_q^n)$, in the tangent space $T^*\mathcal{M}_q^n$. We will consider these further in the next section.

2.2.1.3 The Metric Tensor: Mapping p-Forms to p-Vectors

One important property of differential forms and vectors defined in the general setting of differentiable manifolds is that they are expressed in a manner that is not dependent on a particular coordinate system. (To demonstrate this formally requires observing how each behaves under the action of a transition map between overlapping coordinate patches, which we do not reproduce here, but see Ref. [142] for a full treatment.) As alluded to above in the discussion following Example 2.1, a fundamental distinction between vectors and forms is how they react to a change of coordinates to preserve their coordinate invariance. Specifically, elements of the tangent space $T\mathcal{M}^n$ behave as **contravariant tensors** in that they vary in the "opposite" direction with a change of coordinates, and elements of the cotangent space $T^*\mathcal{M}^n$ behave as **covariant tensors** in that they vary in the "same" direction with a change of coordinates.[4]

Because both contravariant and covariant tensors are implicated in differentiation and can comprise differential operators, it is often confusing when trying to distinguish whether a particular quantity is contravariant or covariant. Intuitively, contravariant quantities include coordinate vectors that represent the position of an object or and derivative of position with respect to time (such as velocity or acceleration); and covariant quantities include the derivatives of some function over space, such as the gradient of a scalar field, in which the quantity arises by "dividing by the coordinates" as in the gradient of a scalar function f, $\nabla f = (\partial f / \partial x) \, \bar{e}_x + (\partial f / \partial y) \, \bar{e}_y$. Furthermore, because of the skew-symmetry imparted on p-vectors and p-forms by the exterior product, p-vectors and p-forms are **antisymmetric tensors** of a contravariant and covariant kind, respectively. (For a

---

[4] As a simple, illustrative example, consider a one-dimensional coordinate system defined in units of centimeters with a position vector at the origin extending to the location at 500 centimeters. If the unit of measurement were to change from centimeters to meters, the basis vectors will become longer and our position vector, if it physically represents the same position, will now extend only 5 meters. Therefore, the coordinate increments are larger in the new system, but the value of the coefficient describing the contravariant vector is smaller.

deeper discussion of the relationship between the theories of differential forms and tensor analysis, see, e.g., Refs. [142, 335].)

The linear mapping provided by a p-form of a p-vector into a scalar does not require a metric. However, if a metric is available then it can be used to establish an isomorphism between p-vectors and p-forms—that is, the metric can provide a mapping of a p-vector into a corresponding p-form.

We begin with an n-dimensional vector space V equipped with an **inner product** that is a mapping of a pair of (1-)vectors $\bar{v}, \bar{w} \in V$, denoted $\langle \bar{v}, \bar{w} \rangle$, into the real numbers, i.e., $\langle \bar{v}, \bar{w} \rangle : V \times V \to \mathbb{R}$. The inner product is symmetric (i.e., $\langle \bar{v}, \bar{w} \rangle = \langle \bar{w}, \bar{v} \rangle$), bilinear, non-degenerate such that if $\langle \bar{v}, \bar{w} \rangle = 0$ for all $\bar{w}$ then $\bar{v} = 0$, and positive definite such that $\|\bar{v}\|^2 = \langle \bar{v}, \bar{v} \rangle$ is positive when $\bar{v} \neq 0$. Under the standard basis $\bar{e}_i$, the bi-linearity of the inner product implies

$$\langle \bar{v}, \bar{w} \rangle = \left\langle \sum_i \bar{e}_i v^i, \sum_j \bar{e}_j w^j \right\rangle \tag{2.33}$$

$$= \sum_i \sum_j v^i w^j \langle \bar{e}_i, \bar{e}_j \rangle \tag{2.34}$$

thus the inner product of two vectors can be evaluated by a weighted linear combination of the inner product of the basis elements. Typically the inner product of the basis elements is expressed by an $n \times n$ matrix (which is symmetric by the definition of the inner product) with entries

$$g_{ij} = \langle \bar{e}_i, \bar{e}_j \rangle \tag{2.35}$$

so that

$$\langle \bar{v}, \bar{w} \rangle = \sum_i \sum_j v^i w^j g_{ij}. \tag{2.36}$$

If the components are assembled into matrices, the inner product calculation can be expressed in terms of standard linear algebra, i.e.,

$$\langle \bar{v}, \bar{w} \rangle = \mathbf{v}^\mathsf{T} \mathbf{G} \mathbf{w} \tag{2.37}$$

where $\mathbf{v} = [v^1, \ldots, v^n]^\mathsf{T}$, $\mathbf{w} = [w^1, \ldots, w^n]^\mathsf{T}$, and $[g_{ij}] = \mathbf{G}$. Thus, under a Euclidean or "flat" metric and Cartesian coordinates the inner product of two vectors is equivalent to the dot product of their components.

Since the inner product $\langle \bar{v}, \bar{w} \rangle$ is a linear function of $\bar{w}$ when $\bar{v}$ is held fixed, the function $\tilde{\alpha}$ defined by

$$\tilde{\alpha}(\bar{w}) = \langle \bar{v}, \bar{w} \rangle \tag{2.38}$$

is a linear functional on V, i.e., $\tilde{\alpha} \in V^*$. Through this construction, every vector $\bar{v}$ in the inner product space V can be associated with a form $\tilde{\alpha}$ that is called the **covariant version** of $\bar{v}$. The inner product, by establishing a scalar product of two 1-vectors, enables a mapping between 1-forms and 1-vectors.

2.2 Differential Forms

This mapping can be exploited to define an inner product on forms. Consider the 1-form $\tilde{\alpha}$ represented by the standard dual basis, $\tilde{\alpha} = \sum_j \alpha_j \tilde{\sigma}^j$. Then, by invoking the mapping between forms and vectors, we see that

$$\tilde{\alpha} = \sum_j \tilde{\alpha}(\bar{e}_j)\tilde{\sigma}^j = \sum_j \langle \bar{v}, \bar{e}_j \rangle \tilde{\sigma}^j = \sum_j \left\langle \sum_i v^i \bar{e}_i, \bar{e}_j \right\rangle \tilde{\sigma}^j$$

$$= \sum_j \left(\sum_i v^i g_{ij} \right) \tilde{\sigma}^j \tag{2.39}$$

and therefore the components of $\tilde{\alpha}$ are expressible as

$$\alpha_j = \sum_i v^i g_{ij}. \tag{2.40}$$

In the special case of a Euclidean metric, $\alpha_j = v^j$.

In the finite-dimensional inner product space V, each linear functional $\tilde{\alpha}$ is the covariant version of *some* vector $\bar{v}$. Therefore we may compute the **contravariant version** of $\tilde{\alpha}$ by solving (2.40) for the components v^i. Since the inner product is defined to be non-degenerate, the inverse matrix $\mathbf{G}^{-1}$ exists and is symmetric. Conventionally, the entries of the inverse matrix are denoted g^{ij}. The contravariant version of $\tilde{\alpha}$ is therefore

$$v^i = \sum_j \alpha_j g^{ij} \tag{2.41}$$

and it is straightforward to show that $\mathbf{G}^{-1}$ provides a inner product on V^* that is equivalent to the original inner product on V. Thus for a pair of 1-forms $\tilde{\alpha}, \tilde{\beta} \in V^*$,

$$\langle \tilde{\alpha}, \tilde{\beta} \rangle = \sum_i \sum_j \alpha_i \beta_j g^{ij} = \mathbf{a}\mathbf{G}^{-1}\mathbf{b}^\mathsf{T} \tag{2.42}$$

where $\mathbf{a} = [\alpha^1, \ldots, \alpha^n]$ and $\mathbf{b} = [\beta^1, \ldots, \beta^n]$.

The matrix $\mathbf{G}$ defined above in (2.35) is known as the **metric tensor**. Although it arises here from a general inner product space, a metric tensor can also be defined for the tangent space $T\mathcal{M}_q^n$ at each point q of a manifold $\mathcal{M}^n$. If we want the metric tensors for the tangent space and cotangent space to be compatible, the metric tensor for the cotangent space $T^*\mathcal{M}_q^n$ is the matrix *inverse* of the metric tensor matrix on the tangent space $T\mathcal{M}_q^n$ (to avoid confusion the metric tensor on the tangent space is referred to as the **primal metric tensor** and the corresponding metric tensor on the cotangent space is referred to as the **dual metric tensor**).[5] Note that, in contrast to p-forms and p-vectors, the metric tensor is a *symmetric* covariant tensor and is

---

[5] A *Riemannian metric* on a manifold $\mathcal{M}^n$ is a family of positive definite inner products defined at each $\mathcal{M}_q^n$ that is differentiable over all $q \in \mathcal{M}^n$. A manifold equipped with a Riemannian metric is called a *Riemannian manifold*.

always rank 2, and the dual metric tensor is a rank 2 symmetric contravariant tensor [335]. In the literature the process of converting a contravariant vector to a covariant vector is referred to as *index lowering* while converting in the opposite direction from covariant to contravariant is referred to as *index raising*, most often in the context of tensor analysis.[6]

In some applications, a metric is provided without explicit identification as to whether the metric applies to contravariant quantities or covariant quantities, and a decision must be made to interpret the metric as primal or dual. We shall consider this issue further below as it arises often in the discrete context of graph theory.

For completeness, we define a **norm** on vectors using the inner product defined by the metric tensor. The norm of any 1-vector $\bar{v}$ is given by $\|\bar{v}\| \equiv \sqrt{\langle \bar{v}, \bar{v} \rangle}$. This norm can be shown to satisfy the triangle inequality and it, as well as the underlying inner product, satisfies the Schwartz inequality. Finally a **metric** can be defined that establishes the distance between two vectors at a point q: for a pair of vectors $\bar{u}$ and $\bar{v}$ in $T\mathcal{M}_q^n$, the metric is defined as $\rho(\bar{u}, \bar{v}) = \|\bar{u} - \bar{v}\|$.

The metric tensor provides a unique mapping between 1-forms and 1-vectors, but requires the specification of a metric on the underlying vector space in the form of a bilinear pairing. This pairing is fundamental in that once it is established, lengths of vectors can be computed, from which one can then compute lengths of curves, angles, and perform parallel translation. In the next section we will consider integration, which is another such pairing that is purely topological and thus does not require a metric.

2.2.2 Differentiation and Integration of Forms

So far we have considered the exterior algebra of both p-vectors and differential p-forms and the mechanism by which the metric tensor can provide a one-to-one mapping between forms and vectors. We now consider the operations of differentiation and integration defined for forms in order to assemble the tools that are needed to use differential forms for calculus in arbitrary dimensions.

2.2.2.1 The Exterior Derivative

The exterior derivative operator extends the notion of a derivative to differential forms in a way that is invariant under coordinate transformations and does not require the specification of a metric. We define the **exterior derivative** as the operator d that maps p-forms to $(p+1)$-forms, $\mathrm{d} : \bigwedge^p(T^*\mathcal{M}^n) \to \bigwedge^{p+1}(T^*\mathcal{M}^n)$. Just as

---

[6]In the differential forms literature this same process is sometimes described by the so-called *musical isomorphisms*, ♯ and ♭, where $\tilde{\alpha}^\sharp$ is the contravariant version of the form $\tilde{\alpha}$, and $\bar{v}^\flat$ is the covariant version of the vector $\bar{v}$.

2.2 Differential Forms

the exterior product builds higher-order forms out of lower-order forms, the exterior derivative also acts to increase the degree of the form. The exterior derivative is uniquely defined as the operator with the following properties, given an arbitrary p-form $\tilde{\alpha}$ and a q-form $\tilde{\beta}$.

1. $d(\tilde{\alpha} + \tilde{\beta}) = d\tilde{\alpha} + d\tilde{\beta}$
2. $d(\tilde{\alpha} \wedge \tilde{\beta}) = (d\tilde{\alpha}) \wedge \tilde{\beta} + (-1)^p \tilde{\alpha} \wedge d\tilde{\beta}$ (Leibniz's Law or the chain rule)
3. $d(d\tilde{\alpha}) = 0$
4. The 1-form $df \in \bigwedge^1$ for any function $f \in \bigwedge^0$ is the usual differential of the function f,

$$df = \sum_i \frac{\partial f}{\partial x^i} dx^i. \tag{2.43}$$

Note that a derivative with property 2 is called an **antiderivation** [335].

Property (2.43) defines the operation of the exterior derivative on a differentiable scalar-valued function or a 0-form, which results in a 1-form whose components are clearly identifiable as the components of the *gradient* of a scalar field as it is defined in standard vector calculus. The exterior derivative on a general degree p differential form is simply the operation defined in (2.43) applied to the scalar-valued components of the form. For the p-form $\tilde{\alpha}$ expressed in components as $\tilde{\alpha} = \sum_i \alpha_i \, dx^i$, the exterior derivative operation yields the $(p+1)$-form

$$d\tilde{\alpha} = \sum_i (d\alpha_i) \wedge dx^i. \tag{2.44}$$

The differential operators in vector calculus can be seen as a special case of the exterior derivative acting on a form of a particular degree. Here we consider two examples showing how the curl and divergence operators may arise as special cases of the exterior derivative.

Example 2.2 (The curl of a scalar field) If we consider a 1-form $\tilde{\omega} = P \, dx + Q \, dy + R \, dz$, then

$$d\tilde{\omega} = dP \wedge dx + dQ \wedge dy + dR \wedge dz$$

$$= \left(\frac{\partial P}{\partial x} dx + \frac{\partial P}{\partial y} dy + \frac{\partial P}{\partial z} dz \right) \wedge dx$$

$$+ \left(\frac{\partial Q}{\partial x} dx + \frac{\partial Q}{\partial y} dy + \frac{\partial Q}{\partial z} dz \right) \wedge dy$$

$$+ \left(\frac{\partial R}{\partial x} dx + \frac{\partial R}{\partial y} dy + \frac{\partial R}{\partial z} dz \right) \wedge dz,$$

thus, by collecting terms and the antisymmetry property of the wedge product,

$$d\tilde{\omega} = \left(\frac{\partial R}{\partial y} - \frac{\partial Q}{\partial z} \right) dy \wedge dz + \left(\frac{\partial P}{\partial z} - \frac{\partial R}{\partial x} \right) dz \wedge dx + \left(\frac{\partial Q}{\partial x} - \frac{\partial P}{\partial y} \right) dx \wedge dy.$$

Example 2.3 (The divergence of a vector field) If we consider a 2-form $\tilde{\alpha} = A\,dy \wedge dz + B\,dz \wedge dx + C\,dx \wedge dy$, then

$$d\tilde{\alpha} = \left(\frac{\partial A}{\partial x}\,dx + \frac{\partial A}{\partial y}\,dy + \frac{\partial A}{\partial z}\,dz\right) \wedge dy \wedge dz$$
$$+ \left(\frac{\partial B}{\partial x}\,dx + \frac{\partial B}{\partial y}\,dy + \frac{\partial B}{\partial z}\,dz\right) \wedge dz \wedge dx$$
$$+ \left(\frac{\partial C}{\partial x}\,dx + \frac{\partial C}{\partial y}\,dy + \frac{\partial C}{\partial z}\,dz\right) \wedge dx \wedge dy$$

and so

$$d\tilde{\alpha} = \left(\frac{\partial A}{\partial x} + \frac{\partial B}{\partial y} + \frac{\partial C}{\partial z}\right) dx \wedge dy \wedge dz.$$

As a brief aside, the previous two examples highlight an important distinction between standard vector calculus and the language of differential forms. In standard vector calculus one deals only with *vectors* (which are typically conceptualized as "1-vectors" in the terminology of this chapter), yet in many applications of vector calculus in physics, some of the fields that require representation as vectors behave quite differently—some "vector fields" are better represented by 1-forms while others by 2-forms [395]. Often it is absolutely critical to find the appropriate representation of a field in order to solve a problem at hand, and this is especially the case in numerical modeling (see Refs. [47, 178], for examples).

Based on these examples, the exterior derivative operator appears able to substitute for the standard vector calculus operators—the exterior derivative applied to a 0-form performs an operation similar to a gradient, the exterior derivative applied to a 1-form performs an operation similar to a curl, and the exterior derivative applied to a 2-form performs an operation similar to a divergence. However there are some details required to precisely translate between the settings of vector calculus and differential forms. For example, in conventional vector calculus, the gradient of a function f is given by the gradient operator and produces a vector field, ∇f, with the same components as in (2.43). Therefore, we may examine the relationship between the 1-form df and the vector field ∇f. This relationship may be characterized by considering that df is a covariant quantity and ∇f is contravariant, meaning that the metric tensor is likely to play a role in distinguishing these quantities. Specifically, given a vector $\bar{v}$, df and ∇f are related via

$$df(\bar{v}) = \langle \nabla f, \bar{v} \rangle \tag{2.45}$$

i.e., $df(\bar{v})$ is the directional derivative of f along $\bar{v}$, as defined above in (2.28) and (2.29). Therefore the components of ∇f can be calculated with the metric tensor from the components of df in order to produce the vector indicating the direction of steepest descent along the scalar field f. Similarly, the exterior derivative of higher-order forms can be converted into the vector fields corresponding to the curl

2.2 Differential Forms

and divergence operators from vector calculus only when a metric is supplied. These examples will be considered in a later section.

There are two classes of differential forms that are of particular interest which are defined using the exterior derivative. A form $\tilde{\omega}$ is called a **closed form** if its exterior derivative is zero, i.e., if $d\tilde{\omega} = 0$. An **exact form** is a form that is the exterior derivative of a form, so if $\tilde{\alpha} = d\tilde{\beta}$, $\tilde{\alpha}$ is an exact form. It is therefore said that a closed form is in the *kernel* or nullspace of the exterior derivative operator, and that an exact form is in the *image* or range of the exterior derivative operator. Note that one can form an interesting equivalence relation between forms that differ by a closed form. With respect to the exterior derivative, two p-forms $\tilde{\alpha}$ and $\tilde{\beta}$ are said to be equivalent, $\tilde{\alpha} \sim \tilde{\beta}$, or **cohomologous**, if they differ by an exact form, i.e., if there exists some exact form $\tilde{\eta}$ such that $\tilde{\alpha} = \tilde{\beta} + \tilde{\eta}$, in which case $d\tilde{\alpha} = d\tilde{\beta}$ but $\tilde{\alpha} \neq \tilde{\beta}$.

The geometric nature of the exterior derivative is a natural extension of the differential operator in one dimension [99]. Similar to the exterior product which (in Euclidean space) expresses the volume of the enclosed parallelepiped, the geometric interpretation of the exterior derivative is related to characterizing the expansion of a form outwards in all directions.

> The *exterior derivative* measures the variation of a p-form simultaneously in each of the p directions of a p-dimensional parallelepiped, and is therefore the natural generalization of the one-dimensional differential operator d/dt.

Although the exterior derivative operator is denoted plainly as d regardless of the degree of the form it operates on, the character of the exterior derivative is somewhat distinct in different dimensions. The distinction will be made apparent in Sect. 2.3 when the exterior derivative is defined in the discrete setting.

The topological nature of the exterior derivative can be seen through its relationship to integration that is summarized by the **Generalized Stokes' Theorem**, otherwise known as the generalized Fundamental Theorem of Calculus, which is expressed as

$$\int_S d\tilde{\omega} = \int_{\partial S} \tilde{\omega} \tag{2.46}$$

where S is a p-dimensional submanifold of $\mathcal{M}^n$ with boundary ∂S and ω is a differential $(p-1)$-form in the cotangent bundle $T^*\mathcal{M}^n$. In the context of integration, the domain of integration of a p-form is often referred to as a p-**domain**. The proof of this generalization requires invoking the Fundamental Theorem of Calculus [353] and will not be reproduced here. It is assumed that the boundary ∂S of the domain S consists of a finite number of closed curves each of which is positively oriented with respect to the orientation of S. If we define a curve $\mathcal{C}$ that is oriented in the opposite direction of ∂S, then by convention

$$\int_{\mathcal{C}} \tilde{\omega} = -\int_{\partial S} \tilde{\omega}. \tag{2.47}$$

In cases where the domain $\mathcal{S}$ is a closed manifold and thus without boundary, e.g., if $\mathcal{S}$ is a sphere, $\partial \mathcal{S} = 0$ and the therefore

$$\int_{\mathcal{S}} d\tilde{\omega} = 0 \quad \text{(if $\mathcal{S}$ is closed)}.$$

While the operation of integration can be seen as the inverse of the exterior derivative operator, the domain of the integration also plays an important role. As with the special cases of the Fundamental Theorem reviewed above—including the Gradient Theorem, Green's Theorem, Stokes' Theorem, Gauss's Theorem, and integration by parts—the general expression in terms of differential forms makes clear the interrelationship between the derivative operator and the boundary operator. If we express the bilinear pairing of forms and regions using a notation similar to the inner product, e.g., if the Fundamental Theorem is denoted as

$$[\![d\tilde{\omega}, \mathcal{S}]\!] = [\![\tilde{\omega}, \partial \mathcal{S}]\!] \qquad (2.48)$$

then we see that the relationship between the boundary operator ∂ and the exterior derivative operator d satisfies an *adjointness relation*. Due to this relationship, the exterior derivative operator is often referred to as the **coboundary operator**, especially in the context of algebraic topology. We will now consider some of the topological implications of the Fundamental Theorem.

2.2.2.2 The Poincaré Lemma

It is a basic topological fact that the boundary of any manifold that is itself a boundary of a higher-dimensional manifold must always be empty. For example, the boundary of the unit ball in three dimensions is the unit sphere, and the unit sphere is a closed manifold without boundary. Algebraically, this means that the image of the boundary operator is in the kernel of the operator, or $\partial \partial = 0$. One can see that the Fundamental Theorem of Calculus provides an intuitive proof for why, given that "the boundary of a boundary is zero", then

$$\int_{\partial \mathcal{S}} d\tilde{\omega} = \int_{\mathcal{S}} d(d\tilde{\omega}) = \int_{\partial(\partial \mathcal{S})} \tilde{\omega} = 0, \qquad (2.49)$$

and therefore $d(d\tilde{\omega}) = 0$. Thus both the boundary operator and the coboundary operator exhibit the property that the image of the operator is a subset of its kernel, i.e., $\partial\partial = 0$ and $dd = 0$. This is a fundamental property of both of these operators. In the case of the exterior derivative, this identity is equivalent to the condition that mixed partial second derivatives are equal.

2.2 Differential Forms

> The boundary of a boundary is always zero, and thus the exterior derivative of an exterior derivative is always zero, which can be seen with the Generalized Stokes' Theorem.

Returning to vector calculus, the property that $dd = 0$ is equivalent to familiar expressions: $\nabla \times \nabla = 0$ (the curl of the gradient vanishes) and $\nabla \cdot \nabla \times = 0$ (the divergence of the curl vanishes). The Fundamental Theorem of Calculus therefore highlights how a topological property of the boundary operator relates to the *topology* of vector fields. A natural question then arises: If the exterior derivative of a differential form is zero, then is that form the exterior derivative of another form? or, When is a closed form also exact? These identities of the differential operators often lead to commonly accepted results in vector calculus that are *not* strictly correct, such as: (i) if $\nabla f = 0$ then f is constant; (ii) if $\nabla \times \vec{A} = 0$ then there is a function f such that $\vec{A} = \nabla f$; and (iii) if $\nabla \cdot \vec{B} = 0$ then there is a vector field $\vec{A}$ such that $\nabla \times \vec{A} = \vec{B}$. Indeed, it is not always the case that a closed form is exact.

The canonical example of a differential form that is closed but not exact within a defined region is the closed 1-form $\tilde{\omega}$ defined in a set $\mathcal{U} \subset \mathbb{R}^2$ as

$$\tilde{\omega} = \frac{-y \, dx + x \, dy}{x^2 + y^2} \tag{2.50}$$

in terms of the two-dimensional Cartesian coordinates x and y. Often $\tilde{\omega}$ is expressed as "$d\theta$" since it arises from the derivative of the polar coordinate $\theta = \operatorname{atan}(y/x)$ for a region $\mathcal{V}$ in $\mathbb{R}^2$ that does not include the origin, $\mathcal{V} = \mathbb{R}^2 \backslash \{\mathbf{0}\}$. Unfortunately θ is not a single-valued continuous function in $\mathcal{V}$,[7] and thus $\tilde{\omega}$ is locally equivalent to $\tilde{\theta}$ but not globally. Another example, set in $\mathbb{R}^3$ but otherwise identical, is the closed form $\tilde{\tau}$ defined as

$$\tilde{\tau} = \frac{x \, dy \wedge dz + y \, dz \wedge dx + z \, dx \wedge dy}{(x^2 + y^2 + z^2)^{3/2}}, \tag{2.51}$$

which known as the "solid angle form" (see Ref. [178]). We see in these examples that if a form is *locally* exact it does not follow that it is *globally* exact, and in general it is far more difficult to determine when a closed form is exact globally.

Locally, on a domain that can be continuously shrunk down to a single point (i.e., a "star-shaped" domain), every closed form is exact. This fact is often referred to as the **Poincaré lemma**, and expresses a sort of integrability conditions on forms. While the Poincaré lemma holds for topologically "simple" regions, it does not hold in general or globally over more topologically "interesting" manifolds. The study of the topological properties that govern whether a closed form is exact is known as **cohomology theory**. It is closely related to **homology theory** which seeks to

---

[7] Note that subsets of $\mathcal{V}$ exist on which $\tilde{\omega}$ is exact. For example, within the unit disk with the point at the origin removed and a "cut" along the positive x axis θ is single-valued. Changing the topology of the region in this way is analogous to *branch cuts* that are defined in complex analysis and the theory of Riemann surfaces.

answer analogous questions regarding whether a closed *region* is the boundary of something. For our purposes, it suffices to highlight that there is a close coupling between the topology of a region and the character of the differential forms that can be defined on that region. In the next section we take one small step closer to the study of cohomology as we review Hodge theory and the Hodge decomposition, which requires the definition of the last essential concept for understanding differential forms: The Hodge star operator.

2.2.3 The Hodge Star Operator

Although the exterior product and the exterior derivative are defined without the use of a metric, some operations exist that require a metric. The Hodge star operator both helps to clarify the relationship between the wedge product and the cross product of standard vector calculus and establishes a new linear pairing for forms, but requires the specification of a metric and an orientation to do so.

The **Hodge star operator** is a linear mapping from p-forms to $(n-p)$-forms, denoted as $\star : \bigwedge^p \to \bigwedge^{n-p}$. Because the dimension of $\bigwedge^p$ and $\bigwedge^{n-p}$ are the same, since $\binom{n}{p} = \binom{n}{n-p}$, the mapping is an isomorphism between the two spaces.[8] First we consider the action of the Hodge star on the $r = \binom{n}{p}$ orthonormal basis elements of $\bigwedge^p$. The Hodge star transforms each subspace spanned by $\tilde{\sigma}^{i_1} \wedge \cdots \wedge \tilde{\sigma}^{i_p}$ to an element of the orthogonal subspace, i.e.,

$$\star(\tilde{\sigma}^{i_1} \wedge \cdots \wedge \tilde{\sigma}^{i_p}) = \tilde{\sigma}^{i_{(p+1)}} \wedge \cdots \wedge \tilde{\sigma}^{i_n}, \tag{2.52}$$

where the ordering of indices from $\{i_1, \ldots, i_k, i_{k+1}, \ldots, i_n\}$ is an even permutation of the integers $1, \ldots, n$. So for example, if we consider the basic 1-forms dx^i in $\mathbb{R}^3$, with the standard right-handed orientation, then $\star(dx^1 \wedge dx^2) = dx^3$ and $\star(dx^2 \wedge dx^3) = dx^1$ but $\star(dx^1 \wedge dx^3) = -dx^2$.

The effect of the Hodge star on a general p-form can be understood in terms its effect on each covariant component. If we consider a p-form $\tilde{\omega}$ expressed as $\tilde{\omega} = \sum_{i=1}^r w_i \tilde{\sigma}^i$, then

$$\star\tilde{\omega} = \sqrt{|g|} \sum_{i=1}^r w_i^\star (\star\tilde{\sigma}^i) \tag{2.53}$$

---

[8] When dealing with complicated, non-orientable spaces, it is often noted that the Hodge star operator maps p-forms to *pseudo*-$(n-p)$-forms. Pseudoforms are forms who change sign whenever the orientation specified for the underlying manifold reverses [142]. We will not make the distinction between forms and pseudoforms here, although the concept of pseudoforms may help conceptualize the operation of the discrete Hodge star that maps between the primal and dual complexes (see Sect. 2.3.4). Note that the distinction between forms and pseudoforms is closely related to that of straight and twisted forms, polar and axial vectors, and across and through variables [48].

2.2 Differential Forms

where $|g|$ represents the (primal) metric tensor determinant and the effect on each covariant coefficient w_i is to convert it to its *contravariant* version w_i^* via the (dual) metric tensor.[9] The $(n - p)$-form $\star\tilde{\omega}$ is said to be the **Hodge dual** of the p-form $\tilde{\omega}$. To make the operation of the Hodge star clear, we consider a simple example.

Example 2.4 If $\tilde{\beta}$ is a 1-form in three dimensions, $\tilde{\beta} = b_1 \mathrm{d}x^1 + b_2 \mathrm{d}x^2 + b_3 \mathrm{d}x^3$, and the metric tensor g is simplified such that only the diagonal entries of the dual metric tensor, $g^{ii} = 1/g_{ii}$, are non-zero, then the Hodge dual of $\tilde{\beta}$ is

$$\star\tilde{\beta} = \frac{1}{\sqrt{g^{11}g^{22}g^{33}}}(b_1 g^{11} \mathrm{d}x^2 \wedge \mathrm{d}x^3 + b_2 g^{22} \mathrm{d}x^3 \wedge \mathrm{d}x^1 + b_3 g^{33} \mathrm{d}x^1 \wedge \mathrm{d}x^2).$$

The Hodge star operator is linear and has the following properties (that are straightforward to verify), given two p-forms $\tilde{\alpha}$ and $\tilde{\beta}$ and a scalar function f.

1. $\star\tilde{\alpha} \wedge \tilde{\beta} = \star\tilde{\beta} \wedge \tilde{\alpha}$
2. $\star(f\tilde{\alpha}) = f \star \tilde{\alpha}$
3. $\star(\star\tilde{\alpha}) = (-1)^{p(n-p)}\tilde{\alpha}$
4. $\star \star \star \star \tilde{\alpha} = \tilde{\alpha}$
5. $\tilde{\alpha} \wedge \star\tilde{\alpha} = 0$ if and only if $\tilde{\alpha} = 0$ (non-degeneracy of the inner product).

The Hodge star is useful because it provides an inner product on p-forms. If we consider two p-forms, $\tilde{\alpha}$ and $\tilde{\beta}$, if the Hodge star is applied to one of the forms, say $\tilde{\beta}$, it becomes an $(n - p)$-form $\star\tilde{\beta}$ and the wedge product of $\tilde{\alpha}$ and $\star\tilde{\beta}$ is an n-form—or a scalar. This construction yields a pointwise inner product (over the manifold $\mathcal{M}^n$) of the two differential p forms given by

$$\tilde{\alpha} \wedge \star\tilde{\beta} = \langle\tilde{\alpha}, \tilde{\beta}\rangle \, \mathrm{vol}^n \tag{2.54}$$

where the inner product on forms is provided by the (dual) metric tensor,[10] and the **canonical volume element** (or the *metric volume element*) vol^n is an n-form scaled by the metric tensor determinant, or

$$\mathrm{vol}^n = \sqrt{|g|}\,\mathrm{d}x^1 \wedge \cdots \wedge \mathrm{d}x^n. \tag{2.55}$$

The scaling by $\sqrt{|g|}$ ensures that vol^n has unit norm, which is partly why this scale factor appears in the Hodge star definition provided in (2.53). Thus the volume

---

[9] Here the Hodge star is defined as operating only on p-forms, however the Hodge star operator has been recently extended to operate also on p-vectors [189], with several interesting implications.

[10] It is also possible to phrase the inner product between two arbitrary p-forms expressed as the exterior product of p individual 1-forms in terms of the inner product defined on 1-forms [99] with the expression

$$\langle \tilde{u}^1 \wedge \cdots \wedge \tilde{u}^p, \tilde{v}^1 \wedge \cdots \wedge \tilde{v}^p \rangle_p = \det\left(\begin{bmatrix} \langle \tilde{u}^1, \tilde{v}^1 \rangle & \cdots & \langle \tilde{u}^1, \tilde{v}^p \rangle \\ \vdots & \ddots & \vdots \\ \langle \tilde{u}^p, \tilde{v}^1 \rangle & \cdots & \langle \tilde{u}^p, \tilde{v}^p \rangle \end{bmatrix}\right).$$

element can be employed to measure the volume V of an n-dimensional region $\mathcal{R}$ through integration, i.e., $V = \int_{\mathcal{R}} \text{vol}^n$. The canonical volume element arises from a special case of the Hodge star operator, since the Hodge dual of the constant function $f = 1$ is

$$\star 1 = \text{vol}^n. \tag{2.56}$$

The inner product provided by the Hodge star operator can be extended over a compact manifold $\mathcal{M}^n$ to produce a *global scalar product* of two forms given by

$$(\tilde{\boldsymbol{\alpha}}, \tilde{\boldsymbol{\beta}}) \equiv \int_{\mathcal{M}} \tilde{\boldsymbol{\alpha}} \wedge \star \tilde{\boldsymbol{\beta}} = \int_{\mathcal{M}} \langle \tilde{\boldsymbol{\alpha}}, \tilde{\boldsymbol{\beta}} \rangle \text{vol}^n. \tag{2.57}$$

Note also that if the orientation changes, vol^n reverses sign, and hence the sign of $\star \tilde{\boldsymbol{\beta}}$ also changes.

Example 2.5 Continuing from Example 2.4, if $\tilde{\boldsymbol{\alpha}}$ is a 1-form given by $\tilde{\boldsymbol{\alpha}} = a_1 dx^1 + a_2 dx^2 + a_3 dx^3$, $\tilde{\boldsymbol{\beta}}$ is a 1-form given by $\tilde{\boldsymbol{\beta}} = b_1 dx^1 + b_2 dx^2 + b_3 dx^3$, and the metric tensor g is simplified such that only the diagonal entries g^{ii} are non-zeros, then

$$\tilde{\boldsymbol{\alpha}} \wedge \star \tilde{\boldsymbol{\beta}} = \frac{1}{\sqrt{g^{11} g^{22} g^{33}}} (a_1 b_1 g^{11} + a_2 b_2 g^{22} + a_3 b_3 g^{33}) dx^1 \wedge dx^2 \wedge dx^3$$

$$= \frac{1}{\sqrt{g^{11} g^{22} g^{33}}} (\mathbf{a} \mathbf{G}^{-1} \mathbf{b}^{\mathsf{T}}) dx^1 \wedge dx^2 \wedge dx^3,$$

where $\mathbf{a} = [a^1, a^2, a^3]$ and $\mathbf{b} = [b^1, b^2, b^3]$.

If we consider the special case of $\mathbb{R}^3$, the Hodge star operator and the wedge product combined provide several common vector identities from traditional vector calculus. If we assume a Euclidean metric so that any contravariant vector $\bar{\boldsymbol{v}}$ can be associated with a covector $\tilde{\boldsymbol{v}}$ with identical components, then it is straightforward to demonstrate that for contravariant vectors $\bar{\boldsymbol{u}}$, $\bar{\boldsymbol{v}}$, and $\bar{\boldsymbol{w}}$,

$$\bar{\boldsymbol{u}} \cdot \bar{\boldsymbol{v}} = \star(\tilde{\boldsymbol{u}} \wedge \star \tilde{\boldsymbol{v}}) = \star(\tilde{\boldsymbol{v}} \wedge \star \tilde{\boldsymbol{u}}), \tag{2.58a}$$

$$\bar{\boldsymbol{u}} \times \bar{\boldsymbol{v}} = \star(\tilde{\boldsymbol{u}} \wedge \tilde{\boldsymbol{v}}), \tag{2.58b}$$

and

$$\bar{\boldsymbol{u}} \cdot (\bar{\boldsymbol{v}} \times \bar{\boldsymbol{w}}) = \star(\tilde{\boldsymbol{u}} \wedge \tilde{\boldsymbol{v}} \wedge \tilde{\boldsymbol{w}}). \tag{2.58c}$$

Similarly, with the exterior derivative and Hodge star operators, we may produce operations identical to the divergence and curl operators. The key to translating the exterior derivative into an operation identical to these familiar operators is to note that, in standard vector calculus, there are only 1-vectors! In this setting, the curl operator maps a 1-vector into a 1-vector, and the divergence operator maps a 1-vector into a 0-vector. As we saw above in Examples 2.2 and 2.3, the curl arises from

2.2 Differential Forms

the exterior derivative of a 1-form and therefore produces a 2-form, whereas the divergence arises from the exterior derivative of a 2-form and therefore produces a 3-form. Thus the Hodge star is required to generate the standard vector calculus versions of these operators.[11] For a 1-form $\tilde{\alpha}$ in $\mathbb{R}^3$, the curl operator is then given by

$$\star d\tilde{\alpha} \Leftrightarrow \nabla \times \bar{\alpha} \tag{2.59a}$$

and the divergence operator is given by

$$\star d \star \tilde{\alpha} \Leftrightarrow \nabla \cdot \bar{\alpha}. \tag{2.59b}$$

The first expression results in a 1-vector and the second expressions results in a 0-vector or a scalar. Note that the inclusion of the Hodge star operator in these expressions for curl and divergence does *not* imply that these two operators require metric information, but rather the Hodge star provides the necessary conversion into the precise definitions in standard vector calculus. As before, to compute either the 1-form that is the "gradient" of a 0-form, the 2-form that is the "curl" of a 1-form, or the 3-form that is the "divergence" of a 2-form requires only an application of the exterior derivative, therefore these operations are purely topological in nature and require no metric information.

The Hodge star thus incorporates geometric information, via the metric tensor, into the language of differential forms, extending the framework beyond the purely algebraic and topological operations provided by the wedge product and exterior derivative. In the context of physics, this metric information is typically embodied by the *constitutive laws* appropriate to the physical domain. These laws, which are interpreted as expressing the metric tensor in each domain, will be discussed further in the next section.

> The *Hodge star operator* $\star$ is a linear mapping from p-forms to $(n-p)$-forms that requires a *metric tensor* and an *orientation* on the underlying manifold. It provides a scalar product between forms, and is necessary to complete the correspondence between standard vector calculus and the exterior calculus of differential forms.

The composite operator of the Hodge star and the exterior derivative, $\star d\star$, used in the above expression for divergence has interesting properties with respect to the global scalar product on forms, and will now be explored in greater detail.

---

[11] Here the notation $\Leftrightarrow$ is meant to convey that these forms and vectors correspond. Oftentimes the vectorfield corresponding to a form is termed a "proxy" field [48]. Of course it is possible to map a form to its corresponding vector using the metric tensor, but for this discussion we do not require this level of detail.

2.2.3.1 The Codifferential Operator and the Laplace–de Rham Operator

The exterior derivative operator, discussed above, maps p-forms to $(p+1)$-forms. The Hodge star enables the definition of a new differential operator that maps in the opposite direction. The **codifferential operator** $d^* : \bigwedge^p \to \bigwedge^{p-1}$ (sometimes denoted as δ) is defined as

$$d^* \equiv (-1)^{n(p+1)+1} \star d \star . \tag{2.60}$$

It appeared in the expression for the vector divergence given above in (2.59b). It can be shown that this operator is the adjoint of the exterior derivative with respect to the global scalar product defined above (up to a boundary term), i.e., for a $(p-1)$-form $\tilde{\alpha}$ and p-form $\tilde{\beta}$,

$$(d\tilde{\alpha}, \tilde{\beta}) = (\tilde{\alpha}, d^*\tilde{\beta}). \tag{2.61}$$

As with the exterior derivative, a p-form $\tilde{\beta}$ is said to be *coclosed* if it is in the kernel of the codifferential operator, i.e., if $d^*\tilde{\beta} = 0$, and *coexact* if it is the codifferential of a $(p+1)$-form $\tilde{\alpha}$, i.e., if $\tilde{\beta} = d^*\tilde{\alpha}$. Also, the Poincaré lemma applies to the codifferential operator thus $d^*d^* = 0$.

Example 2.6 (Adjointness of grad and $-\mathrm{div}$ in two dimensions) Consider a 0-form f and a 1-form $\tilde{\omega}$ defined on a manifold $\mathcal{M} \subset \mathbb{R}^2$, and the 2-form given by

$$d(f \wedge \star\tilde{\omega}) = (df) \wedge \star\tilde{\omega} + f \wedge (d \star \tilde{\omega})$$

at a point $p \in \mathcal{M}$. If this expression is integrated over all of $\mathcal{M}$, we see that

$$\int_{\partial \mathcal{M}} f \wedge \star\tilde{\omega} = \int_{\mathcal{M}} d(f \wedge \star\tilde{\omega})$$

$$= \int_{\mathcal{M}} (df) \wedge \star\tilde{\omega} + \int_{\mathcal{M}} f \wedge (d \star \tilde{\omega})$$

$$= (df, \tilde{\omega}) + (f, \star\star\star d \star \tilde{\omega})$$

$$= (df, \tilde{\omega}) - (f, \star d \star \tilde{\omega}).$$

If $\mathcal{M}$ is closed such that it is boundaryless, i.e., $\partial \mathcal{M} = 0$, then

$$(df, \tilde{\omega}) = (f, \star d \star \tilde{\omega}) = (f, d^*\tilde{\omega})$$

which expresses the adjointness of d and d^* with respect to this scalar product. It also demonstrates that the adjointness of the vector calculus operators grad and $-\mathrm{div}$ is a special case of the adjointness of d and d^*.

Using the codifferential we may define an operator on forms that generalizes the Laplacian operator from vector calculus. The **Laplace–de Rham operator** is a

2.2 Differential Forms

mapping from p-forms to p-forms, denoted as $\mathbf{\Delta} : \bigwedge^p \to \bigwedge^p$, and formally defined as

$$\mathbf{\Delta} \equiv dd^* + d^*d = (d+d^*)^2 \tag{2.62}$$

in terms of the exterior derivative and codifferential operators.[12] The Laplace–de Rham operator is thus self-adjoint with respect to the global scalar product, i.e., $\mathbf{\Delta} = \mathbf{\Delta}^*$. In $\mathbb{R}^3$, the Laplace–de Rham operator on 0-forms is identical up to a sign to the Laplacian operator on scalars, i.e.,

$$\mathbf{\Delta} f = -\nabla^2 f, \tag{2.63}$$

and to the Laplacian operator on vectors, i.e., if we assume a Euclidean metric and assign $\tilde{\tau} \Leftrightarrow \vec{\tau}$,

$$\mathbf{\Delta}\tilde{\tau} \Leftrightarrow -\vec{\nabla}^2\vec{\tau} = -(\nabla\nabla\cdot\vec{\tau} - \nabla\times\nabla\times\vec{\tau}). \tag{2.64}$$

Through the Laplace–de Rham operator, several concepts of potential theory carry over to the domain of differential forms. A **harmonic form** is a p-form $\tilde{\omega}$ that is in the nullspace of the Laplace–de Rham operator such that $\mathbf{\Delta}\tilde{\omega} = 0$. Note that on a closed (i.e., boundaryless) manifold $\mathbf{\Delta}\tilde{\omega} = 0$ if and only if $d\tilde{\omega} = 0$ and $d^*\tilde{\omega} = 0$, thus harmonic forms on a closed manifold are both closed and coclosed.

The Laplace–de Rham operator plays a central role in Hodge theory, which is described in detail in Appendix C. In summary, a result of **Hodge's Theorem** is that any p-form on a closed manifold can be written as the sum of an exact form, a coexact form, and a harmonic form, and the decomposition of the p-form into these three components is known as its **Hodge decomposition**. This decomposition is a generalization of the Helmholtz decomposition of vector fields, which is also discussed in detail in Appendix C.

2.2.4 Differential Forms and Linear Pairings

To summarize the brief review of differential forms, we have:

- established a duality between contravariant p-vectors and covariant p-forms;
- defined an exterior derivative operator using only the topological structure of the manifold;
- shown that a coupling exists between differential forms and the p-domains over which they are integrated with the Generalized Stokes' Theorem; and
- have seen how the Hodge star, through exploiting a metric structure imposed on a manifold, can provide a correspondence between standard tools in traditional vector calculus with the framework of differential forms.

---

[12] The *Laplace–Beltrami operator* is a special case of the Laplace–de Rham operator restricted to $p = 0$ and defined on Riemannian manifolds.

One theme of this review has been to establish linear pairings between objects—including the contraction of a p-vector with a p-form, the integration of a form over a domain, and the global scalar product of p-forms—and the adjointness relations of d and ∂ and of d and d* that arise from these pairings. These relations will play a larger role in the discrete implementation of exterior calculus, which is the topic of the next section.

2.3 Discrete Calculus

Now that we have reviewed the essential components of the generalization of standard vector calculus provided by the exterior calculus, we are positioned to develop the analogous concepts in a purely discrete setting. The broad concept of a discrete calculus has been approached by many different authors in a wide variety of contexts (e.g., [40, 47, 59, 103, 105, 111, 118, 178, 200, 232, 275, 307, 323, 360, 364, 379, 388, 400, 403]), and the formulation that we present in this section has been assembled from several of these sources. Given the finite-dimensional vector spaces forming the substrate of discrete manifolds, the discrete differential form can be defined just as it is in the continuous setting. Here we emphasize that a distinction must be drawn between formulating an inherently *discrete* version of differential forms and an alternative approach in which discrete representations of continuous differential forms are defined—an approach that we term the *discretized* version. While the discretized approach is extremely useful for solving many practical problems in computational physics [47, 178], it views the discrete representations as being samples of underlying continuous fields, whereas the viewpoint here and in the rest of this book is that the fields are *intrinsically* discrete—meaning there is no underlying continuous field or continuous operations that we seek to approximate, but rather the discrete fields themselves are the main objects of interest.

The discrete formulation of differential forms retains all of the features outlined above, including the coordinate-free expression of differential forms and the exterior derivative as well as the independence of the exterior calculus from a metric. One strong advantage of the discrete formulation is that the purely topological nature of the exterior derivative—and the tight relationship between the derivative and the boundary operator—is made far more apparent and concrete.

We begin the translation of the exterior calculus on differential forms into the discrete setting by reconsidering the components of the Generalized Stokes' Theorem,

$$\int_S d\tilde{\omega} = \int_{\partial S} \tilde{\omega},$$

which lays the foundation of calculus.

Recall that the Generalized Stokes' Theorem is a generalization of the Fundamental Theorem of Calculus—which is typically phrased in terms of 0-forms or functions—to higher-order forms and domains of integration. The discrete calculus must reproduce all aspects of this Theorem if we are to construct a formally analogous calculus. Notice that this expression contains several components, including: the domain of integration S; the boundary operator ∂; an orientation on the domain S; the form $\tilde{\omega}$; the derivative operator d; and the bilinear pairing given by the integral itself. In addition to these basic ingredients (which, through their definition, will provide the algebraic and topological structure of the discrete calculus), we will also require a discrete version of the Hodge star operator and a metric tensor to enable a complete discrete calculus on manifolds.

A description of discrete calculus can be provided in terms of chains, cochains and the boundary and coboundary operators, all of which are defined below. However, the goal of the following discussion is to simultaneously introduce the *representations* of the relevant ingredients in terms of linear algebra—namely, matrix operators and vector representations. Although it is not surprising that operators in such discrete systems may be considered as systems of linear equations, the matrix formulation of discrete calculus enables powerful tools for evaluating operations and for solving large problems. Because these representations arise so naturally in discrete calculus, throughout this discussion we will refer to the more abstract discrete operators and their concrete representations and data structures interchangeably.

2.3.1 Discrete Domains

We begin by considering the discrete analogue of the domain. A discrete domain will be represented by a **cell complex**, which is comprised of a collection of finite-dimensional vector spaces of p-cells. A **p-cell**, σ^p is defined as a set of points which is homeomorphic to a closed, unit p-ball $\mathcal{B}_p$, i.e.,

$$\mathcal{B}_p = \{x \in \mathbb{R}^p | \|x\| \leq 1\}. \tag{2.65}$$

The *boundary*, $\partial \sigma^p$, of the p-cell σ^p is the portion of the cell which is mapped by the homeomorphism to the boundary of the unit ball, i.e.,

$$\partial \mathcal{B}_p = \{x \in \mathbb{R}^p | \|x\| = 1\}. \tag{2.66}$$

A p-cell may be represented by an ordered set of vertices comprising a convex p-polytope. We denote the set of all p-cells as S_p. The simplest, low-order p-cells are the 0-cells or **nodes**, 1-cells or **edges**, and 2-cells or **faces**. Because these three types of cells will play the most prominent role in the applications discussed in later chapters, we denote these sets with the special notation: $\mathcal{V}$ represents the set of nodes or "vertices", $\mathcal{E}$ the set of edges, and $\mathcal{F}$ the set of faces. The ith element of these sets will be denoted by $v_i \in \mathcal{V}$, $e_i \in \mathcal{E}$, and $f_i \in \mathcal{F}$. The notation e_{ij} will also be used to describe an (oriented) edge that connects nodes v_i and v_j.

A **graph** is a special case of a cell complex that is comprised of only nodes (i.e., 0-cells) and edges (i.e., 1-cells), and is therefore a 1-complex. Thus, the framework of discrete calculus developed here applies to graphs as well as more general

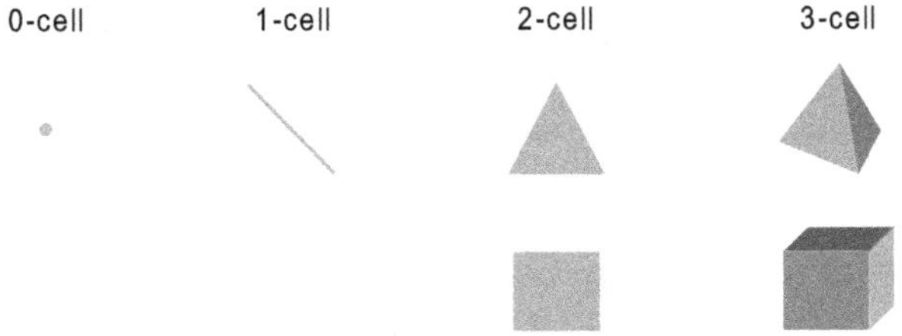

Fig. 2.2 Examples of cells of order 0 through 3. The *top row* examples are all simplices, while the *bottom row* examples also include more general p-cells

p-complexes. This provides the possibility of exploiting the many results and tools from the rich fields of algebraic and computational graph theory to solve many sophisticated problems in the present context of calculus.

The boundary of a p-polytope consists of a set of $(p-1)$-polytopes (defined for $p > 0$), thus at least $p + 1$ vertices are required to define a p-polytope. When a p-cell consists of *exactly* $p + 1$ vertices, it is called a p-**simplex**. Note how this definition of a cell and its boundary excludes the possibility for the cell's boundary to self-intersect. Therefore, a 1-cell for which the endpoints are the same 0-cell (i.e., a self-loop) is excluded from the definition and therefore not allowed as a cell. Consequently, none of the discrete domains that we consider will possess self-loops. In graph theory, a fully-connected set of nodes is called a **clique**,[13] where a p-clique indicates a clique comprised of p nodes. Similarly, a p-simplex is also fully connected with $p + 1$ nodes.

The terminology for various p-cells varies widely across disciplines. The basic building blocks of all p-cells are 0-cells. From a geometric viewpoint, the 0-cells are commonly called *vertices* or *points*, while in graph theory the 0-cells are typically called *nodes*. More abstractly, the 0-cells can be thought of as any set consisting of one single element. The 1-cells consist of pairs of 0-cells and are variously referred to as *edges* (in the context of graph theory) or *branches* (in the context of circuit theory). A 2-cell is composed of at least three vertices and is variously referred to as a *face* or *facet* (in the context of geometry), *cycle* (in graph theory), and *mesh* or *loop* (in the context of circuit theory). Since the terms "mesh" and "loop" appear with different definitions in different fields, we will generally avoid these terms. Higher order p-cells are generally referred to simply as p-cells. We will refer to p as the *order* of the p-cell. Figure 2.2 provides various examples of cells, both simplices and non-simplices.

A collection of cells defines a **cell complex** if the collection satisfies the following two requirements.

1. The boundary of each p-cell (for $p > 0$) is comprised of the union of lower-order p-cells.[14]

---

[13] Cliques play an important role in Markov Random Fields (due to the Hammersley–Clifford Theorem [31]) and the identification of cliques with simplices allows us to consider p-cliques as geometric objects with dimension p.

[14] These boundary cells are sometimes referred to as the *faces* of the p-cell, which can be confused with the usage of the term *face* to describe a 2-cell. We shall reserve the term "face" for a 2-cell only.

2.3 Discrete Calculus

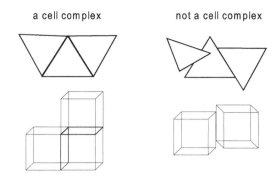

Fig. 2.3 A cell complex is a collection of cells which are joined together to satisfy certain rules. The *left column* shows a 2-complex and a 3-complex, while the *right column* show cells which were improperly glued together and therefore do not comprise a cell complex

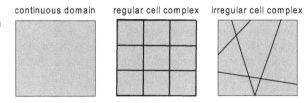

Fig. 2.4 A cell complex may be considered as a tessellation of a continuous domain, where the tessellation may be either regular or irregular

2. The intersection of any two cells is either empty or a boundary element of both cells.

Figure 2.3 depicts examples of cells joined to create a valid cell complex, and cells which were joined so as to violate the conditions for a valid cell complex. A complex is referred to as a p-complex if p is the maximum order of all cells in the complex.

From a discretization viewpoint, a p-complex may be viewed as a tessellation of a p-manifold, as depicted in Fig. 2.4. Throughout this and later chapters, we assume that the complex under consideration is connected unless stated otherwise. When the complex is not connected then, without loss of generality, each connected complex may be treated separately.

Typically, the 0-cells of a complex are considered as vertices of a discrete manifold and as such are thought of as being *embedded* in some extrinsic embedding space, i.e., to each vertex is assigned a coordinate in n dimensions. This embedding of the vertices allows one to define other features of the complex, like orientation and duality (discussed below), in terms of the ambient space into which the complex is embedded. However, major applications for graph theory and discrete calculus have no natural geometric embedding (e.g., communication networks, social networks, gene regulatory networks, and the World Wide Web). Since our goal is to provide the tools of discrete calculus so that they may be applied to the broadest class of real-world problems, our approach will be to avoid any dependence on an embedding in our definitions and exposition.

2.3.1.1 Orientation

We will require that each p-cell is *oriented*, and the orientation is specified by the ordering of nodes used to represent the p-cell. Any p-cell (regardless of the order p) may have one of two possible orientations. Intuitively, an orientation establishes a default direction on a cell which is considered positive or negative. For example, flow passing through an oriented edge in the same direction as its orientation is considered positive while flow passing through the edge in the opposite direction is considered negative. Note however that an *oriented graph is different from a directed graph*. An orientation on an edge defines which direction of flow is considered positive or negative, while a *directed* edge only permits flow in one direction. This distinction is very important since all of the edges in this book will be *oriented*, but unless otherwise stated the edges will be considered *undirected*.

We begin by defining orientation for a p-simplex and then show how this definition may be extended to general p-cells. When a p-cell is a p-simplex, we may provide a formal definition of orientation. A p-simplex consists of $p+1$ nodes and the orientation may be defined by an *ordering* of those nodes as the list

$$\tau_p = \{\sigma_0^0, \sigma_1^0, \ldots, \sigma_{p+1}^0\}. \tag{2.67}$$

If we exchange the position of any two nodes in the ordering an odd number of times (an odd permutation), then we say that the orientation has changed with respect to the initial orientation. However, an even number of exchanges (even permutation) keeps the orientation the same as the initial orientation. We consider two orientations to be the same (called *coherent*) if one can be obtained from the other by an even permutation. For completeness, a 0-cell is considered to have two orientations ("sourceness" and "sinkness") although it is defined by only a single node. Conventionally, all nodes are given the same orientation, "sourceness", meaning that the negative (sink) end of an edge will not be coherent with the orientation of a node, while the positive (source) end of an edge will be coherent with the orientation of a node.

We can see how this purely combinatorial specification of orientation defines orientation in an intuitive way by considering Fig. 2.5. An edge may be considered as pointing from node A to node B or from node B to node A. If the edge points from node A to node B, then flow from A to B is considered *positive* and flow from B to A is considered *negative*. Similarly, a triangle (i.e., a face or a 2-simplex) may take a *clockwise* or a *counterclockwise* orientation. If the triangle is oriented clockwise, then a clockwise rotation around the triangle is considered *positive* and a counterclockwise rotation is considered *negative*. The clockwise orientation is represented by listing the triangle's points in the order (A, B, C) and the counterclockwise orientation may be obtained by exchanging points A and C to give the ordering (C, B, A). If a second exchange were made on this ordering, by exchanging C and B, then the ordering (B, C, A) would again represent a clockwise orientation. The same concepts apply to the orientation of a 3-simplex in which the orientation may be viewed intuitively as the concept of "screw-sense" about the tetrahedron embodied in the

2.3 Discrete Calculus

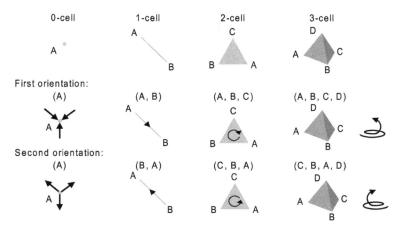

Fig. 2.5 A simplex is given an orientation by ordering its vertices. Given a reference ordering, all orderings obtained by an even permutation are considered to be the same as the reference orientation, while all orderings obtained by an odd permutation are considered to be opposite to the reference orientation. The orientation of a 1-simplex corresponds to a notion of direction, while the orientation of a 2-simplex corresponds to a notion of clockwise/counterclockwise and the orientation of a 3-simplex corresponds to the screw sense of the volume. For completeness, 0-simplices are considered to have two orientations, "sinkness" and "sourceness"

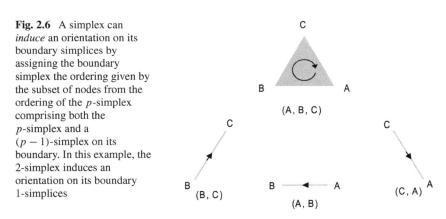

Fig. 2.6 A simplex can *induce* an orientation on its boundary simplices by assigning the boundary simplex the ordering given by the subset of nodes from the ordering of the p-simplex comprising both the p-simplex and a $(p-1)$-simplex on its boundary. In this example, the 2-simplex induces an orientation on its boundary 1-simplices

familiar "right-hand rule" from physics. Therefore, the opposite orientation on a 3-simplex would be defined by the "left-hand rule".

An oriented simplex is said to *induce* an orientation on a boundary simplex by assigning the boundary simplex the ordering given by the subset of nodes from the ordering of the p-simplex comprising both the p-simplex and a $(p-1)$-simplex on its boundary. For example, Fig. 2.6 depicts the induced orientation of a triangle (2-cell) on the edges (1-cells) comprising its boundary. Therefore, in Fig. 2.7 the induced orientation of the cycle on the edges is coherent with the reference orientation for edges e_1 and e_2, while the induced orientation is not coherent for edges e_3 and e_4.

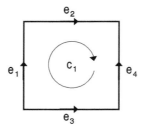

Fig. 2.7 Every p-cell has a boundary comprised by p-cells. In this example, the boundary of this 2-cell is comprised of four 1-cells. The induced orientation of the cycle on the edges is coherent with the reference orientation for edges e_1 and e_2, while the induced orientation is not coherent for edges e_3 and e_4

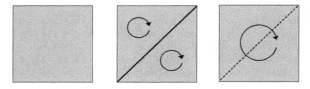

Fig. 2.8 The ordering of a cell which is not a simplex may be produced by decomposing the cell into simplices and orienting the simplices to agree on the internal boundaries. In this example, two 2-simplices are used to orient a 2-cell

Two p-cells may be merged to comprise a single oriented p-cell when the following conditions are satisfied:

1. The two p-cells share exactly one $(p-1)$-cell boundary element.
2. The shared $(p-1)$-cell boundary element is not a boundary element of any other p-cell in the complex.
3. The two p-cells induce opposite orientations on the shared $(p-1)$-cell boundary element.

Consequently, a general p-cell may be oriented by subdividing the cell into internal p-simplices (a **simplicial decomposition**). By orienting a single internal simplex, the third condition above effectively forces all of the other internal simplices to assume an orientation and therefore the orientation propagates to the entire cell [379]. Figure 2.8 illustrates how this simplicial decomposition is used to orient a cell.

2.3.1.2 The Incidence Matrix

Given these definitions of the orientation of p-cells, then we may represent the structure of the complex algebraically via an **incidence matrix**. The incidence matrix $\mathbf{N}_p^T$, which encodes which p-cells are incident to which $(p-1)$-cells in the n-complex,

2.3 Discrete Calculus

is defined as

$$\mathbf{N}_p(i,j) = \begin{cases} 0 & \text{if } \sigma_j^{p-1} \text{ is not on the boundary of } \sigma_i^p, \\ +1 & \text{if } \sigma_j^{p-1} \text{ is coherent with the induced orientation of } \sigma_i^p, \\ -1 & \text{if } \sigma_j^{p-1} \text{ is not coherent with the induced orientation of } \sigma_i^p. \end{cases}$$
(2.68)

Therefore, the collection of n incidence matrices fully describe the boundary and orientation structure of an n-complex. The p-incidence matrix $\mathbf{N}_p^\mathsf{T}$, defined for $0 < p \leq n$, thus consists of $n_{p-1} \times n_p$ elements, where $n_p = |\mathcal{S}_p|$ and $\mathcal{S}_p$ is the set of p-cells in the complex. For a graph (a 1-complex), this definition agrees with the standard definition of *the* incidence matrix when $p = 1$, which is an $n \times m$ matrix (where $n = |\mathcal{V}|$ and $m = |\mathcal{E}|$) whose rows index into the node set and whose columns index into the edge set (i.e., each row corresponds to a node and each column corresponds to an edge). Since we consider more general complexes, we will refer to the $p = 1$ incidence matrix as the *node–edge* incidence matrix and the $p = 2$ incidence matrix as the *edge–face* incidence matrix. Since these two matrices play a dominant role in this book, we use special notation for these matrices. Following Strang [360], we let $\mathbf{N}_1^\mathsf{T} = \mathbf{A}^\mathsf{T}$ represent the *node–edge* incidence matrix and $\mathbf{N}_2^\mathsf{T} = \mathbf{B}^\mathsf{T}$ represent the *edge–face* incidence matrix.[15]

In all of the subsequent discussion, we assume that we are treating an oriented n-complex defined with corresponding incidence matrices. In other words, the cells are all assigned a unique, arbitrary orientation which acts as a reference orientation for subsequent quantities.

2.3.1.3 Chains

The incidence matrix moves us from the subject of topology toward algebra since the matrix represents the topological structure of the complex algebraically. In order to define a *domain of integration* analogous to the concept used in conventional calculus, we must take a second step toward algebra by defining a p-chain.

A p-**chain** is an n_p-tuple of scalars which assigns a coefficient to each p-cell, where n_p is the number of distinct p-cells in the complex. Because the space of p-cells is finite-dimensional (since there is only a finite number of p-cells in a complex), a p-chain can alternately be viewed as a *vector* in this space, and each component of a p-chain is the coefficient of a single p-cell. In this context, the collection of p-cells comprises the basis set, and thus each p-cell is referred to as a **basic p-chain**, and denote the vector space of p-chains as C_p. Typically a p-chain is represented as a one-dimensional array or a column vector, $\boldsymbol{\tau}_p$, that is always an $n_p \times 1$ vector with zeros in the entries corresponding to p-cells not included in the chain.

---

[15] We define the incidence matrix as the transpose of $\mathbf{A}$ and $\mathbf{B}$ since we will see in the next section that $\mathbf{A}$ and $\mathbf{B}$ play a more prominent role than the incidence matrices $\mathbf{A}^\mathsf{T}$ and $\mathbf{B}^\mathsf{T}$.

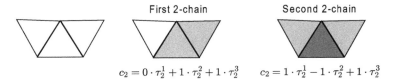

Fig. 2.9 A *chain* defines an oriented domain of integration in an oriented complex. Given the 2-complex on the *left*, the first 2-chain indicates the rightmost two 2-cells for integration (shaded in *blue*), while the second 2-chain indicates all of the cells for integration (with the middle cell integrated in reverse orientation, shaded in *red*)

The p-chain is intended to represent some portion of the p-complex (analogous to the relationship between a domain of integration and the entire domain). Therefore each coefficient of the p-chain can be considered as representing the inclusion of the p-cell in the domain of integration and its orientation within the p-chain (relative to its reference orientation). In other words, a p-chain may be viewed as an (oriented) **indicator vector** representing a set of p-cells, in which values greater than 1 represent greater multiplicity in the set. A given p-chain τ_p can therefore be expressed as

$$\tau_p = \sum_{i=1}^{n_p} a^i \sigma_i, \quad \sigma_i \in C_p, \ a^i \in \mathbb{R}. \tag{2.69}$$

Additionally, the sign of each coefficient a^i indicates the orientation, i.e.,

$$\tau_p(\sigma_i) = -\tau_p(\sigma_i'), \quad \text{if } \sigma_i \text{ and } \sigma_i' \text{ are opposite orientations of a } p\text{-cell.} \tag{2.70}$$

Two p-chains may be added together by adding the coefficients of each p-cell basis element. The resulting group may be called the group of oriented p-chains [290]. Figure 2.9 gives examples of 2-chains on a small 2-complex. From the standpoint of calculus, these chains may be considered as defining the (oriented) domain of integration, and we may identify a p-chain with the p-vectors described above in the theory of differential forms.

2.3.1.4 The Discrete Boundary Operator

Any p-chain may be represented as an $n_p \times 1$ column vector, and the incidence matrix is an $n_{p-1} \times n_p$ matrix that stores the incidence relations between p-cells and $(p-1)$-cells. The incidence matrix therefore naturally maps p-chains into their corresponding boundary elements. In other words, when the incidence matrix $\mathbf{N}_p^\mathsf{T}$ is applied to a p-chain, the result is a $(p-1)$-chain, i.e.,

$$\tau_{p-1} = \mathbf{N}_p^\mathsf{T} \tau_p. \tag{2.71}$$

Remarkably, this chain τ_{p-1} represents the oriented set of $(p-1)$-cells on the *boundary* of the cells represented by the chain τ_p. Consequently, the incidence matrix is the *matrix representation* of the **discrete boundary operator**, i.e., $\mathbf{N}_p^\mathsf{T} : C_p \rightarrow$

2.3 Discrete Calculus

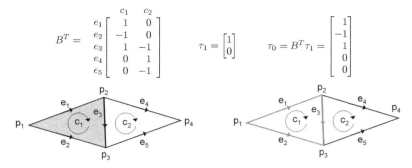

Fig. 2.10 The incidence matrix may be used as an operator to find the boundary of a chain. Therefore, the incidence matrix defines the *boundary* operator on a complex. In this example, the chain τ_1 consists of one cycle in the complex, c_1. The application of the boundary operator $\mathbf{B}^T$ to τ_1 yields τ_0, which represents the oriented edges $\{e_1, e_3, -e_2\}$ comprising the boundary of the cycle c_1

C_{p-1}. The incidence matrix provides us with a purely algebraic representation of the boundary—the boundary of a chain is defined algebraically as the result of applying the incidence relations captured by the incidence matrix to the chain. In other words, the incidence matrix $\mathbf{N}_p^T$ is a natural representation of the boundary operator ∂_p on p-chains. Figure 2.10 demonstrates this effect of the incidence matrix as boundary operator on a small 2-complex using the edge–face incidence matrix. As before, the boundary of a boundary is algebraically zero, i.e., $\mathbf{N}_{p-1}^T \mathbf{N}_p^T = 0$. Also, a p-chain $\boldsymbol{\beta}_p$ in the nullspace of the boundary operator, such that $\mathbf{N}_p^T \boldsymbol{\beta}_p = 0$, is called a *cycle*, and a $(p-1)$-chain $\boldsymbol{\alpha}_{p-1}$ that is in the range of the boundary operator, i.e, such that $\boldsymbol{\alpha}_{p-1} = \mathbf{N}_p^T \boldsymbol{\beta}_p$ for some p-chain $\boldsymbol{\beta}_p$, is called a *boundary*.

> The incidence matrix $\mathbf{N}_p^T$ provides both (i) a representation (or data structure) of the *topology* of the discrete manifold and (ii) a representation of the *boundary operator* ∂_p.

This conception of the incidence matrix as the boundary operator allows us to offer a intuitive understanding of the algebraic properties of the incidence matrix, which are used throughout the book. These same results are achieved elsewhere through more algebraic means [34, 36]. Consider the node–edge boundary operator $\mathbf{A}^T$. This matrix is of size $m \times n$ where $m = |\mathcal{E}|$ and $n = |\mathcal{V}|$ (using the conventional notation in graph theory). The right nullspace of $\mathbf{A}^T$ is of rank $m - n + 1$. To understand this fact, first consider that a 1-chain τ_1 which represents a cycle has no boundary, i.e., $\mathbf{A}^T \tau_1 = 0$. By definition, a graph which is a **spanning tree** has no cycles and connects every node. The number of edges required to connect every node without cycles is $n - 1$. Therefore, when a graph contains no cycles, $m = n - 1$ and the rank of the right nullspace of $\mathbf{A}^T$ is of rank 0. For every edge added to a spanning tree, a new linearly independent cycle is defined, which adds to the dimension of the right nullspace of $\mathbf{A}^\mathsf{T}$. Therefore, the difference between m and $n-1$ determines the number of cycles forming a basis for the nullspace of the boundary operator.

An important aspect of the node–edge incidence matrix $\mathbf{A}^\mathsf{T}$ is the property of **total unimodularity**. A matrix is totally unimodular if the determinant of every submatrix is either zero or ± 1. Total unimodularity is an extremely important property in linear programming, since a totally unimodular constraint matrix (and integer constraints) guarantees that the solution to the linear programming problem will be integer [304]. Consequently, it is possible to solve integer programming problems using efficient linear programming solvers when the constraint matrix is totally unimodular. Nemhauser and Woolsley [291] go further to state that every totally unimodular matrix is the edge–node incidence matrix of some graph with only two exceptions.

We may similarly examine the face–edge incidence matrix $\mathbf{B}^\mathsf{T}$. Since $\mathbf{A}^\mathsf{T}\mathbf{B}^\mathsf{T} = 0$, then each of the column vectors for $\mathbf{B}^\mathsf{T}$ represent cycles. When $\mathbf{B}^\mathsf{T}$ contains $m-n+1$ independent columns, then we may view the complex as *simply connected*. If a complex contains a handle, then its set $\mathcal{F}$ is missing an independent cycle and therefore $\mathbf{B}^\mathsf{T}$ will contain less than $m-n+1$ independent columns. As the boundary operator, the right nullspace of $\mathbf{B}^\mathsf{T}$ is spanned by the space of boundary-less 2-chains. A 2-chain has no boundary when it encloses a volume.

These definitions provide us with an algebraic foundation for representing discrete domains: The oriented cell complex represents the entire domain, the chain represents an oriented subdomain (e.g., as the domain of integration) and the incidence matrix represents the boundary operator. In order to complete Stokes' Theorem on a discrete domain, we now need to define functions on the domain and a derivative operator.

2.3.2 Discrete Forms and the Coboundary Operator

As in the continuous case, we may define a vector space of linear functionals that locally map p-chains to scalars at each p-cell in a complex. In the discrete setting, these linear functionals are analogous to differential forms and are thus called p-**cochains**. Intuitively, a p-cochain may be viewed as a function defined on the domain of p-cells. In other words, just as a form is a linear functional that maps vectors to scalars, a cochain is a linear functional that maps chains to scalars. Because of the finite number of basic p-chains, the vector space of p-cochains C^p is also finite-dimensional and each p-cochain can be expressed in terms of the p-cochain basis σ^i as

$$\mathbf{c}^p = \sum_{i=1}^{n_p} a_i \sigma^i, \quad \sigma^i \in C^p,\ a_i \in \mathbb{R}. \tag{2.72}$$

We represent a p-cochain as an $n_p \times 1$ column vector, $\mathbf{c}^p$. As with a p-chain, the sign of each coefficient a_i indicates the orientation, i.e.,

$$\mathbf{c}^p(\sigma^i) = -\mathbf{c}^p(\sigma^{i'}) \tag{2.73}$$

2.3 Discrete Calculus

if σ^i and $\sigma^{i'}$ are opposite orientations of the same p-cell. Thus there is a single basic p-chain and a single basic p-cochain defined at each p-cell of the complex, and therefore the basis sets of p-cochains and p-chains are biorthogonal. For this reason, the two vector spaces are isomorphic.[16]

Armed with these definitions of a discrete domain and a linear functional on that domain, we can define integration as the pairing of a p-chain with a p-cochain,

$$[\![\mathbf{c}^p, \boldsymbol{\tau}_p]\!] = \sum_{i=1}^{n_p} \tau_p(\sigma_i) \mathbf{c}^p(\sigma^i) = \mathbf{c}^p \boldsymbol{\tau}_p, \qquad (2.74)$$

producing a scalar quantity. This pairing can be thought of as applying the p-cochain (i.e., a linear functional) to the p-chain at each p-cell of the complex then adding up the resulting scalars. Because both the chain and cochain are represented as vectors, integration is therefore defined as the sum of the element-by-element product of the vector components.[17] We may then consider including the boundary operator $\mathbf{N}_p^\mathsf{T}$ into this pairing and consider the adjoint of the boundary operator, $(\mathbf{N}_p^\mathsf{T})^*$, defined through the expression

$$[\![(\mathbf{N}_p^\mathsf{T})^* \mathbf{c}^{p-1}, \boldsymbol{\tau}_p]\!] = [\![\mathbf{c}^{p-1}, \mathbf{N}_p^\mathsf{T} \boldsymbol{\tau}_p]\!]. \qquad (2.75)$$

Note that this adjointness relation is simply an expression of the Generalized Stokes' Theorem! We may then define the discrete version of the exterior derivative on p-cochains, d_p, as the **coboundary operator**, which is the adjoint of the boundary operator, or $d_p = (\partial_{p+1})^*$, and with a natural matrix representation: $d_p \equiv (\mathbf{N}_{p+1}^\mathsf{T})^*$. Since the boundary operator is represented by a real matrix, and this pairing is simply the scalar product of a chain and cochain, its adjoint is simply the transpose, thus the discrete coboundary operator is given by $\mathbf{N}_p : C^{p-1} \to C^p$.

As with the boundary operator, it is instructive to consider the algebraic properties of the coboundary operator applied to 0-cochains and 1-cochains. If we consider the coboundary operator on 0-cochains $\mathbf{A}$ (which is the transpose of the node–edge incidence matrix $\mathbf{A}^\mathsf{T}$) the right nullspace of $\mathbf{A}$ is dimension r where r indicates the number of connected components in the graph. To understand this fact, consider which sets of nodes have no coboundary (i.e., edges connecting the nodes to other nodes). If the 0-cochain $\mathbf{c}^0(v_i) = 1$ for all $v_i \in \mathcal{V}$, then $\mathbf{Ac}^0 = 0$ since there is no boundary for the entire domain (i.e., no edges connect the entire set of nodes to any extraneous nodes). When $\mathcal{V}$ is composed of two connected components, $\mathcal{V}_1$ and $\mathcal{V}_2$ such that $\mathcal{V}_1 \cap \mathcal{V}_2 = \emptyset$ and $\mathcal{V}_1 \cup \mathcal{V}_2 = \mathcal{V}$, then the 0-cochain $\mathbf{c}^0(v_i) = 1, \forall v_i \in \mathcal{V}_1$ and $\mathbf{c}^0(v_i) = 0, \forall v_i \in \mathcal{V}_2$ will also evaluate to $\mathbf{Ac}^0 = 0$ since there is no coboundary (outgoing edges) from the set represented by the 0-cochain $\mathbf{c}^0$. Therefore, for each

---

[16] In the remainder of the book, we will adopt Strang's [360] notation and adopt **x** to represent 0-cochains while using **y** to represent 1-cochains. However, in this chapter we will continue to use $\mathbf{c}^p$ to represent a p-cochain.

[17] Recall that integration is a pairing of chains and cochains that does not require any metric information whatsoever.

connected component there is a 0-cochain in the nullspace of **A**. If any cochain represents a node set that is *not* a disconnected component then, by definition, there is at least one edge leaving the set. Consequently, $\mathbf{Ac}^0 \neq 0$ unless $\mathbf{c}^0$ is an indicator vector representing a component which is disconnected from the remaining graph. We may also consider the properties of the coboundary operator on 1-cochains **B** (which is the transpose of the edge–face incidence matrix $\mathbf{B}^T$). Similarly, because $\mathbf{BA} = 0$, the right nullspace of **B** is spanned by the $n = |\mathcal{V}|$ independent columns of **A**, corresponding to the nodes. A common, alternative, interpretation of the nullspace of **B** is that it is spanned by the set of all possible *cuts* in the graph (i.e., oriented sets of edges which disconnect the graph). More treatment of these properties may be found in [34, 36].

2.3.3 Primal and Dual Complexes

The dual of a cell complex is an old concept with different meanings and interpretations in the literature. However, the basic idea is that, from a given cell complex (the *primal*), we can construct a related cell complex called the *dual*. The dual of an n-complex is generally viewed geometrically—in the sense that it is typically constructed by placing a node or 0-cell of the dual complex at the barycenter of each n-cell of the primal complex, and then connecting two dual nodes with an edge if the two primal n-cells share an $(n-1)$-cell as a border. Once this identification has been made between the n-cells of the primal complex and the 0-cells of the dual, maintenance of the incidence structure forces every primal p-cell to map to a dual $(n-p)$-cell.

We will define the notion of duality employed in this book by the following condition: If the primal n-complex is constructed such that each $(n-1)$-cell is the boundary of exactly two n-cells (and such that each of these incident n-cells induce an opposite orientation on the shared $(n-1)$-cell), then we can define a dual complex by identifying each p-cell in the primal complex with a unique $(n-p)$-cell in the dual complex. Note that this requirement effectively indicates that the p-complex defines a *combinatorial manifold* [284] that, analogous to a continuous manifold, is everywhere locally homeomorphic to Euclidean space.

Under this definition of the dual, the incidence structure of the primal complex is preserved in the dual. That is, if the duality condition above is true, then *the primal complex incidence matrix $\mathbf{N}_n^T$ is the transpose of the node–edge incidence matrix of the dual complex*, i.e., $\mathbf{N}_n^T = \mathbf{M}_1$. This can be extended to link the incidence matrices between the primal and dual complexes for all degrees of chains. Since

$$\mathbf{N}_n^T \mathbf{N}_{n-1}^T = 0, \tag{2.76}$$

then, if we take a transpose of this expression we have

$$\mathbf{N}_{n-1}\mathbf{N}_n = \mathbf{N}_{n-1}\mathbf{M}_1^T = 0. \tag{2.77}$$

This suggests that $\mathbf{N}_{n-1} = \mathbf{M}_2^\mathsf{T}$, and this construction can be extended for each cell degree of the complex. Therefore, given a primal n-complex, the boundary operator on primal p-cells is the transpose of the boundary operator on dual $(n-p+1)$-cells, or

$$\mathbf{N}_p^\mathsf{T} = \mathbf{M}_{(n-p+1)}, \tag{2.78}$$

and, equivalently, we may associate each p-cell in the primal with an $(n-p)$-cell in the dual. Constructing the dual complex in this way, the algebraic structure of the primal complex is preserved. Furthermore, this notion of duality is completely independent of the embedding and therefore does not require any knowledge of the complex beyond its connectivity. This purely topological statement embodies the concept of **Poincaré duality**, which is generally made in terms of the homology/cohomology groups of a complex (e.g., [276]) and defines a combinatorial manifold [284].

Note that the incidence matrices of the dual complex $\mathbf{M}_p$ naturally provide a coboundary operator, given that their structure is identical to that of the primal complex. However, we have not yet defined cochains on the dual complex or how they are related to cochains on the primal complex. This will be considered in the next section.

In order to build a dual complex corresponding to a given primal n-complex, we must identify $(n-1)$-cells that form the boundary of exactly two adjacent n-cells and that exhibit opposing induced orientation as a single $(n-1)$-cell. Often, when we draw a finite complex there are a series of border $(n-1)$-cells which do not appear to have a second n-cell on which they are incident. In these cases, we note that we can include an "outside cell" in order to provide the second n-cell for these border cells. For example, if we draw a finite planar graph on the plane, then the **outside cell** may be considered to be the face attached to all of the border cells that extends "to infinity". The inclusion of this outside cell can be motivated by considering the graph embedded on the surface of a topological sphere, which is a manifold in the strict sense of the term (i.e., not a manifold with boundary) and therefore a *closed surface*. The requirement of the outside cell can be viewed as requiring the complex to be a strict combinatorial manifold in order to define a dual complex. Figure 2.11 gives an example of a graph defining the surface of a cube. When the graph is embedded on the plane, we seem to lose a face of the cube unless the outside face is included in the set of 2-cells.

We note that the definition of the dual complex presented above is a generalization of the standard conception of duality in graph theory, which states that two graphs are dual if there is an isomorphism between the cycles of the primal graph and the cuts of the dual graph. Given this definition of a dual graph, it was shown by Whitney [402] that a graph has a dual if and only if the graph is planar. However, when the primal graph is planar, the standard graph theory notion of duality and the definition given here agree, since for a planar graph there exists an oriented cycle set

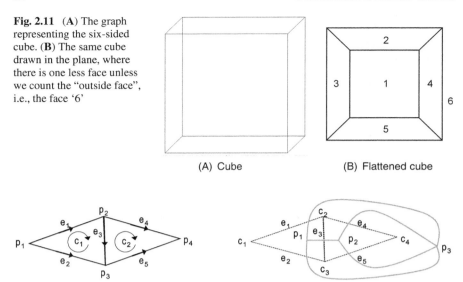

Fig. 2.11 (A) The graph representing the six-sided cube. (B) The same cube drawn in the plane, where there is one less face unless we count the "outside face", i.e., the face '6'

Fig. 2.12 A planar graph and its dual. Each shaded contour represents a dual edge

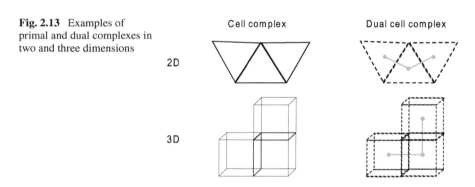

Fig. 2.13 Examples of primal and dual complexes in two and three dimensions

which satisfies the condition that each edge is the boundary of exactly two cycles with opposite orientation (see Fig. 2.12).[18]

One consequence of this definition of the duality of a n-complex is that the dual *changes with the value of n*. Figure 2.13 illustrates this concept for two small complexes and Fig. 2.14 illustrates this concept by considering the dual of a 4-connected lattice in two dimensions and the dual of a 6-connected lattice in three dimensions. In the two-dimensional case, nodes are dual to faces, edges are dual to edges and faces are dual to nodes. However, in three dimensions, nodes are dual to volumes, edges are dual to cycles, cycles are dual to edges and volumes are dual to nodes.

---

[18] The (oriented) cycle double cover conjecture [340] postulates that every bridgeless graph has a cycle set which admits the type of duality discussed here. See Sect. 4.1.2.1 for more discussion.

2.3 Discrete Calculus 53

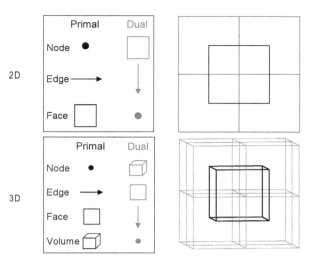

Fig. 2.14 The duality relationships between primal and dual complexes depend on the order of the complex. For example, primal edges correspond to dual edges in two dimensions, whereas primal edges correspond to dual faces in three dimensions

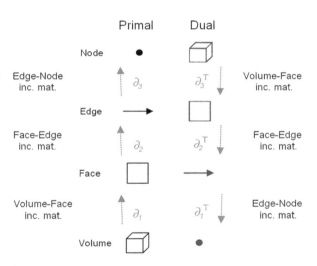

Fig. 2.15 Correspondence between chains in primal and dual complexes in three dimensions. The *boundary* operator for a primal p-cell is the *coboundary* operator of a dual $(p-1)$-cell

Figure 2.15 shows how the boundary operators in the primal complex relate to the coboundary operators in the dual complex.

We defined an orientation for each cell in purely algebraic terms by using the ordering of the vertices of each cell. However, since each cell in the dual complex is identified with a cell in the primal complex, then we may use the orientation on a primal cell to define the orientation of a dual cell. The combinatorial orientation of a primal cell has been termed *intrinsic* (since the orientation is defined by the cell nodes) while the orientation of a dual cell has been termed *extrinsic* (since this orientation is defined by the primal cell nodes) [275, 380]. Defining the extrinsic orientation of a p-cell in terms of the dual $(n-p)$-cell has the consequence that the

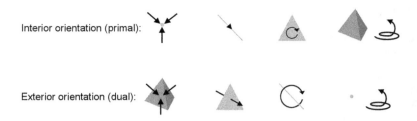

Fig. 2.16 Interior orientation prescribed on a primal complex induces exterior orientation on the corresponding dual complexes. Graphical representations of this mapping between dual simplices in a 3-complex are demonstrated

extrinsic orientation *depends on n* for a primal *n*-complex. Figure 2.16 illustrates the relationship between interior and exterior orientation for a 3-complex.

2.3.4 The Role of a Metric: the Metric Tensor, the Discrete Hodge Star Operator, and Weighted Complexes

2.3.4.1 The Metric Tensor and the Associated Inner Product

As in the general setting described above in Sect. 2.2.1.3, in order to convert chains into cochains and vice versa we require the definition of an inner product specified by a metric tensor. We begin by defining an inner product in each chain space C_p. Then, given a set of basic chains $\sigma_i \in C_p$ we can define a *metric tensor*

$$g_{ij}^p = \langle \sigma_i, \sigma_j \rangle. \tag{2.79}$$

Typically in the discrete setting, the basis set of p-cells is defined to be orthogonal such that $g_{ij}^p = 0$ for $i \neq j$. This is a direct result of the coarse, sparse topology of a complex. For this reason the role of the metric is dramatically simplified. The (primal) metric tensor is thus represented by the $n_p \times n_p$ diagonal matrix $\mathbf{G}_p$ where $\mathbf{G}_p = \text{diag}(g_{ii}^p)$ for $1 \leq i \leq n_p$. Thus converting a p-chain into its equivalent p-cochain consists of a simple scaling of the chain coefficients for each basic cochain, i.e., for a given p-chain $\boldsymbol{\tau}_p$ and p-cochain $\mathbf{c}^p$,

$$\mathbf{c}^p = \mathbf{G}_p \boldsymbol{\tau}_p. \tag{2.80}$$

The reverse mapping from p-cochains to p-chains can be computed from the dual metric tensor using the inverse of the metric tensor matrix, $\mathbf{G}_p^{-1}$. The specification of a metric tensor provides a means to compute the inner product between pairs of p-chains or pairs of p-cochains. As expected, the inner product between p-chains $\boldsymbol{\mu}$ and $\boldsymbol{\nu}$ is simply $\langle \boldsymbol{\mu}, \boldsymbol{\nu} \rangle = \boldsymbol{\mu}^T \mathbf{G} \boldsymbol{\nu}$, and the inner product between p-cochains $\mathbf{u}$ and $\mathbf{v}$ is $\langle \mathbf{u}, \mathbf{v} \rangle = \mathbf{u}^T \mathbf{G}^{-1} \mathbf{v}$. As in the continuous setting, the mapping provided by the metric tensor preserves the inner product of a pair of cochains when each cochain is converted into their corresponding chain, and vice versa. It is straightforward to show that if $\mathbf{u} = \mathbf{G}\boldsymbol{\mu}$ and $\mathbf{v} = \mathbf{G}\boldsymbol{\nu}$, then $\langle \mathbf{u}, \mathbf{v} \rangle = \langle \boldsymbol{\mu}, \boldsymbol{\nu} \rangle$.

In the special case where $g_{ii}^p = 1$ for all n_p entries, the metric is analogous to a Euclidean metric, in which case there is a natural correspondence between 0-chains and 0-cochains—in this case there is no distinction between them. Note that because the inner product is defined to be positive definite at each basic p-chain, it provides a Riemannian metric over the complex.

2.3.4.2 The Discrete Hodge Star Operator

Given this definition of an inner product and corresponding metric tensor, we may consider the Hodge star operator articulated in the discrete setting. Recall that the Hodge star operator is a mapping from p-forms to $(n-p)$-forms in n-dimensions that depends on both the metric tensor and the prescribed orientation of space. This construction was made possible by the matching dimensions of the space of p-forms and $(n-p)$-forms (each of which have dimension $\binom{n}{p}$ in the continuous setting), thus the Hodge star operator was an isomorphism between these two spaces. A similar isomorphism exists in the discrete setting between each space of p-chains on the primal complex and the space of $(n-p)$-chains on the dual complex, which is a manifestation of Poincaré duality. The discrete Hodge star operator therefore simply transfers p-cochains on the primal complex, represented with respect to the basic p-cochains on the primal complex (or, equivalently in this setting, the basic p-chains), into a representation in terms of the $(n-p)$-cochains on the dual complex scaled by the inverse metric tensor. We can therefore view the discrete Hodge star construction as a combination of Poincaré duality, which is a purely topological notion, with the metric information provided by the metric tensor. Thus in the case of an n-complex, any given p-cochain $\mathbf{x}$ on the primal complex is mapped to its corresponding $(n-p)$-cochain $\mathbf{x}^*$ on the dual complex such that

$$\mathbf{x}^* \equiv \star\mathbf{x} = \mathbf{G}_p^{-1}\mathbf{x}. \tag{2.81}$$

Due to the finite-dimensional cochain bases and the simple form of the metric tensor, the Hodge star operator is straightforward to implement in the discrete setting.[19] Note, however, that in the construction of a dual complex requires a prescribed orientation on the primal complex, therefore if the orientation of the primal complex were to reverse direction the orientation of the dual complex and its basic p-cochains would change, resulting in a sign change in the Hodge dual $\star\mathbf{x}$.

---

[19] In the context of finite elements, the search for a definitive discrete Hodge star operator has been elusive [47, 178, 199, 248]. In other discrete calculus formulations—in which the goal is to provide a *discretized* version of calculus (e.g., [102])—several formulations of the Hodge star operator have been suggested, including the Galerkin–Hodge star operator that has many advantageous properties [248]. However a fundamental difficulty arises in defining an operator that converges in the limit of finer mesh sizes to the continuum operator. In the present framework (i.e., the *discrete* formulation of calculus) the definition of the Hodge star is straightforward and is provided by the well-defined inner product on any cochain basis, all of which are finite-dimensional.

Oftentimes it is desirable to be able to compute the inner product between a pair of $(n-p)$-cochains on the dual complex in such a way that it is equivalent to the inner product between the corresponding pair of p-cochains on the primal complex. That is, we wish to define the inner product on two $(n-p)$-cochains $\mathbf{u}^*$ and $\mathbf{v}^*$ such that

$$\langle\langle \mathbf{u}^*, \mathbf{v}^* \rangle\rangle = \langle \mathbf{u}, \mathbf{v} \rangle, \tag{2.82}$$

where, to avoid confusion, we have denoted the inner product on the dual complex as $\langle\langle \cdot, \cdot \rangle\rangle$. From the definition of the Hodge star operator, the inner product on the dual complex is

$$\langle\langle \mathbf{u}^*, \mathbf{v}^* \rangle\rangle = (\star\mathbf{u})^\mathsf{T} \mathbf{G}^*_{n-p} (\star\mathbf{v}) = \mathbf{u}^\mathsf{T} (\mathbf{G}_p^{-1})^\mathsf{T} \mathbf{G}^*_{n-p} \mathbf{G}_p^{-1} \mathbf{v} \tag{2.83}$$

then to satisfy the requirement that the inner products are equivalent implies that

$$\mathbf{u}^\mathsf{T} (\mathbf{G}_p^{-1})^\mathsf{T} \mathbf{G}^*_{n-p} \mathbf{G}_p^{-1} \mathbf{v} = \langle \mathbf{u}, \mathbf{v} \rangle = \mathbf{u}^\mathsf{T} \mathbf{G}_p^{-1} \mathbf{v}, \tag{2.84}$$

and therefore, since the metric tensor matrix is symmetric,

$$\mathbf{G}^*_{n-p} = \mathbf{G}_p. \tag{2.85}$$

Thus the inner product of a pair of $(n-p)$-cochains on the dual complex—that is compatible with the inner product of the corresponding pair of p-cochains on the primal complex—is given by the quadratic form using the *primal* metric tensor on p-chains $\mathbf{G}_p$. This implies that, in order to maintain a compatible inner product between pairs of p-cochains on the primal complex and their corresponding pairs of $(n-p)$-cochains on the dual complex, the mapping from an $(n-p)$-cochain on the dual complex $\mathbf{y}^*$ into the appropriate p-cochain on the primal complex $\mathbf{y}$ must be

$$\mathbf{y} = \star \mathbf{y}^* = \mathbf{G}^*_{n-p} \mathbf{y}^* = \mathbf{G}_p \mathbf{y}^*. \tag{2.86}$$

Note that the standard notation for the Hodge star operator, $\star$, does not distinguish between these two distinct mappings: the $\star$ operator performs both the mapping from a p-cochain on the primal complex to an $(n-p)$-cochain on the dual complex and its inverse.[20] Therefore one must always be mindful of what is the operand of the Hodge star operator in order to properly interpret its action.

For a general n-complex, metric tensors can be defined for all p-cells for each value $0 \leq p \leq n$, forming a collection of matrices $\mathbf{G}_p$. While higher-order metric tensors can be constructed from lower-order ones, as discussed in the previous section on the Hodge star operator, in several branches of physics an independent metric tensor is more appropriate for each class of p-cochains. In the context of physics, the discrete Hodge star incorporates metric information into the system and represents

---

[20] So as to not deviate completely from the literature, we forgo cumbersome but explicit notation, e.g., $\star_p$ and $\star^{-1}_{n-p}$, that could be employed to help limit the ambiguity of this overloaded operator.

2.3 Discrete Calculus

a class of laws known as **constitutive laws**. These laws embody the material properties of a system and its substances and can be identified in virtually all branches of physics.

Fortunately, because of the isomorphism between p-cochains on the primal complex and $(n - p)$-cochains on the dual complex, the transition between the two complexes is practically unnoticeable. As in the continuous setting, tracking which elements of the complex represent the fields that we are studying can help provide insight into the geometry and topology of the fields and how they interact. As discussed above, it is also critical to represent various fields on the appropriate discrete structures to faithfully capture their geometric character in the representation.

2.3.4.3 Weights

Since the metric tensor is fully described by a single scale factor at each basic p-chain or each p-cell of the complex, conventionally the metric tensor is prescribed by assigning a **weight** to each p-cell. For example, a **weighted graph** is a 1-complex for which each edge possesses a weight, and this weight specifies the metric tensor for each edge. In this book, we assume that $w_i > 0$ with finite value, unless otherwise noted. With the weights playing the role of the metric tensor, many geometric quantities can be readily computed. See Chap. 8 for examples.

2.3.4.4 The Volume Cochain

As in the continuous setting, the Hodge star operator allows us to define at each basic n-chain in an n-complex a canonical volume element or **volume cochain** as $\star 1 = \text{vol}^n$, which in the discrete case is given by the n-cochain

$$\mathbf{vol}^n = \mathbf{G1} = \mathbf{g} \tag{2.87}$$

which is simply the (column) vector of n-chain weights and is therefore defined over the entire complex. (Note that the discrete Hodge star in this definition is sensitive to changes in orientation, thus the volume cochain also changes sign with a change in orientation of the primal complex.) Under this definition for the volume cochain, the volume of a specific n-chain $\tau_n = \sum a^i \sigma_i$ is simply the sum (i.e., the integral) of the volume cochain over the chain, or

$$\text{volume}(\tau_n) = \mathbf{g}^\mathsf{T} \tau_n. \tag{2.88}$$

Therefore the volume of a chain is simply the sum of the weights of the basic n-chains comprising the chain scaled by the chain coefficients a^i.

Similarly, we can define the "volume of a cochain" **c** using a chain of all ones, via

$$\mathbf{vol}_n = \mathbf{G}^{-1}\mathbf{1} = \mathbf{g}^{-1}, \tag{2.89}$$

and
$$\text{volume}^*(\mathbf{c}^n) = (\mathbf{g}^{-1})^T \mathbf{c}^n. \tag{2.90}$$

One consequence of the above definitions is an explanation of the common usage of edge weights in the computational literature to represent both *distances* and *affinities*. For example, if a graph represents a road network between cities, the weights are used to encode the length of a road connecting two cities. This **distance weight** could then be used to calculate the shortest path between two cities (e.g., using Dijkstra's algorithm). If we view the shortest-path calculation as finding a 1-chain τ_1 of road edges that connect the two cities of interest by a path of minimum distance, then the volume (length) of the chain is given by $\mathbf{1}^T \mathbf{G}_1 \tau_1 = \mathbf{g}^T \tau_1$ meaning that the weights of the roads in the shortest path are summed to produce a distance. A distance weight on an edge between two nodes is small if the two nodes are close together and large if the two nodes are further apart. In contrast, other applications and algorithms treat weights as *affinities*, meaning that an edge connecting two nodes has a large **affinity weight** if the two nodes are similar and a small affinity weight if the two nodes are less similar. For example, the flow capacities in a max-flow/min-cut problem are high if two nodes are well-connected and low if two nodes are poorly connected. In this case, each edge capacity is used to control the amount of flow through an edge. Since edge flow is a 1-cochain, $\mathbf{c}^1$, then the total flow is measured by $\mathbf{1}^T \mathbf{G}_1^{-1} \mathbf{c}^1$. Therefore, we see that if we knew distances between two nodes, then the corresponding *affinity edge weight is the reciprocal of the distance edge weight*. For example, in the context of circuit theory, a branch resistance is the *reciprocal* of the branch conductance. Therefore, throughout the book, we will refer to the edge weights used to measure distances between nodes (and chains) as *distance weights* and the edge weights used to measure affinities (and cochains) as *affinity weights*, with the understanding that when the two sets of weights are compatible on a particular complex the values of these weights are reciprocal to each other. Although edge weights will play the most prominent role in this book, we will also use the terms *distance weights* and *affinity weights* to describe volume and inverse weights on p-cells. This reciprocal relationship between distance weights and affinity weights has been noted previously in the literature (e.g., [54]).

Note that we will use the notation $\mathbf{G}_p$ to represent the diagonal matrix of distance weights for the p-cells. This constitutive matrix is represented by many other authors using different notation (including Strang [360], with which our notation mostly aligns). When $\mathbf{G}$ appears without a subscript, then it is assumed that $p = 1$, since edge weights play the most prominent role in this work.

It was shown above that distances in primal space are measured the same as affinities in dual space and vice versa. This connection between the weights of primal graphs and dual graphs has been occasionally exploited in the literature. For example it has been noted that the higher-complexity max-flow/min-cut algorithm can be replaced by the low-complexity Dijkstra's shortest path algorithm run on the *dual* space [191]. Figure 2.17 gives an example of a weighted primal graph, a dual graph and the equivalence of the min-cut in the primal graph and the shortest path in the dual graph when calculated with the same weights.

2.3 Discrete Calculus

Fig. 2.17 The max-flow/min-cut problem on a planar graph is equivalent to a minimum path problem on the dual graph *using the same weights*. The max-flow/min-cut problem solves for a cochain and therefore employs affinity weights. The conversion to a shortest path problem changes the variable to a chain, inducing distance weights (reciprocal), but the chain is on the dual graph which brings back the original affinity weights (by taking a second reciprocal)

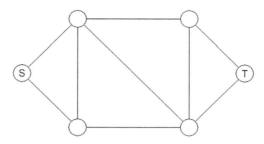

(A) Max-flow/min cut problem on a planar graph

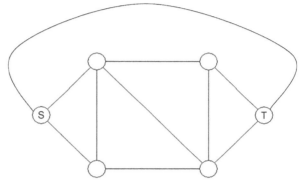

(B) Connect source and terminal with an edge

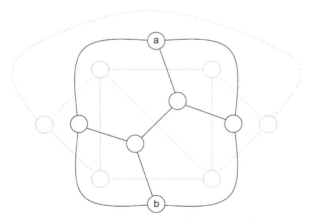

(C) Solve shortest path problem from *a* to *b*

2.3.5 The Dual Coboundary Operator

Armed with a discrete Hodge star operator we may define the discrete version of a codifferential operator. The **dual coboundary operator**, $\mathbf{N}_p^*$, maps p-cochains into

($p-1$)-cochains—via a brief mapping into the dual complex—and is defined as

$$\mathbf{N}_p^* \equiv \star \mathbf{M}_{n-p+1} \star . \qquad (2.91)$$

For clarity, we will consider each step in this composite operator. The first discrete Hodge star operator maps the given p-cochain on the primal complex into an $(n-p)$-cochain on the dual complex. Then, the incidence matrix on the dual graph is employed as the coboundary operator that maps the $(n-p)$-cochain on the dual complex into an $(n-p+1)$-cochain on the dual complex. Finally, the second discrete Hodge star operator maps the $(n-p+1)$-cochain on the dual complex into a $(p-1)$-cochain in the primal complex. In essence, this states that the dual coboundary operator on p-cochains defined on the primal complex can be phrased in terms of the $(n-p+1)$-coboundary operator on the dual complex with the discrete Hodge star acting to shuffle cochains back and forth between the complexes.

Because of the relationship between the incidence matrices in the primal and dual complexes (i.e., Poincaré duality) discussed in the previous section, summarized in (2.78), this operator can be phrased entirely in terms of the incidence structure of the primal complex as

$$\mathbf{N}_p^* = \star \mathbf{N}_p^\mathsf{T} \star . \qquad (2.92)$$

In the discrete setting, the Hodge star operator is simply a diagonal matrix, so the dual coboundary operator may be equivalently expressed as

$$\mathbf{N}_p^* = \mathbf{G}_{n-p+1}^* \mathbf{N}_p^\mathsf{T} \mathbf{G}_p^{-1} = \mathbf{G}_{p-1} \mathbf{N}_p^\mathsf{T} \mathbf{G}_p^{-1} \qquad (2.93)$$

since $\mathbf{G}_{n-p+1}^* = \mathbf{G}_{p-1}$ by (2.85). Thus the dual coboundary operator is simply a weighted version of the primal coboundary operator.

In the special case in which the metric is Euclidean at all locations of the complex on both p- and $(p-1)$-cochains (i.e., $\mathbf{G}_p = \mathbf{I}$ and $\mathbf{G}_{p-1} = \mathbf{I}$), the dual coboundary operator matrix is the transpose of the coboundary operator matrix $\mathbf{N}_p$, i.e.,

$$\mathbf{N}_p^* = \mathbf{M}_{n-p+1} = \mathbf{N}_p^\mathsf{T} \quad \text{when the metrics are Euclidean.} \qquad (2.94)$$

In this case the discrete instantiation of the codifferential operator can be phrased entirely in terms of the incidence matrix of the primal complex. Otherwise the dual coboundary operator and the boundary operator differ by an element-by-element scaling imposed by the p and $p-1$ metric tensor matrices.

2.3.6 The Discrete Laplace–de Rham Operator

A remarkable aspect of the above discussion is that the boundary operator and the exterior derivate are *defined by the structure of the complex itself*. In other words, the structure of the derivative operator depends on the topological structure of space— in a sense, the graph *is* the operator. That the incidence matrix is simultaneously

2.3 Discrete Calculus

used as a data structure representing the topology *and* the coboundary operator implies that this is quite literally the case! After some consideration, we might not be so surprised by this realization since the same dependence occurs in conventional calculus. The gradient operator behaves differently when applied to functions on the surface of different manifolds. In this section, we revisit the Laplace–de Rham operator in the setting of discrete calculus.

Recall the definition of the *Laplace–de Rham operator* in terms of the exterior derivative and the codifferential, $\Delta \equiv dd^* + d^*d$. We may apply this definition to the p-complex in which the incidence matrix and its transpose play the role of the codifferential and exterior derivative to give

$$\mathbf{L}_p = \mathbf{N}_p \mathbf{N}_p^* + \mathbf{N}_{p+1}^* \mathbf{N}_{p+1}. \tag{2.95}$$

Therefore, the special case of the Laplace–de Rham operator on 0-cochains, which is identical to the Laplacian from standard vector calculus, may be written as $\mathbf{L}_0 = \mathbf{A}^T\mathbf{A}$. Within graph theory, this matrix is referred to as the **Laplacian matrix**. The Laplacian matrix is size $|\mathcal{V}| \times |\mathcal{V}|$ and symmetric. Since the Laplacian matrix is composed of the matrix $\mathbf{A}$, its nullspace is also rank r (where r is the number of connected components). When the graph is connected, the nullspace of $\mathbf{L}$ is spanned by a constant vector. Since the Laplacian matrix is composed of $\mathbf{A}^T\mathbf{A}$, the Laplacian matrix is positive semi-definite.

The Laplacian matrix also represents the connectivity of the graph structure, but not the orientation of the edges. Given any graph then the Laplacian matrix may be defined directly by

$$L_{ij} = \begin{cases} d_i & \text{if } i = j, \\ -1 & \text{if } e_{ij} \in \mathcal{E}, \\ 0 & \text{otherwise}, \end{cases} \tag{2.96}$$

where L_{ij} is indexed by vertices v_i, v_j and d_i represents the number of edges incident on node v_i, called the node **degree**. Note that this definition of the term *degree* is different from the conventional use of the word *degree* to describe a differential form. This discrete calculus analogue of the Laplacian operator first appeared in [118]. When the underlying graph is a 4-connected grid, this Laplacian matrix is identical to the Laplacian matrix used in finite differences which is obtained from discretization with a 5-point stencil.

The Laplacian matrix is also related to the **adjacency matrix**, $\mathbf{W}$, typically employed in graph theory. Define the adjacency matrix as

$$W_{ij} = \begin{cases} 1 & \text{if } e_{ij} \in \mathcal{E}, \\ 0 & \text{otherwise}. \end{cases} \tag{2.97}$$

Like the Laplacian matrix, the adjacency matrix encodes the connectivity information of the graph but not the edge orientations. All of these matrices which describe the structure of the graph are related by the expression

$$\mathbf{L} = \mathbf{D} - \mathbf{W} = \mathbf{A}^T \mathbf{A}, \tag{2.98}$$

where **D** represents a diagonal matrix of the node degree, $D_{ii} = d_i$. Note however, that only the incidence matrix preserves the information about edge orientation information. Therefore, we may consider the incidence matrix more fundamental, since the other matrices may be easily derived from it.

The Laplacian operator changes in the presence of a nontrivial metric. Let $\mathbf{G}_0$ represent the diagonal matrix of node (distance) weights and let $\mathbf{G}_1$ represent the diagonal matrix of edge (distance) weights. Therefore, for a weighted graph, the Laplacian operator is defined as

$$\mathbf{L}_0 = \mathbf{N}_0^* \mathbf{N}_0 + \mathbf{N}_1^* \mathbf{N}_1 = \mathbf{N}_1^* \mathbf{N}_1 = \mathbf{G}_0 \mathbf{A}^\mathsf{T} \mathbf{G}_1^{-1} \mathbf{A}, \tag{2.99}$$

since $\mathbf{N}_0$ is defined as the null operator and matrix $\mathbf{A}$ represents the special case of the node–edge incidence matrix, as usual. Therefore, the Laplacian operator for a weighted graph takes the form

$$L_{ij} = \begin{cases} \frac{1}{w_i} d_i & \text{if } i = j, \\ -\frac{1}{w_i} w_{ij} & \text{if } e_{ij} \in \mathcal{E}, \\ 0 & \text{otherwise,} \end{cases} \tag{2.100}$$

where $d_i = \sum_{e_{ij} \in \mathcal{E}} w_{ij}$ is the (weighted) degree of each node. Note that in the above expression we avoid the reciprocal notation needed for the inverse metric tensor by identifying each edge weight w_{ij} (and node weight w_i) with the *affinity weights*, which is conventional for definition of the Laplacian matrix.

The form of the unweighted Laplacian that we have been considering, $\mathbf{L}_0 = \mathbf{A}^\mathsf{T} \mathbf{A}$, is therefore a special case of this more general weighted scalar Laplacian where $\mathbf{G}_0^{-1} = \mathbf{I}$ and $\mathbf{G}_1^{-1} = \mathbf{I}$. The choice of $\mathbf{G}_0 = \mathbf{D}^{-1}$ is also sometimes employed in the literature, where $\mathbf{D}$ is the diagonal matrix with $D_{ii} = d_i$. The Laplacian matrix defined as $\mathbf{L}_0 = \mathbf{D}^{-1} \mathbf{A}^\mathsf{T} \mathbf{G}_1^{-1} \mathbf{A}$ is sometimes referred to as the *normalized Laplacian* or the *random walk Laplacian*. The normalized Laplacian is often symmetrized to produce $\tilde{\mathbf{L}}_0 = \mathbf{D}^{-\frac{1}{2}} \mathbf{A}^\mathsf{T} \mathbf{G}_1^{-1} \mathbf{A} \mathbf{D}^{-\frac{1}{2}}$. The normalized Laplacian fits naturally into the study of random walks [115], has some attractive theoretical properties [81, 112] and has been preferred in some applications [23, 345].

We may now consider the extension of the Laplacian operator to 1-cochains. From the general definition of the Laplacian matrix in (2.95), we can express the Laplacian matrix operator on 1-cochains in terms of the node–edge incidence matrix **A**, the edge–face incidence matrix **B**, and the corresponding metric tensor matrices as

$$\mathbf{L}_1 = \mathbf{N}_1 \mathbf{N}_1^* + \mathbf{N}_2^* \mathbf{N}_2 = \mathbf{A} \mathbf{G}_0 \mathbf{A}^\mathsf{T} \mathbf{G}_1^{-1} + \mathbf{G}_1 \mathbf{B}^\mathsf{T} \mathbf{G}_2^{-1} \mathbf{B}, \tag{2.101}$$

which we may consider to be the discrete calculus analogue of the *vector Laplacian*, which we will refer to as the **edge Laplacian**. Unlike the scalar Laplacian, the edge Laplacian *does* preserve orientation information. The unweighted edge Laplacian matrix (i.e., $\mathbf{G}_0 = \mathbf{G}_1 = \mathbf{G}_2 = \mathbf{I}$) has appeared in the literature for purposes of analyzing networks [1, 80, 288].

2.4 Structure of Discrete Physical Laws

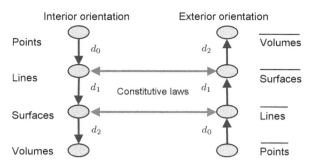

Fig. 2.18 A Tonti diagram expressing the structure of primal and dual complexes, their associated coboundary operators, and the constitutive laws that map primal cochains into dual cochains for a 3-complex. This general structure is common to many branches of physics

However, there are several cases in which non-trivial higher-order weights can be quite useful. In particular, the cycle weights $\mathbf{G}_2^{-1}$ and node weights $\mathbf{G}_0$ in the edge Laplacian reflect two aspects of the *viscosity* of a flow variable (represented by a 1-cochain). The proper assignment and interpretation of these weights for the edge Laplacian is given in Sect. 4.3.

2.4 Structure of Discrete Physical Laws

We end this discussion on conventional and discrete calculus with a description of the geometric structure of physics that is generated by this theoretical construct. Since the discrete Hodge star operator provides a mapping from p-cochains on the primal complex to $(n - p)$-cochains on the dual complex, the primal coboundary operator maps p-cochains into $(p+1)$-cochains on the primal complex, and the dual coboundary operator maps p-cochains into $(p-1)$-cochains on the dual complex, a natural structure to the calculus begins to emerge. This structure can be summarized schematically by means of a **Tonti diagram**. An example Tonti diagram is given for 3-complexes in Fig. 2.18.

A **Roth diagram** is similar to a Tonti diagram, except that it is generally used to describe a discrete calculus setting and often explicitly accounts for the image and nullspace of the boundary operator. For example, the Roth diagram for electrical circuits is given in Fig. 2.19. Many authors have discussed these relationships and the striking order that can be seen to exist across multiple disciplines. The interested reader is strongly encouraged to refer to Refs. [59, 178, 323, 359, 379] for more details.

2.5 Examples of Discrete Calculus

We now have in place all of the machinery of conventional calculus phrased on a discrete space. It is worthwhile at this point to consider the implications of these definitions. One important practical difference between the standard vector calculus and this discrete calculus is in the distinction between scalar fields and vector fields. In conventional calculus, scalar fields *and* vector fields may be evaluated at

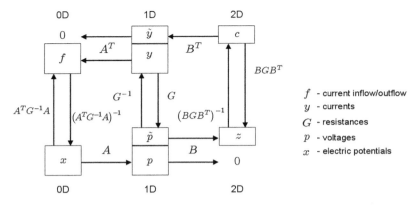

Fig. 2.19 A Roth diagram expressing the geometric structure of electrical circuits. The "1D" boxes are separated into two compartments to denote the range of **A**, $\hat{\mathbf{y}}$, and the nullspace of **A**, $\tilde{\mathbf{y}}$

a point. However, the same is not true in the discrete calculus setting. For example, a scalar field may be viewed as a 0-cochain, $\mathbf{c}^0$, which assigns a value to each node. However, if we apply a gradient operator, **A**, to this function, we produce

$$\mathbf{c}^1 = \mathbf{A}\mathbf{c}^0. \tag{2.102}$$

In other words, by applying the gradient operator to a 0-cochain scalar function we generate a 1-cochain edge function. Therefore, the "vector field" produced by application of a gradient operator to a "scalar field" is *identified with edges*. The theory of differential forms also supports this concept that the application of the exterior derivative to a 0-form will produce a 1-form. However, in continuous space, the concept of an infinitesimal allows us to shrink, in the limit, a one-dimensional vector into a zero-dimensional location. Therefore, in the conventional calculus a vector field is generally indistinguishable from an *n-tuple* of scalar fields (e.g., heat, electrical charge and chemical concentration) since they are both evaluated at point locations in space.[21] In contrast, the discrete calculus makes these situations very distinct in the sense that multiple scalar fields would be represented by multiple 0-cochains while a vector field would be represented by a 1-cochain. Similarly, a 2-form in three dimensions is represented by three coefficients, which can make it difficult to distinguish a 2-form from 1-form. For example, application of the curl operator to a vector field is generally conceived of as returning a vector field, even though in the language of differential forms the application of the exterior derivative to a 1-form yields a 2-form. In contrast, the discrete calculus distinguishes between these fields explicitly, since a 1-cochain is a function assigning values to edges while a 2-cochain is a function assigning values to cycles.

---

[21] Throughout the applications sections of the book, multiple scalar (data) fields associated with a node will be referred to as a *tuple* and denoted as $\tilde{x}$. In contrast, the word *vector* will be reserved for either conventional (continuous) vectors denoted as $\vec{x}$ or column vectors denoted as **x**.

2.5 Examples of Discrete Calculus

In the remainder of this chapter we demonstrate how these definitions for discrete calculus allow us to preserve the structure of conventional calculus. In this section we give several illustrations of how theory from standard calculus translates directly to the discrete calculus developed above.

2.5.1 Fundamental Theorem of Calculus and the Generalized Stokes' Theorem

As reviewed above, the Fundamental Theorem of Calculus as it is typically expressed is a special case of the Generalized Stokes' Theorem on 0-forms or functions. In this section, we provide examples of how the Generalized Stokes' Theorem holds on a 1-complex, a 2-complex and a 3-complex. Additionally, we show that the Generalized Stokes' Theorem may *not* hold in the same way for other standard discretizations of continuous operators, specifically the central differences approximation.

Recall that the Generalized Stokes' Theorem is given by

$$\int_{\mathcal{S}} \mathrm{d}\tilde{\omega} = \int_{\partial \mathcal{S}} \tilde{\omega}, \tag{2.103}$$

and the corresponding discrete calculus version is

$$[\![\mathbf{N}_p \mathbf{c}^{p-1}, \boldsymbol{\tau}_p]\!] = [\![\mathbf{c}^{p-1}, \mathbf{N}_p^\mathsf{T} \boldsymbol{\tau}_p]\!], \tag{2.104}$$

where $\mathbf{N}_p$ is the incidence matrix (representing the coboundary operator) applied to the $(p-1)$-cochain, $\mathbf{c}^{p-1}$ and $\mathbf{N}_p^\mathsf{T}$ represents the boundary operator applied to p-chain $\boldsymbol{\tau}_p$ for any value p, $0 < p \leq n$. Expressed in this discrete manner using matrix representations of the boundary and coboundary operators and vector representations of the chains and cochains, the Generalized Stokes' Theorem simply states that, with respect to the bilinear pairing of integration, $\mathbf{N}_p$ is adjoint to its transpose $\mathbf{N}_p^\mathsf{T}$.

2.5.1.1 Generalized Stokes' Theorem on a 1-Complex

The Fundamental Theorem of Calculus states that

$$\int_a^b f(t)\, \mathrm{d}t = F \bigg|_a^b = F(b) - F(a), \tag{2.105}$$

for some differentiable function F and interval of evaluation between a and b. The discrete calculus instantiation of this theorem is

$$[\![\mathbf{A}\mathbf{c}^0, \boldsymbol{\tau}_1]\!] = [\![\mathbf{c}^0, \mathbf{A}^\mathsf{T} \boldsymbol{\tau}_1]\!], \tag{2.106}$$

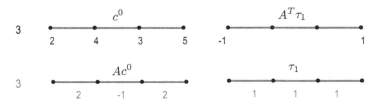

Fig. 2.20 Example of the Generalized Stokes' Theorem on a 1-complex. The *top row* depicts the pairing of a 0-cochain c^0 and 0-chain $A^T \tau_1$, with coefficients on the nodes represented in *blue*. The *bottom row* depicts the pairing of a 1-cochain Ac^0 with a 1-chain τ_1, with coefficients on the edges represented in *red*. In this example, both pairings result in a value of '3'

for edge–node incidence matrix A, 0-cochain c^0 (i.e., a function assigned to the nodes) and 1-chain τ_1 (i.e., an indicator vector of edges for evaluation of the integral). This relationship consists of the pairing of a 1-cochain Ac^0 with a 1-chain τ_1 on the left-hand side of (2.106) to yield a scalar, and a pairing of a 0-cochain c^0 with a 0-chain $A^T \tau_1$ on the right-hand side to yield a scalar, and these two resulting scalars are equivalent. The discrete calculus version of the Fundamental Theorem of Calculus can be seen as an instance of the generalized Stokes' Theorem when $p = 1$, in which case $N_1 = A$.

We first apply the discrete version of the Fundamental Theorem of Calculus to an example involving a very simple graph, illustrated in Fig. 2.20. In this example,

$$c^0 = \begin{bmatrix} 2 \\ 4 \\ 3 \\ 5 \end{bmatrix}, \tag{2.107}$$

$$A = \begin{bmatrix} 0 & 0 & -1 & 1 \\ 0 & -1 & 1 & 0 \\ -1 & 1 & 0 & 0 \end{bmatrix}, \tag{2.108}$$

and

$$\tau_1 = \begin{bmatrix} 1 \\ 1 \\ 1 \end{bmatrix}. \tag{2.109}$$

The reader may verify that the discrete Fundamental Theorem of Calculus given holds for this case. However, it is instructive to examine each side of (2.106) separately. On the left side, the gradient of c^0, assigned to each edge, is summed over all of the edges represented in τ_1. On the right side, $A^T \tau_1$ yields zeros on the interior of the line and just a '1' and '−1' on the endpoints. Consequently, the product of $[\![c^0, A^T \tau_1]\!]$ evaluates c^0 only on the endpoints of the line, as in the right hand side of the familiar form in (2.105). For both the left and right sides, the inner product evaluates to the quantity '3'.

2.5 Examples of Discrete Calculus

Writing out the Fundamental Theorem of Calculus on this simple graph may appear very familiar in the sense that it seems to represent nothing more than a forward differences approximation to the continuous gradient operator evaluated on a line discretized at four locations. While it is true that the our first example coincides with this more familiar case, we now proceed to show that the Fundamental Theorem of Calculus holds for more complicated graphs for which an interpretation as a finite difference approximation becomes less clear. Consider the graph in Fig. 2.21 which is exactly the same as before, except that a single edge and node have each been added. In this example,

$$\mathbf{c}^0 = \begin{bmatrix} 2 \\ 4 \\ 3 \\ 5 \\ 7 \end{bmatrix}, \tag{2.110}$$

$$\mathbf{A} = \begin{bmatrix} 0 & 0 & -1 & 1 & 0 \\ 0 & -1 & 1 & 0 & 0 \\ -1 & 1 & 0 & 0 & 0 \\ 0 & 0 & -1 & 0 & 1 \end{bmatrix}, \tag{2.111}$$

and

$$\boldsymbol{\tau}_1 = \begin{bmatrix} 1 \\ 1 \\ 1 \\ 1 \end{bmatrix}. \tag{2.112}$$

As before, the discrete version of the Fundamental Theorem of Calculus given in (2.106) holds. The left side of (2.106) is straightforward in the sense that we are again summing the differences of the 0-cochain given by $\mathbf{c}^0$. However, the right hand side no longer represents a simple evaluation of the difference of $\mathbf{c}^0$ at two locations. In this example of a tree, the right hand side of (2.106) represents the difference of $\mathbf{c}^0$ at the endpoints from the previous example *and* the difference of endpoint of the new branch with the node joining the new branch. The traditional Fundamental Theorem of Calculus summarizes the integral of the function derivative by the evaluation of the function at two points, due to the internal cancellations along the interior of the evaluation interval. In this example with a tree, the summation of the derivatives are summarized by evaluation of the function at all *terminals and joins*, while any other "internal" edges cancel. However it is important to point out that this 1-complex does not represent a combinatorial manifold (either with or without boundary) as it is defined in Sect. 2.3.3: The join node of this tree represents a non-manifold node and in this case appears to be in the "boundary" of the 1-complex as defined by the boundary operator $\mathbf{A}$. However, the discrete version of the Fundamental Theorem of Calculus still holds. While there are possible interpretations of this example that could lead to a continuous analogue, it is clear that using a tree as a domain of integration is fully possible within the discrete calculus framework yet the finite

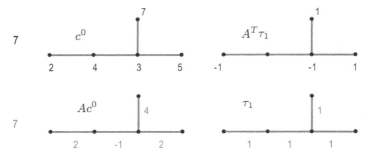

Fig. 2.21 A second example of the discrete calculus form of the Fundamental Theorem of Calculus in the same format as Fig. 2.20. In this example, the additional edge in the tree means that c^0 is no longer evaluated at just the endpoints, but rather the endpoints *and* the join point. This 1-complex is not a combinatorial manifold yet the Fundamental Theorem of Calculus still holds. In this case, both pairings result in a value of '7'

differences method cannot account for such a domain. This example highlights some of the flexibility of the discrete calculus approach.

If we apply the Fundamental Theorem of Calculus to the evaluation of a *line integral* in two dimensions, then it is well known that integration of a single-valued function around a closed loop evaluates to zero, since the beginning and end points are the same (i.e., $a = b$) and thus the loop has no boundary. We can test the same property of the discrete version of the Fundamental Theorem of Calculus by considering the example given by Fig. 2.22. This example shows a graph consisting of four nodes joined in a cycle by four edges. In this example,

$$\mathbf{c}^0 = \begin{bmatrix} 2 \\ 4 \\ 7 \\ 3 \end{bmatrix}, \tag{2.113}$$

$$\mathbf{A} = \begin{bmatrix} 0 & 0 & -1 & 1 \\ 0 & -1 & 1 & 0 \\ -1 & 1 & 0 & 0 \\ 1 & 0 & 0 & -1 \end{bmatrix}, \tag{2.114}$$

and

$$\boldsymbol{\tau}_1 = \begin{bmatrix} 1 \\ 1 \\ 1 \\ 1 \end{bmatrix}. \tag{2.115}$$

Clearly, $\mathbf{A}^\mathsf{T} \boldsymbol{\tau}_1 = \mathbf{0}$, so the pairing of the 0-chain and 0-cochain on right hand side of (2.106) evaluates to zero—regardless of the values of $\mathbf{c}^0$. However, the reader may verify that the pairing of the 1-chain and 1-cochain on left hand side of (2.106) also evaluates to zero for this example. In fact, $\mathbf{A}^\mathsf{T} \boldsymbol{\tau}_1 = \mathbf{0}$ whenever $\boldsymbol{\tau}_1$ represents a cycle,

2.5 Examples of Discrete Calculus

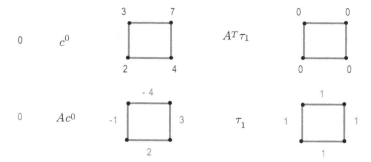

Fig. 2.22 Example of the Fundamental Theorem of Calculus evaluated on a cycle. Both the pairing of the 0-chain and 0-cochain depicted in the *top row* and the pairing of the 1-chain and 1-cochain depicted on the *bottom row* evaluate to zero in this example

since cycles are always in the nullspace of $\mathbf{A}^\mathsf{T}$. Therefore, any line integral around a cycle in a graph will always evaluate to zero, as anticipated from the familiar setting of standard calculus in which a closed line integral in a gradient field would also evaluate to zero.

2.5.1.2 Comparison with Finite Differences Operators

A reader with a numerical analysis background may view this example of integrating around a cycle as a very coarse discretization of the two-dimensional plane (consisting of only four points) using a 5-point stencil. In this case, we may examine a slightly larger graph consisting of a 4×4 lattice for which our previous example comprises only the central four points. Figure 2.23 illustrates this new example in which the values associated with each node are seen again in the central four nodes, but now additional values are assigned to the new nodes. If we were to evaluate the discrete version of the Fundamental Theorem of Calculus on this new graph, we would find that each side still evaluates to zero when τ_1 represents any cycle.

The reader familiar with numerical analysis might casually regard $\mathbf{A}$ as simply a forward difference approximation to the continuous gradient and question the importance being given to the operator in this book. However, it is important to note that $\mathbf{A}$ is a discrete analogue of the continuous gradient operator, while a finite differences discretization of the gradient must *adapt the numerical implementation of the gradient to the computational task*. In fact, if we try to use a finite differences gradient as a discretization of the gradient *operator* then we are led into problems not encountered by the discrete calculus analogue of the gradient. For example, one property of the gradient operator is that it can be used to produce a gradient field from a scalar field. A closed line integral within this gradient field should yield a zero value of the integral, regardless of the path of the line integral. In our example we will compare the properties of the gradient field produced by central differences (treated as the discretized gradient *operator*) with the gradient field produced by the discrete calculus gradient operator $\mathbf{A}$.

> The method of finite differences provides a means to *discretize* the gradient operator, and is useful in some applications, but does not preserve the topological properties of the exterior derivative and thus does not yield relations that satisfy the Fundamental Theorem of Calculus. Specifically, the "gradient field" produced by this finite differences procedure integrated around a closed contour does not evaluate to zero. In contrast, the discrete calculus gradient defined as a special case of the coboundary operator does exhibit the desirable topological behavior.

We showed in Fig. 2.22 that the gradient field obtained by applying the discrete calculus gradient operator to a scalar field, i.e., an exact form, always integrates to zero around a closed loop. We now demonstrate that applying the same steps for the central differences operator will not produce a zero integration around a closed loop. This example is illustrated in Fig. 2.23 by displaying the central differences approximation to the gradient *at the nodes* for each of the central points. Recall that the central differences gradient approximation of the two-dimensional function $f(x, y)$ at location (x, y) is given by

$$\nabla f(x, y) = \begin{bmatrix} \frac{f(x-1,y)+f(x+1,y)}{2} \\ \frac{f(x,y-1)+f(x,y+1)}{2} \end{bmatrix}. \quad (2.116)$$

Using $\mathbf{c}^0$ as $f(x, y)$, we see from Fig. 2.23 that summing these vectors around the *nodes* in a cycle (since central differences only produces vectors at the nodes) gives

$$\begin{bmatrix} 2 \\ 3 \end{bmatrix} + \begin{bmatrix} 2 \\ 2 \end{bmatrix} + \begin{bmatrix} 3 \\ 1 \end{bmatrix} + \begin{bmatrix} -2 \\ -2 \end{bmatrix} = \begin{bmatrix} 5 \\ 4 \end{bmatrix} \neq \begin{bmatrix} 0 \\ 0 \end{bmatrix}. \quad (2.117)$$

This example illustrates that it is important to distinguish the discrete calculus operators from the discretization strategies common in numerical analysis. Because the gradient operation is enacted by the coboundary operator, the gradient of a 0-cochain defined on nodes is a 1-cochain on the edges. If the gradients at these edges are summed in any cycle, then the sum will be zero, precisely because the incidence matrix $\mathbf{A}^\mathsf{T}$ represents the *boundary operator* on a graph, because cycles have no boundary. In contrast, traditional numerical analysis associates gradients (and vectors in general) with nodes and the central differences operator does *not* represent a boundary operator. In summary, the central differences operator may represent a useful *discretization* of the continuous gradient, but the analogous discrete calculus operator is designed to capture the theoretical properties of the gradient in a discrete setting, such as the behavior of the gradient as a boundary operator.

2.5 Examples of Discrete Calculus

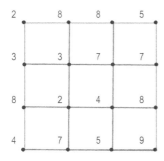
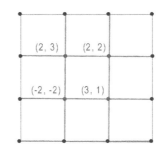

Fig. 2.23 Demonstration of the violation of the Fundamental Theorem of Calculus that is made when employing the finite differences approximation to the gradient operator (as a substitute for the coboundary operator). In this example a graph lattice is considered as an approximation to the two-dimensional plane, and a scalar field is defined on the nodes (shown in *blue*). The central differences method provides an approximation to the gradient the central region of the graph, resulting in a tuple of components at each node (shown in *green*). If we integrate these resulting values around a closed loop, the sum does not evaluate to zero

2.5.1.3 Generalized Stokes' Theorem on a 2-Complex

Letting $p = 2$, the discrete version of the Generalized Stokes' Theorem in (2.104) becomes

$$[\![\mathbf{Bc}^1, \tau_2]\!] = [\![\mathbf{c}^1, \mathbf{B}^T \tau_2]\!]. \qquad (2.118)$$

In this case, the edge–face incidence matrix **B** represents the coboundary operator and $\mathbf{B}^T$ represents the boundary operator, $\mathbf{c}^1$ a 1-cochain (establishing values at each edge), and τ_2 a 2-chain over which to perform the integration. This discrete version of the Generalized Stokes' Theorem corresponds to the traditional conception of Stokes' Theorem

$$\iint_S (\nabla \times \vec{H}) \cdot d\vec{S} = \int_{\partial S} \vec{H} \cdot d\vec{r}, \qquad (2.119)$$

in which **B** plays the role of $\nabla \times$, the vector field $\vec{H}$ corresponds to the flow field[22] $\mathbf{c}^1$ and the domain of integration S is represented by τ_2.

---

[22] We employ the term *flow* field to represent a 1-cochain throughout the text. This term is used because flows are common 1-cochains (e.g., the maximum flow problem optimizes a 1-cochain) and the term instills a sense of *direction* for the flow through each edge.

Figure 2.24 gives an example of the Generalized Stokes' Theorem on edges. In this example,

$$\mathbf{c}^1 = \begin{bmatrix} 2 \\ 6 \\ 3 \\ 6 \\ 4 \\ 5 \\ 4 \\ 3 \\ 7 \end{bmatrix}, \qquad (2.120)$$

$$\mathbf{B} = \begin{bmatrix} 1 & 1 & -1 & 0 & 0 & 0 & 0 & 0 & 0 \\ 0 & 0 & 0 & 1 & -1 & 0 & 0 & 1 & 0 \\ 0 & 0 & 1 & 0 & 1 & 1 & 0 & 0 & 0 \\ 0 & 0 & 0 & 0 & 0 & -1 & 1 & 0 & 1 \end{bmatrix}, \qquad (2.121)$$

and

$$\boldsymbol{\tau}_2 = \begin{bmatrix} 1 \\ 1 \\ 1 \\ 1 \end{bmatrix}. \qquad (2.122)$$

We can see in Fig. 2.24 that the left hand side of the discrete version of the Generalized Stokes' Theorem calculates the circulation (i.e., the curl) around each cycle which is then summed over all cycles, while the right hand side sums the edge flows *on the boundary of the chain* since the interior circulations cancel each other (as commonly seen in the traditional proofs of the standard Stokes' Theorem). In this example, both pairings evaluate to 28.

2.5.1.4 Generalized Stokes' Theorem on a 3-Complex

The manifestation of the Generalized Stokes' Theorem in three dimensions is given by the classical Divergence Theorem (or Gauss's Theorem)

$$\iiint_V \nabla \cdot \vec{H} \, dV = \int_{\partial V} \vec{H} \cdot d\vec{S}, \qquad (2.123)$$

and its discrete analogue

$$[\![\mathbf{N}_3 \mathbf{c}^2, \boldsymbol{\tau}_3]\!] = [\![\mathbf{c}^2, \mathbf{N}_3^\mathsf{T} \boldsymbol{\tau}_3]\!], \qquad (2.124)$$

where in this case the volume-face incidence matrix $\mathbf{N}_3$ represents the coboundary operator and $\mathbf{N}_3^\mathsf{T}$ represents the boundary operator, $\mathbf{c}^2$ is a 2-cochain (establishing values on each face) and $\boldsymbol{\tau}_3$ is a 3-chain over which to perform the integration.

2.5 Examples of Discrete Calculus

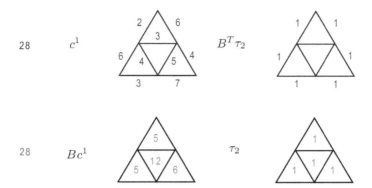

Fig. 2.24 Example of Generalized Stokes' Theorem on a 2-complex. The *top row* depicts the pairing of a 1-cochain $\mathbf{c}^1$ with a 1-chain $\mathbf{B}^T\tau_2$, with coefficients on the edges represented in *blue*. This pairing sums the edge flows *on the boundary of the chain* since the interior circulations cancel each other. The *bottom row* depicts the pairing of a 2-cochain $\mathbf{Bc}^1$ with a 2-chain τ_2, with coefficients on the faces represented in *red*. This pairing computes the circulation or curl around each face and sums over all faces in the 2-chain. In this example, both pairings evaluate to '28'

Although the Divergence Theorem is naturally expressed in three dimensions, it is more difficult to adequately draw an example. In Sect. 2.3.4 we saw that, in the discrete setting, each p-cochain is dual to an $(n - p)$-cochain, where n is the ambient dimension. Therefore, we may express the *dual* form for the divergence theorem

$$[\![\mathbf{A}^T\mathbf{c}^1, \tau_0]\!] = [\![\mathbf{c}^1, \mathbf{A}\tau_0]\!]. \tag{2.125}$$

The interpretation of this dual version of the Divergence Theorem states that the sum of the divergence of the flow field $\mathbf{c}^1$, given by $\mathbf{A}^T\mathbf{c}^1$, integrated over a region of nodes τ_0 equals the flow out of the region subtracted from the flow into the region. Therefore, this dual formulation of the Divergence Theorem corresponds more closely to the intuition of the conventional Divergence Theorem.

Since the dual variant of the combinatorial Divergence Theorem is the most intuitive, it is important to note that this theorem also applies to a 2-complex, for which the dual is also a 2-complex. Figure 2.25 illustrates this case, in which two units of edge flow pass into the central three nodes and two units of flow exit these three nodes. Consequently, the total divergence inside the three nodes is zero, since all of the flow entering these nodes also leaves the nodes.

2.5.2 The Helmholtz Decomposition

The conventional Helmholtz Theorem states that each sufficiently smooth, rapidly decaying vector field may be decomposed into the sum of two orthogonal components: an irrotational (curl-free) and a solenoidal (divergence-free) vector fields.

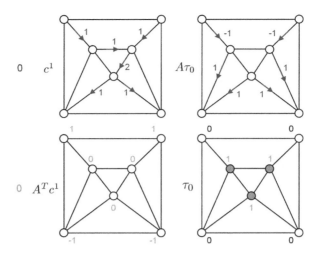

Fig. 2.25 Example of the (dual) Divergence Theorem on a 2-complex. The *top row* depicts the pairing of a 1-cochain $\mathbf{c}^1$ with a 1-chain $\mathbf{A}\tau_0$, with coefficients on the edges represented in *blue*. Any edge not labeled is assumed to have zero flow. This pairing sums the edge flows *into* the central nodes and subtracts the edge flow *leaving* the central nodes, to produce zero (i.e., all flow entering also leaves). The *bottom row* depicts the pairing of a 0-cochain $\mathbf{A}^\mathsf{T}\mathbf{c}^1$ with a 0-chain τ_0, with coefficients on the nodes represented in *red*. This pairing shows that because each of the central nodes have zero divergence, then all of the flow entering the nodes must also leave

Formally, a vector field $\vec{F}$ may be decomposed into

$$\vec{F} = \vec{F}_{\text{irrot}} + \vec{F}_{\text{sol}}, \tag{2.126}$$

where

$$\vec{F}_{\text{irrot}} = \nabla\phi, \tag{2.127}$$

$$\nabla \cdot \vec{F}_{\text{sol}} = 0, \tag{2.128}$$

for some scalar potential field ϕ, and these two components are orthogonal, i.e.,

$$\iiint_V (\vec{F}_{\text{irrot}} \cdot \vec{F}_{\text{sol}})\, dV = 0. \tag{2.129}$$

We now prove that the discrete analogue of the Helmholtz Theorem is also true. We begin by associating the conventional vector field $\vec{F}$ with a 1-cochain (representing edge flows) $\mathbf{c}^1$. Therefore, we can define $\mathbf{c}^1_{\text{irrot}}$ and $\mathbf{c}^1_{\text{sol}}$ as any vectors satisfying

$$\mathbf{c}^1_{\text{irrot}} = \mathbf{A}\mathbf{c}^0, \tag{2.130}$$

$$\mathbf{A}^\mathsf{T}\mathbf{c}^1_{\text{sol}} = \mathbf{0}, \tag{2.131}$$

2.5 Examples of Discrete Calculus

for some 0-cochain $\mathbf{c}^0$ or, alternately,

$$\mathbf{B}^\mathsf{T}\mathbf{c}^1_{\text{irrot}} = \mathbf{0}, \tag{2.132}$$

$$\mathbf{c}^1_{\text{sol}} = \mathbf{B}^\mathsf{T}\mathbf{c}^2, \tag{2.133}$$

for some 2-cochain $\mathbf{c}^2$. Therefore, the name "irrotational" to describe the $\mathbf{c}^1_{\text{irrot}}$ component can be seen from the fact that $\mathbf{B}$ is the discrete analogue of the curl operator (i.e., the curl of $\mathbf{c}^1_{\text{irrot}}$ is zero). Note that $\mathbf{B}$ must represent $|\mathcal{F}| = |\mathcal{E}| - |\mathcal{V}| + 1$ cycles, such that its columns provide a basis for the nullspace of $\mathbf{A}$ (see Chap. 4).

Given these definitions, it is straightforward to show that the two components are orthogonal.

Theorem 2.1

$$\langle \mathbf{c}^1_{\text{sol}}, \mathbf{c}^1_{\text{irrot}} \rangle = 0. \tag{2.134}$$

Proof This statement is true by

$$\langle \mathbf{c}^1_{\text{sol}}, \mathbf{c}^1_{\text{irrot}} \rangle = 0, \tag{2.135}$$

$$(\mathbf{c}^1_{\text{sol}})^\mathsf{T}\mathbf{c}^1_{\text{irrot}} = 0, \tag{2.136}$$

$$\mathbf{c}^{0\mathsf{T}}\mathbf{A}^\mathsf{T}\mathbf{c}^1_{\text{irrot}} = 0, \tag{2.137}$$

$$\mathbf{c}^{0\mathsf{T}}\mathbf{0} = 0. \tag{2.138}$$

□

With this preliminary result, we can state the discrete calculus analogue of the Helmholtz Theorem.

Theorem 2.2 *Every $\mathbf{c}^1$ may be decomposed into*

$$\mathbf{c}^1 = \mathbf{c}^1_{\text{irrot}} + \mathbf{c}^1_{\text{sol}}. \tag{2.139}$$

Proof By (2.131) and (2.132) we can rewrite the discrete calculus Helmholtz decomposition in (2.139) as

$$\mathbf{c}^1 = \mathbf{A}\mathbf{c}^0 + \mathbf{B}^\mathsf{T}\mathbf{c}^2, \tag{2.140}$$

$$\mathbf{c}^1 = \begin{bmatrix} \mathbf{A} & \mathbf{B}^\mathsf{T} \end{bmatrix} \begin{bmatrix} \mathbf{c}^0 \\ \mathbf{c}^2 \end{bmatrix}. \tag{2.141}$$

Since $\mathbf{A}$ is of size $m \times n$ where $m = |\mathcal{E}|$ and $n = |\mathcal{V}|$, $\mathbf{B}^\mathsf{T}$ is of size $m \times (m - n + 1)$, the combined matrix in (2.141) is of size $m \times (m + 1)$. However, since the rank of $\mathbf{A}$ is $(n - 1)$, the rank of $\mathbf{B}$ is $(m - n + 1)$ and $\mathbf{A}$ is orthogonal to $\mathbf{B}^\mathsf{T}$, the combined matrix has rank equal to $(m - n + 1) + (n - 1) = m$. Consequently, the combined matrix in (2.141) provides a basis for the entire range space, meaning that any vector $\mathbf{c}^1$ is represented. □

Intuitively, the discrete Helmholtz decomposition states that each flow field on the edges can be decomposed into a component which represents the gradient of some scalar (node) potential field and a cyclic component. In broader generality, the Helmholtz decomposition represents a special case of the Hodge decomposition, outlined above in Sect. 2.2. The precise relationship between these two decompositions is provided in detail in Appendix C.

We now consider how to compute a Helmholtz decomposition for some flow field. Since the Helmholtz decomposition is not unique, the method we present produces only one of many possible Helmholtz decompositions. If the decomposition is to be used for a particular purpose, some regularization might be employed to find a unique decomposition which is well-suited to the problem. There has been little treatment of this topic in the existing literature (although there is some recent activity [24, 25, 219]). However, the method we present for computing a Helmholtz decomposition is straightforward. We now consider a simple algorithm for computing the Helmholtz decomposition of a flow field on a graph.

2.5.2.1 Algorithm for Computing a Helmholtz Decomposition of a Flow Field

In order to produce a Helmholtz decomposition of $\mathbf{c}^1$, our goal is to find a $\mathbf{c}^0$ that can be used to generate $\mathbf{c}^1_{\text{irrot}}$ via (2.130). Once we have $\mathbf{c}^1_{\text{irrot}}$, then $\mathbf{c}^1_{\text{sol}} = \mathbf{c}^1 - \mathbf{c}^1_{\text{irrot}}$. In order for $\mathbf{A}^T \mathbf{c}^1_{\text{sol}} = 0$ to be satisfied, then the condition for $\mathbf{c}^0$ becomes

$$\mathbf{A}^T \mathbf{c}^1_{\text{sol}} = 0, \tag{2.142}$$

$$\mathbf{A}^T (\mathbf{c}^1 - \mathbf{A}\mathbf{c}^0) = 0, \tag{2.143}$$

$$\mathbf{A}^T \mathbf{A} \mathbf{c}^0 = \mathbf{L} \mathbf{c}^0 = \mathbf{A}^T \mathbf{c}^1. \tag{2.144}$$

Consequently, if we simply solve (2.144) for $\mathbf{c}^0$, then this condition is automatically satisfied. However, we know that the Laplacian matrix $\mathbf{L}$ of a connected graph is missing one rank because $\mathbf{A}\mathbf{1} = 0$. Fortunately, $\mathbf{A}^T \mathbf{c}^1$ is orthogonal to $\mathbf{1}$, since $\mathbf{1}^T \mathbf{A}^T \mathbf{c}^1 = 0$. Consequently, we can arbitrarily fix the value of $\mathbf{c}^0$ at some node v_0 to be zero, $v_0 = 0$ and solve

$$\mathbf{L}_0 \mathbf{c}^0_0 = (\mathbf{A}^T \mathbf{c}^1)_0, \tag{2.145}$$

in which the zero subscript is used to indicate the removal of the row corresponding to v_0 (and the column as well for $\mathbf{L}_0$).

The algorithm for computing a Helmholtz decomposition of a flow field is then given by the following steps.

1. Arbitrarily identify some node v_0.
2. Solve (2.145) for $\mathbf{c}^0_0$.
3. Form $\mathbf{c}^0(v_0) = 0$, $\mathbf{c}^0(v_i) = \mathbf{c}^0_0(v_i)$, $\forall v_i \neq v_0$.

4. Set $\mathbf{c}_{\text{irrot}}^1 = \mathbf{A}\mathbf{c}^0$.
5. Set $\mathbf{c}_{\text{sol}}^1 = \mathbf{c}^1 - \mathbf{c}_{\text{irrot}}^1$.

The primary computational burden of this algorithm is the linear system solve required by solving (2.145). Although the linear system is generally sparse, the linear system solution will have a super-linear time complexity for most linear solver algorithms. Therefore, this algorithm represents the most simple approach to computing a Helmholtz decomposition, but it may not have the lowest computational complexity.

2.5.3 Matrix Representation of Discrete Calculus Identities

In this section, we examine the discrete analogues of identities from conventional vector calculus. Many of these vector identities are memorable largely because they are surprising. In contrast, we will see that the discrete calculus analogues of these identities are often more straightforward, but we include them here because these identities are familiar tools from continuous analysis that may also be applied in the discrete setting.

2.5.3.1 Integration by Parts

Integration by parts is a key technique to produce a closed-form expression for a complicated integral. In discrete calculus, we are generally less concerned with finding closed-form expressions (since computers can be used to find solutions as needed), but it is still interesting to examine the discrete analogue of the integration by parts technique. Conventionally, the integration by parts formula is written as

$$\int_a^b u \frac{dv}{dt} = uv \Big|_a^b - \int_b^a \frac{du}{dt} v. \qquad (2.146)$$

This expression may be generalized to two dimensions, giving the **Gauss–Green Theorem** [360]

$$\iint_{\mathcal{R}} [u(\nabla \cdot \vec{F})] \, dx \, dy = \int_{\mathcal{C}} u\vec{F} \cdot d\vec{S} - \iint_{\mathcal{R}} [(\nabla u) \cdot \vec{F}] \, dx \, dy. \qquad (2.147)$$

The connection between the Gauss–Green Theorem and the Divergence Theorem is often noted, e.g., if u is equal to unity, then the Gauss–Green Theorem (2.147) is equivalent to the Divergence Theorem for $\vec{H}$ (2.123). Therefore, it should be no surprise that the Gauss–Green Theorem can be viewed as a generalization of the

Table 2.1 Relationship between second-order vector identities in the conventional notation and corresponding identities in the discrete setting. Here u represents a scalar field and $\vec{F}$ a vector field. Correspondingly, **u** represents a 0-cochain and **f** a 1-cochain

Conventional	Discrete Calculus
$\nabla \times \nabla u = 0$	$\mathbf{BAu} = 0$
$\nabla \cdot \nabla \times \vec{F} = 0$	$\mathbf{A^T B^T f} = 0$

Divergence Theorem that additionally includes the scalar field u. Recall that the discrete analogue of the Divergence Theorem is

$$\boldsymbol{\tau}_0^T (\mathbf{A^T f}) = (\mathbf{A}\boldsymbol{\tau}_0)^T \mathbf{f}, \tag{2.148}$$

for 0-chain $\boldsymbol{\tau}_0$, edge–node incidence matrix $\mathbf{A}$ and 1-cochain $\mathbf{f}$. We may now introduce the 0-cochain $\mathbf{u}$ to give a discrete analogue of the Gauss–Green formula

$$\mathbf{u}^T \operatorname{diag}(\boldsymbol{\tau}_0)(\mathbf{A^T f}) = (\mathbf{A}\operatorname{diag}(\boldsymbol{\tau}_0)\mathbf{u})^T \mathbf{f}. \tag{2.149}$$

In this expression the left hand side, $\mathbf{u}^T \operatorname{diag}(\boldsymbol{\tau}_0)(\mathbf{A^T f})$, corresponds to the left hand side in the conventional Gauss–Green formula $\iint_{\mathcal{R}} [u(\nabla \cdot \vec{F})]\, dx\, dy$ since $\boldsymbol{\tau}_0$ defines the domain of integration $\mathcal{R}$ and $\mathbf{A^T f}$ corresponds to $-\nabla \cdot \vec{F}$. The right hand side of the discrete calculus analogue in (2.149) corresponds to both terms of the right hand side in the conventional Gauss–Green Theorem. The right hand side of the discrete calculus analogue could also be broken into two terms like the conventional Gauss–Green Theorem by setting

$$(\mathbf{A}\operatorname{diag}(\boldsymbol{\tau}_0)\mathbf{u})^T \mathbf{f} = (\mathbf{A}\operatorname{diag}(\boldsymbol{\tau}_0)\mathbf{u})^T \operatorname{diag}(\boldsymbol{\tau}_{\text{boundary}})\mathbf{f}$$
$$+ (\mathbf{A}\operatorname{diag}(\boldsymbol{\tau}_0)\mathbf{u})^T \operatorname{diag}(\boldsymbol{\tau}_{\text{interior}})\mathbf{f}, \tag{2.150}$$

in which the set of edges connecting two nodes in $\boldsymbol{\tau}_0$ are represented by the 1-chain $\boldsymbol{\tau}_{\text{interior}}$ and the set of edges incident to exactly one node in $\boldsymbol{\tau}_0$ is represented by $\boldsymbol{\tau}_{\text{boundary}}$. As with the conventional Gauss–Green Theorem, the discrete calculus analogue given by (2.149) equals the Divergence Theorem in (2.148) when $\mathbf{u}$ is equal to unity.

2.5.3.2 Other Identities

We close this section by listing a few additional identities from conventional vector calculus and their discrete calculus analogues, using the same notation as in the previous section. Table 2.1 displays identities describing the homology relationships between the conventional vector calculus operators and their discrete calculus counterparts. Table 2.2 lists the conventional forms of the Green's Theorems and the analogous identities in discrete calculus.

The last identity in Table 2.2 appears somewhat asymmetric between the conventional and discrete calculus expressions: The conventional expression includes two

2.5 Examples of Discrete Calculus

Table 2.2 Green's identities [127]. Here u and v represent scalar fields, **u** and **v** represent the corresponding 0-cochains, and τ_0 represents the 0-chain as the domain of integration. Note that here we use the substitution $\mathbf{Y}_0 = \text{diag}(\boldsymbol{\tau}_0)$

Conventional	Discrete Calculus
$\int_{\mathcal{R}} (\nabla^2 u)\, dV = \int_{\partial \mathcal{R}} \nabla u \cdot d\vec{S}$	$\boldsymbol{\tau}_0^T \mathbf{A}^T \mathbf{A} \mathbf{u} = (\mathbf{A}\boldsymbol{\tau}_0)^T \mathbf{A}\mathbf{u}$
$\int_{\mathcal{R}} (\nabla u \cdot \nabla v)\, dV = -\int_{\mathcal{R}} (u\nabla^2 v)\, dV + \int_{\partial \mathcal{R}} u \nabla v \cdot d\vec{S}$	$(\mathbf{u}\mathbf{Y}_0\mathbf{A})^T \mathbf{A}\mathbf{Y}_0 \mathbf{v} = \mathbf{u}^T \mathbf{Y}_0 (\mathbf{A}^T \mathbf{A}\mathbf{Y}_0 \mathbf{v})$

volume integrals and a boundary integral, whereas only two terms appear explicitly in the discrete calculus expression. However it is possible to split this discrete calculus expression up into separate components for the volume integral and the boundary integral, just as in the integration by parts example presented above (see (2.150)).

Note that the two identities in Table 2.2 are sometimes referred to as Green's First Identity, in which case the top entry is a special case of the bottom entry. The general form of this Green's identity is usually expressed as

$$\int_{\mathcal{R}} (u\nabla^2 v - v\nabla^2 u)\, dV = \int_{\partial \mathcal{R}} [u(\nabla v) - v(\nabla u)] \cdot d\vec{S}, \qquad (2.151)$$

which may be seen as two applications of the identity expressed in Table 2.2 in which

$$\int_{\mathcal{R}} (\nabla u \cdot \nabla v)\, dV = -\int_{\mathcal{R}} (u\nabla^2 v)\, dV + \int_{\partial \mathcal{R}} u(\nabla v) \cdot d\vec{S}$$

$$= -\int_{\mathcal{R}} (v\nabla^2 u)\, dV + \int_{\partial \mathcal{R}} v(\nabla u) \cdot d\vec{S}. \qquad (2.152)$$

2.5.4 Elliptic Equations

We begin our treatment of elliptic equations by discussing the properties of harmonic functions on a graph, following the treatments in [36, 74, 81, 111, 115, 161]. In continuous mathematics, a function ϕ is called **harmonic** if it satisfies

$$\nabla^2 \phi = 0, \qquad (2.153)$$

at every location in the domain except at the boundary. The discrete calculus analogue is naturally

$$\mathbf{L}\mathbf{x} = \mathbf{0}. \qquad (2.154)$$

In the absence of boundary conditions, the only solution to (2.154) is when **x** equals a constant value on all nodes. Therefore, we can assume a set of Dirichlet boundary conditions on some set $\mathcal{D} \subset \mathcal{V}$, meaning that x_i is fixed to a boundary value q_i for all $v_i \in \mathcal{D}$. In this section, we can ignore the node weights, since they can be easily removed from (2.154).

Following [161], we can state the **mean-value theorem** for harmonic functions.

Theorem 2.3 *For a harmonic 0-cochain* **x**, *the value of the component x_i for any $v_i \notin \mathcal{D}$ is equal to the weighted mean of the values x_j at its neighbors.*

Proof The proof is trivially seen by writing (2.154) in summation form. In order to satisfy (2.154), each x_i must satisfy

$$x_i = \frac{1}{d_i} \sum_{e_{ij} \in \mathcal{E}} w_{ij} x_j, \qquad (2.155)$$

for node degree d_i. □

Note that the edge weights w_{ij} in the Laplacian represent *affinity* weights. A corollary is the **local maximum principle** and **local minimum principle**

Corollary 2.1 *For a harmonic 0-cochain* **x**, *the component x_i for every $v_i \notin \mathcal{D}$ satisfies*

$$\min_{v_j | e_{ij} \in \mathcal{E}} (x_j) \leq x_i \leq \max_{v_j | e_{ij} \in \mathcal{E}} (x_j). \qquad (2.156)$$

Proof The result is trivially true from (2.155), provided that $w_{ij} > 0$ (as assumed throughout unless otherwise noted). □

The local maximum principle also implies the **strong local maximum principle** (and the strong local minimum principle)

Corollary 2.2 *If $v_i \notin \mathcal{D}$ and $\exists e_{ij} \in \mathcal{E}$ such that $x_i = x_j$, then $x_i = x_j$ for all $e_{ij} \in \mathcal{E}$.*

Proof Equality is achieved in (2.155) only when x_j is a constant for all neighbors of v_i, i.e., $\exists e_{ij} \in \mathcal{E}$. Consequently, if $x_i = x_j$ for any neighbor of v_i, then all neighbors of v_i must have the same value. □

A second corollary is the discrete calculus analogue of the conventional **maximum principle** and **minimum principle** which state that the value of a harmonic function inside a domain is less than the maximum value of the function on the boundary and greater than the minimum value on the boundary. Since the nodes in $\mathcal{D}$ represent the boundary, the discrete calculus analogue is given by

Corollary 2.3 *For a harmonic 0-cochain* **x**, *the component x_i for every $v_i \notin \mathcal{D}$ satisfies*

$$\min_{v_j \in \mathcal{D}} (x_j) \leq x_i \leq \max_{v_j \in \mathcal{D}} (x_j). \qquad (2.157)$$

2.5 Examples of Discrete Calculus

Proof This statement follows directly from the local maximum/minimum principle. □

The analogue of the **strong maximum principle** (and the strong minimum principle) states

Theorem 2.4 *If* $\mathcal{V} - \mathcal{D}$ *is connected and* $x_i = \max_{v_j \in \mathcal{D}}(x_j)$ *for any* $v_i \notin \mathcal{D}$, *then* $x_i = \max_{v_j \in \mathcal{D}}(x_j)$ *for all* $v_i \notin \mathcal{D}$.

Proof This statement follows directly from the strong local maximum/minimum principle and the condition that $\mathcal{V} - \mathcal{D}$ is connected. □

Several discrete calculus analogues of variants on the **Harnack Inequality** have appeared in the literature [81, 111, 112]. Here we treat an analogue of the conventional form of the inequality given by Evans [127]

Theorem 2.5 *For any connected vertex subset* $\mathcal{V}' \subseteq \mathcal{V} - \mathcal{D}$, *there exists a positive constant* C, *depending only on* $\mathcal{V}'$, *such that*

$$\sup_{v_i \in \mathcal{V}'} x_i \leq C \inf_{v_i \in \mathcal{V}'} x_i, \qquad (2.158)$$

for all nonnegative harmonic functions **x** *in* $\mathcal{V}$.

Proof This statement follows directly from the strong local maximum/minimum principle and the condition that $\mathcal{V}'$ is connected. □

The Harnack Inequality therefore implies that within any connected set the values of a nonnegative harmonic function are all comparable in the sense that no x_i for a $v_i \in \mathcal{V}'$ can be too large (small) unless the other x_j for all $v_j \in \mathcal{V}'$ are also large (small). Other inequalities on the spectrum of the Laplacian for which the conventional theory also has discrete calculus analogues, such as **Cheeger's inequality** and **Buser's inequality**, are treated in Chap. 8.

By definition, a harmonic function **x** comprises the solution to the classic elliptical PDE, the **Laplace equation**

$$\mathbf{L}\mathbf{x} = \mathbf{0}. \qquad (2.159)$$

Similarly, the discrete calculus version of the **Poisson equation** is given by

$$\mathbf{L}\mathbf{u} = \mathbf{v}, \qquad (2.160)$$

for some function **v** and a twice differentiable function **u**. Due to the rank deficiency of **L**, these equations do not have a unique solution unless boundary conditions are enforced (see Appendix B).

2.5.4.1 Variational Principles

A common viewpoint on conventional PDEs is that the solutions of the equations represent the *minimum energy* state of a functional. In particular, Dirichlet's Principle states that the solution of the Poisson equation, i.e.,

$$\nabla^2 u = v \quad \text{in } \mathcal{R}, \tag{2.161}$$

$$u = f \quad \text{on } \partial \mathcal{R}, \tag{2.162}$$

minimizes the energy functional

$$\mathcal{Q}[u] = \int_{\mathcal{R}} \frac{1}{2}(\|\nabla u\|^2 + uv) dV, \tag{2.163}$$

among all twice differentiable u that respect the boundary conditions.

Similarly, the solution of the discrete calculus Poisson equation

$$\mathbf{Lu} = \mathbf{v} \quad \text{in } \mathcal{R}, \tag{2.164}$$

$$\mathbf{u} = \mathbf{f} \quad \text{on } \partial \mathcal{R}, \tag{2.165}$$

minimizes the energy functional

$$\mathcal{Q}[\mathbf{u}] = \frac{1}{2}\mathbf{u}^\mathsf{T}\mathbf{Lu} - \mathbf{u}^\mathsf{T}\mathbf{v}, \tag{2.166}$$

among all $\mathbf{u}$ that satisfy the boundary conditions (note the sign change with the conventional formulation because of the identification of $\mathbf{A}^\mathsf{T}$ with $-\nabla \cdot$). The proof of Dirichlet's Principle in the discrete case is much simpler, since it depends only on the fact that the reduced Laplacian matrix $\mathbf{L}$ is positive definite. Note that the reduction occurs as a result of the imposed boundary conditions (see Appendix B for more details).

Energy minimization comprises an important theme of this entire book. In the applications chapters we often motivate an algorithm by the energy minimization that drives the optimization, in which the energy is typically composed of data information and a regularization scheme based on a prior model that helps us extract information from the data. Additionally, the formulation of objective energies in a discrete calculus setting often leads to a more straightforward optimization which can additionally make use of the strong body of combinatorial optimization techniques. Appendix B contains more information specifically on optimization techniques, but examples are spread throughout the book of establishing an objective functional for which the optimum provides a solution to filtering, clustering, etc.

2.5.5 Diffusion

Diffusion is a ubiquitous process in physics which governs many situations such as the spread of chemical concentration in still water through time or the distribution of

2.5 Examples of Discrete Calculus

heat in a solid body through time. Specifically, the **diffusion equation** (also known as the **heat equation**) is given by [127]

$$\frac{\partial u}{\partial t} = \alpha \nabla^2 u, \qquad (2.167)$$

where u is a scalar field representing the distribution of heat (chemical concentration) and α is the *diffusion constant* describing the rate of heat transfer in the body. When α has the same value at all locations in space, we say that the material is *homogeneous* and when α varies with spatial location we say that the material is *inhomogeneous*.

Before presenting a discrete calculus analogue of the diffusion equation, we show how the equation may be derived from conservation of mass (energy) and Fick's Law. Conservation of mass requires that the amount of the substance u in any region must either leave through the boundary or have an external source, i.e.,

$$\frac{\partial}{\partial t} \int_S u \, dV = -\int_{\partial S} \vec{F} \cdot d\vec{S} + \int_S s \, dV, \qquad (2.168)$$

where s is a scalar field of heat sources (a positive value of s indicates a source and a negative value a sink) and $\vec{F}$ is the flux of heat through the boundary. Since S may be arbitrary, then this expression may be rewritten as

$$\frac{\partial u}{\partial t} = -\nabla \cdot \vec{F} + s. \qquad (2.169)$$

In the physical process of diffusion, the expression for the flux $\vec{F}$ is modeled by **Fick's Law**

$$\vec{F} = -\alpha \nabla u, \qquad (2.170)$$

which states that the amount of flux is proportional to the temperature gradient. Putting Fick's Law together with the conservation equation gives

$$\frac{\partial u}{\partial t} = \alpha \nabla^2 u + s, \qquad (2.171)$$

which is equal to the diffusion equation (2.167) in which the source s term is set to zero (i.e., when there are no sources or sinks present).

Following the same derivation for the discrete calculus analogue gives us the expression for conservation of mass

$$\frac{\partial}{\partial t} \tau_0^\mathsf{T} \mathbf{u} = (\mathbf{A}\tau_0)^\mathsf{T} \mathbf{f} + \tau_0^\mathsf{T} \mathbf{s}, \qquad (2.172)$$

where S is represented by the 0-chain τ_0. Note that $\mathbf{u}$ and $\mathbf{s}$ are 0-cochains (i.e., node functions) while $\mathbf{f}$ is a 1-cochain (i.e., an edge function). By restricting S to a single node, we can write the equivalent of (2.169) in the whole domain as

$$\frac{\partial \mathbf{u}}{\partial t} = \mathbf{A}^\mathsf{T} \mathbf{f} + \mathbf{s}. \qquad (2.173)$$

Note that the sign has changed between this expression and (2.169) due to the sign change induced by taking the adjoint of the boundary operation in conventional vector calculus. Similarly, the discrete calculus analogue of Fick's Law is given by

$$\mathbf{f} = -\alpha \mathbf{A}\mathbf{u}. \tag{2.174}$$

Note that the diffusion constant α modifies the flux *through an edge*. Consequently, we may identify α with the edge weight. Putting these expressions together gives us the discrete calculus analogue of the diffusion equation as

$$\frac{\partial \mathbf{u}}{\partial t} = -\mathbf{A}^\mathsf{T} \operatorname{diag}(\alpha) \mathbf{A}\mathbf{u} + \mathbf{s}. \tag{2.175}$$

For simplicity, we also neglect sources, i.e., set $\mathbf{s} = 0$. If our domain is inhomogeneous, then α may vary for different edges. Therefore, we may write the diffusion equation for an inhomogeneous problem, using the diagonal constitutive matrix $\mathbf{G}_1^{-1} = \operatorname{diag}(\alpha)$ to represent edge weights, as

$$\frac{\partial \mathbf{u}}{\partial t} = -\mathbf{A}^\mathsf{T} \mathbf{G}_1^{-1} \mathbf{A}\mathbf{u}. \tag{2.176}$$

In the derivation of Fick's Law, we assumed that the flux across an edge was given by the difference in concentration at the edge endpoints. However, when combined with conservation of mass, this expression for Fick's Law may be interpreted by considering the behavior at a single node—the dynamics distributes the substance u to each neighbor and collects substance u from each neighbor. The difference between this distribution and collection determines whether the value of u at the node increases or decreases. A different model for determining flux is that each node distributes a fraction of its value of u to each of its neighbors, such that the total sum of distribution at each infinitesimal unit of time is equal to its value of u, i.e., the flux is given by

$$\mathbf{f} = -\alpha \mathbf{A} \mathbf{D}^{-1} \mathbf{u}, \tag{2.177}$$

where $\mathbf{D}$ is the diagonal matrix of node degrees. Then, if we change variables to let $\mathbf{q} = \mathbf{D}^{-\frac{1}{2}} \mathbf{u}$ to normalize the substance by node degree, then we can rewrite the diffusion equation in terms of $\mathbf{q}$ as

$$\frac{\partial \mathbf{q}}{\partial t} = -\mathbf{D}^{-\frac{1}{2}} \mathbf{A}^\mathsf{T} \mathbf{G}_1^{-1} \mathbf{A} \mathbf{D}^{-\frac{1}{2}} \mathbf{q}, \tag{2.178}$$

which represents a form of the diffusion equation using the *normalized* Laplacian, which was studied in [81]. Therefore, the distinction between the diffusion equation with the normalized and unnormalized Laplacian operator lies in a different model for flux (Fick's Law) and a change of variables from the concentration $\mathbf{u}$ to the variable $\mathbf{q}$. This subtlety does not typically appear in the conventional equations because each (non-boundary) point in space is considered to have the same neighborhood structure. Note that the distinction between the normalized and unnormalized diffusion equations also disappears in the discrete case when the graph

2.5 Examples of Discrete Calculus

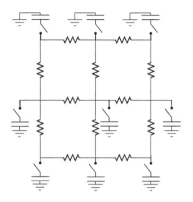

Fig. 2.26 The circuit governed by the discrete calculus analogue of the diffusion equation in (2.179). The electrical potential (charge) at each node is given by **u**, the capacitance by p and diagonal matrix of branch conductances by $\mathbf{G}_1^{-1}$. The switches are considered to be turned on at time zero

is regular (i.e., the degree is equal for every node). For simplicity, we continue our discussion of diffusion using the unnormalized form of the diffusion equation.

The discrete calculus analogue of the diffusion equation also governs physical phenomena. For example, the diffusion equation with inhomogeneous constitutive in (2.176) describes the spread of electrical charge in a DC circuit with capacitors and resistors (an R–C circuit) in which each node is connected to ground via a capacitor and connected to other nodes via resistors. Identifying **u** with the electrical potential at each node, **f** with the current through each branch and α with the conductance of the branch in (2.174), then Kirchhoff's Voltage Law and Ohm's Law combine to give Fick's Law. Together with Kirchhoff's Current Law (representing conservation of charge), we arrive at

$$p\frac{\partial \mathbf{u}}{\partial t} = -\mathbf{A}^T \mathbf{G}_1^{-1} \mathbf{A}\mathbf{u}, \tag{2.179}$$

where p represents the capacitance of each capacitor. Figure 2.26 gives an illustration of the circuit construction governed by this equation. Chapter 3 gives more information on the circuit interpretation of these equations.

One commonality between the analysis of the conventional diffusion equation and the discrete calculus analogue is the utility of Fourier analysis to solve the equation when the underlying domain is shift-invariant. Chapter 5 gives more information about the applicability of Fourier analysis to the diffusion equation.

We now consider how the diffusion changes when the underlying graph is *directed*. For simplicity, we assume that the edge orientation matches the edge direction (i.e., the edge is oriented from positive to negative if flow is permitted from the positive node to the negative node). On a directed graph, the conservation of mass represented by (2.172) is unchanged, but Fick's Law is altered. To see this alteration, we can further deconstruct Fick's Law in terms of the outgoing and incoming flux across each edge:

$$\mathbf{f} = -\alpha(\mathbf{A}^+\mathbf{u} - \mathbf{A}^-\mathbf{u}), \tag{2.180}$$

where the notation $\mathbf{A}^+$ indicates a matrix which contains a '1' in an entry only if **A** contains '1' in the same entry (otherwise zero) and $\mathbf{A}^-$ contains a '1' only if **A** contains a '−1'. Seen this way, the (negative) flux across an edge is composed of the outgoing flow from the positive side node subtracted by the flow coming from the negative side node. However, when the edges are *directed*, flow is permitted in only one direction. Consequently, the expression for flux on a directed graph contains only the positive side part, i.e.,

$$\mathbf{f} = -\alpha \mathbf{A}^+ \mathbf{u}. \tag{2.181}$$

Therefore, diffusion on a directed graph is governed by the equation

$$\frac{\partial \mathbf{u}}{\partial t} = -\mathbf{A}^\mathsf{T} \mathbf{G}^{-1} \mathbf{A}^+ \mathbf{u}. \tag{2.182}$$

Because of its role in the diffusion equation on a directed graph, it is tempting to consider $\mathbf{L} = \mathbf{A}^\mathsf{T} \mathbf{G}^{-1} \mathbf{A}^+$ as the Laplacian operator (matrix) for a directed graph. Indeed, variants of many of the properties of elliptic equations detailed above also hold when this matrix is viewed as the Laplacian matrix for a directed graph. For example, a harmonic function $\mathbf{u}$, satisfying

$$\mathbf{A}^\mathsf{T} \mathbf{G}^{-1} \mathbf{A}^+ \mathbf{u} = 0, \tag{2.183}$$

given a set of boundary conditions, will satisfy a variant of the mean-value property. In this case, the value u_i for node v_i will equal the (weighted) average of its outgoing *neighbors*. The property can be seen by writing (2.183) in summation form for u_i

$$u_i = \frac{1}{d_i} \sum_{e_{ij}} w_{ij} u_j, \tag{2.184}$$

where d_i is the degree of v_i computed using only outgoing edges. Although $\mathbf{A}^\mathsf{T} \mathbf{G}^{-1} \mathbf{A}^+$ behaves in some ways like the Laplacian operator for a directed graph, note that Chung detailed a different (symmetric) conception of the Laplacian operator on a directed graph [79].

In this example of diffusion, we treated *time as continuous*. Although the conception of the discrete calculus employs a discrete *space*, the functions defined on that space (cochains) and time may be real valued. There is no physical reason why we could not have a discrete space and a continuous time (the electrical circuit in Fig. 2.26 is a physical system with exactly these properties). However, time may also be treated discretely in a manner similar to our treatment of space (see [275]). Unless otherwise noted, we use a continuous treatment of time throughout this book.

2.5.6 Advection

When a distribution is actively transported by a flow field, the process is known as **advection**. Advection occurs in many different physical circumstances. For example, consider a chemical dropped into a flowing river. Since the river is flowing (and

2.5 Examples of Discrete Calculus

we neglect diffusion) the change in chemical concentration will occur as a result of the water flow. In a real physical situation, the chemical concentration would change as a result of both advection and diffusion. In this section, we consider the equation describing the advection of a scalar quantity (e.g., chemical concentration) and its discrete calculus analogue.

Given a scalar field u and flow field $\vec{v}$, the **advection equation** (also known as the transport equation) that describes the change in u under $\vec{v}$ is conventionally given by [127]

$$\frac{\partial u}{\partial t} = -\nabla \cdot (\vec{v}u). \tag{2.185}$$

At first, the conventional advection equation seems as though it does not fit well with the discrete calculus framework, since it requires taking the divergence of the multiplication of a scalar field and a vector field. In order to consider how to produce a discrete calculus analogue, we consider the method for deriving the conventional advection equation. As with diffusion, deriving this relation requires combining mass conservation and a definition of flux. The conservation equation is written in the same form as (2.169),

$$\frac{\partial u}{\partial t} = -\nabla \cdot \vec{F} + s. \tag{2.186}$$

The distinction between advection and diffusion lies in the manner in which flux is determined. Fick's Law states that, in a diffusion process, the flux is given by the gradient of the concentration u. However, in an advection process, the flux is given directly by a vector field, i.e., $\vec{F} = \vec{v}u$.

In a discrete calculus analogue of the advection equation, **v** will represent a 1-cochain of flows that will transport node concentrations of the 0-cochain **u**. We can immediately re-use the expression for conservation of mass from the diffusion equation, i.e.,

$$\frac{\partial \mathbf{u}}{\partial t} = \mathbf{A}^\mathsf{T} \mathbf{f} + \mathbf{s}. \tag{2.187}$$

If we adopt a view of the flux through an edge as consisting of the prescribed flow through the edge, **v**, modified by the concentration **u**, then we may define the flux as

$$\mathbf{f} = -\operatorname{diag}(\mathbf{v})\mathbf{A}_\mathbf{v}\mathbf{u}, \tag{2.188}$$

where $\mathbf{A}_\mathbf{v}$ denotes the matrix $(\operatorname{diag}(\operatorname{sign}\{\mathbf{v}\})\mathbf{A})^+$, and the sign operator $\operatorname{sign}\{\cdot\}$ on vectors operates on each element and is defined for each element v_i as

$$\operatorname{sign}\{v_i\} = \begin{cases} +1 & \text{if } v_i > 0, \\ -1 & \text{if } v_i < 0, \\ 0 & \text{if } v_i = 0. \end{cases} \tag{2.189}$$

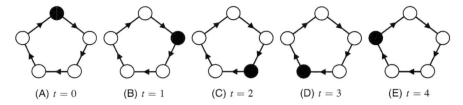

Fig. 2.27 An example of the advection process on a graph. The *arrows* represent a unit flow field cycling around the ring. A node is *filled* if its **u** value is large (in this case, unity) and *empty* if its **u** value is small (zero). Under the process defined by solving the advection equation (2.190) using a forward Euler method ($\Delta t = 1$), the unit flow field transports the initial **u** field around the ring

With this interpretation of the flux, the discrete calculus analogue of the advection equation (in the absence of sources or sinks) is given by

$$\frac{\partial \mathbf{u}}{\partial t} = -\mathbf{A}^\mathsf{T} \operatorname{diag}(\mathbf{v}) \mathbf{A}_\mathbf{v} \mathbf{u}. \tag{2.190}$$

When **v** is a unit flow field which is positive everywhere (i.e., flow is in the same direction as the edge orientation), then (2.190) becomes

$$\frac{\partial \mathbf{u}}{\partial t} = -\mathbf{A}^\mathsf{T} \mathbf{G}^{-1} \mathbf{A}^+ \mathbf{u}, \tag{2.191}$$

which is the same expression as we found for the diffusion equation on a directed graph. This connection will be used later in Chap. 7 to give an interpretation of the PageRank algorithm in terms of an advection equation.

Figure 2.27 gives an illustration of this advection equation on a small cycle graph. In this case, we solved (2.190) using a forward Euler method with time step set to $\Delta t = 1$, i.e., via the iteration

$$\mathbf{u}^{[k+1]} = \mathbf{u}^{[k]} - \Delta t\, \mathbf{A}^\mathsf{T} \operatorname{diag}(\mathbf{v}) \mathbf{A}_\mathbf{v} \mathbf{u}^{[k]}. \tag{2.192}$$

In this case we used unit flow vectors for **v**. If we assume that the orientation is in the same direction as the flow, then for this graph $\operatorname{diag}(\mathbf{v}) \mathbf{A}_\mathbf{v}$ becomes the identity matrix, leading to a particularly simple update. Our initial **u** was zero everywhere except at the first (top) node, which had a unity value. Note how the flow field transports the initial **u** around the cycle without changing it.

2.6 Concluding Remarks

In this chapter we have reviewed the exterior calculus of differential forms as a generalization of standard vector calculus to arbitrary dimensions and, in the process, highlighted the topological character of calculus as well as instances in which a metric is required. We then demonstrated how this more general framework could

2.6 Concluding Remarks

be naturally extended into a discrete domain. The differential operators in this discrete setting were seen to simultaneously represent the topology of the discrete spaces, further demonstrating the inherent interdependence of calculus and topology. Phrasing these operators as matrices opens possibilities for easily representing discrete systems in simple data structures and for solving practical problems in discrete physics with powerful numerical computing software. Indeed, casting discrete physics in a linear algebraic and graph theoretic framework enables results from applied mathematics and graph theory to solve problems with high levels of accuracy.

The universality of this framework across many branches of physics, and our ability to reuse the concepts of differential forms cleanly in a discrete setting, shows that the template outlined in this chapter will be useful across several disciplines and can be applied to solving a wide variety of problems. Given that the foundations of topology are so fundamentally intuitive it is not surprising that the language that mathematicians and physics have developed for describing physical behavior are imbued with easily comprehensible topological relations such as "inside" and "outside". Whether the effectiveness of calculus for modeling many distinct types of physical structures and behavior is due to an underlying commonality between the many compartments of physics or to our human intuition and the natural ways in which we communicate about the world, the discrete calculus appears to carry over to a wide range of disciplines. Just as in other applied sciences and engineering, we can strive to find inspiration from a diverse set of disciplines to help solve challenging problems in any one domain.

In the next chapter we consider a physical domain that readily takes advantage of the tools described here: circuit theory. Not only is this domain a natural fit for this framework but, because of the physical intuition that many practitioners have for voltages and currents, circuit theory often provides a useful set of basic analogies that, once learned, can assist users in bridging into new domains and in translating the tools of discrete calculus to other discrete physical systems.

Chapter 3
Circuit Theory and Other Discrete Physical Models

Abstract In this chapter we present linear electric circuit theory as the central physical model for performing calculus on networks. We adopt circuit theory as the central physical model for applying and understanding the concepts of discrete calculus on graphs for three reasons: because much of the progress in graph theory over the last century was created in the context of circuit theory; because of the early connection made between circuit theory and algebraic topology (Branin Jr. in Proc. of Conf. on Neural Networks, pp. 453–491, 1966; Roth in Proceedings of the National Academy of Sciences of the United States of America 41(7):518–521, 1955); and because circuits are physical, realizable systems which need not be seen as discretizations of an underlying continuous domain but rather as a domain unto themselves. Our focus in this chapter will be to cover the main concepts in linear circuit theory from an algebraic standpoint with a focus on operators. This will prepare the reader with notation and concepts that tie naturally to the previous chapter in which the discrete analogues of differential operators were introduced. We begin with definitions of the physical quantities and corresponding quantities in circuit theory, then proceed to define the laws that relate the quantities to each other, treat methods of solving for the unknowns, and end by connecting circuit theory to other discrete processes.

We recognize that our readers may have varying levels of familiarity with circuit theory, therefore no previous experience with circuit theory is assumed. However, regardless of the reader's familiarity with circuit theory, we believe that this chapter is useful for producing an intuitive understanding of the mathematical devices and discrete analogues of continuous differential operators that were developed in Chap. 2. In order to develop this intuition, our intention with this chapter is to establish notation, provide a physical model for the variables involved in the formalism presented in this work, and to gently introduce concepts that we will encounter again and again in this work. Therefore, we would strongly encourage the reader to at least briefly read this chapter regardless of the reader's knowledge level of circuit theory.

We consider a circuit as a graph consisting of a set of nodes and edges. An edge in a circuit is interpreted physically as a wire that connects two nodes and which has some resistance to the flow of current. In circuit theory, edges are also known as **branches** or **arcs**. A node in a circuit is interpreted physically as any junction between two or more wires (edges) or as a terminal location that receives or sinks energy (more on this concept in a moment).

Viewing electrical circuits from a physics perspective, we are concerned with electrons and their motion through the circuit governed by an energy minimization principle. When there is an unequal distribution of electrons among different nodes in a circuit, then there is an **electric potential** for each node to equilibrate with a reference node, known as the **ground node**. The term *ground* is derived from the safe, neutral potential maintained naturally by the metallic content of the earth. The variable $x_i \in \mathbb{R}$ will be used to represent the electric potential at a single node v_i. The grounded node v_0 has a fixed potential $x_0 = 0$. Electric potential at a node is quantified in units of volts. If a potential x_i is negative, then there are more electrons present at node v_i then at the reference ground and if a potential x_i is positive, then there are fewer electrons present at node v_i then at the reference ground node. We will manipulate the node potentials as a single vector-valued variable $\mathbf{x} \in \mathbb{R}^{|\mathcal{V}|}$, which may be interpreted as a 0-cochain (node function).

The **current** that passes through a wire is a measure of the number of electrons that pass through the wire per unit time. Current is measured in units of amperes (i.e., coulombs per second) and is represented here by the vector $\mathbf{y} \in \mathbb{R}^{|\mathcal{E}|}$. The elements of $\mathbf{y}$, $y_i \in \mathbb{R}$, represent the current passing through the edges of the circuit $e_i \in \mathcal{E}$. Since $\mathbf{y}$ is defined on the edges, then this current vector is a 1-cochain. The value of any current variable y_{ij} through a wire may be positive or negative, and the sign indicates the direction y_i of positive current flow through edge e_i. In order to accommodate this interpretation of the sign of y_i, it is necessary to endow each edge with a sense of orientation (as seen in Chap. 2). The orientation for each edge variable may be assigned arbitrarily, thus a negative value is interpreted as a current flow *in the opposite direction* of the assigned orientation of the edge variable. Therefore, given a set of n nodes $\mathcal{V}$ with elements $v_i \in \mathcal{V}$, we represent each edge, e_{ij}, by the ordered pair $e_{ij} = \{v_i, v_j\}$. If current is flowing from node v_i to v_j the current will be *positive* and if the current is flowing from node v_j to v_i then the current will be *negative*. The designation of edge orientation is essential to solve problems in circuit theory, but the particular choice of e_{ij} compared to e_{ji} has no consequence for the solution to any circuit theory problem (except that the current calculated over edge e_{ij} will be the negative of the current calculated over e_{ji}). Given an edge with orientation e_{ij}, then the **voltage** across e_{ij} is defined by the difference of the potentials at the pair of nodes at the ends of the edge, $p_{ij} = v_i - v_j$.

The vectors of voltages $\mathbf{p}$ and currents $\mathbf{y}$ both take a value for every edge and each component of these two vectors is related to each other via the **resistance** of the edge, r_{ij} (in Sect. 3.3 the edges are generalized to represent capacitors and inductors as well). Resistance is measured in units of Ohms and is physically interpreted as the amount of work required to move electrons through a wire. For our purposes, we will assume that all $r_{ij} > 0$. A wire with higher resistance requires more work to

Fig. 3.1 Conventions for electric circuit theory. (*Left*) An edge in an oriented graph between nodes v_i and v_j with edge weight w_{ij} is equivalent to (*right*) a resistor in a circuit with node potentials x_i and x_j where the edge variable y_{ij} represents a current flow from node v_i to node v_j. The (distance) weight is provided by the resistance r_{ij}

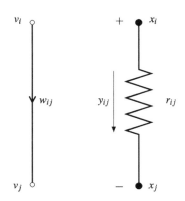

move electrons through the wire and a wire with low resistance requires less work to move electrons through the wire. The resistance of each edge is a material property of each wire. The **conductance** at each edge, w_{ij}, is defined as the reciprocal of the resistance $w_{ij} = 1/r_{ij}$. Wire conductance is measured in units of Siemens and will emerge as the more convenient quantity to use in the theoretical development and applications in this work. In the context of Chap. 2, resistance represents a *distance* weight and conductance represents an *affinity* weight. As seen in the last chapter, the distance weight (resistance) is the reciprocal of the affinity weight (conductance). The collection of nodes, oriented edges, and resistors is the domain upon which all of the circuit laws and dynamics play out. Figure 3.1 summarizes the relationship between a resistive branch in an electric circuit and the equivalent graph edge.

3.1 Circuit Laws

Having defined the variables in the previous section, we now proceed to describe how these variables relate to each other.

Voltage across an edge and current through an edge are related through the conductance via **Ohm's Law** which states that

$$\mathbf{p} = \mathbf{G}\mathbf{y}, \tag{3.1}$$

where **p** represents the voltage each edge, **y** represents the current through each edge and **G** is a diagonal matrix containing the edge resistances along the diagonal, i.e., with $\mathbf{G}_{ii} = r_i$ for some $e_i \in \mathcal{E}$. Ohm's Law states that an edge with a linear resistor induces a linear relationship between voltage across an edge and the current through the edge.

Kirchhoff's Current Law (KCL) states that all of the current flowing into a node also flows out of the node. This law may be represented in matrix form via the equation

$$\mathbf{A}^T \mathbf{y} = \mathbf{0}, \tag{3.2}$$

where **A** is the *edge–node* incidence matrix as defined in Chap. 2.

The relationship between electric potentials at nodes and voltages across edges may also be written in terms of the incidence matrix as

$$\mathbf{Ax} = \mathbf{p}, \tag{3.3}$$

which is known as **Kirchhoff's Voltage Law** (KVL). A consequence of KCL and KVL together is **Tellegen's Theorem** [305] which states that the vector of voltages is orthogonal to the vector of currents, i.e.,

$$\mathbf{p}^\mathsf{T}\mathbf{y} = \mathbf{x}^\mathsf{T}\mathbf{A}^\mathsf{T}\mathbf{y} = \mathbf{0}. \tag{3.4}$$

Tellegen's theorem has been employed several times in the literature as a tool of network analysis, e.g., [301, 302].

These three laws of circuit theory, Ohm's Law, Kirchhoff's Current Law and Kirchhoff's Voltage Law define all of the behavior of linear circuits. These three laws may be composed into one single law which states

$$\mathbf{A}^\mathsf{T}\mathbf{G}^{-1}\mathbf{Ax} = \mathbf{Lx} = \mathbf{0} \tag{3.5}$$

where $\mathbf{L}$ represents the Laplacian matrix from Chap. 2. It was shown in Chap. 2 that the nullspace of the $\mathbf{L} = \mathbf{A}^\mathsf{T}\mathbf{G}^{-1}\mathbf{A}$ matrix is dimension one and spanned by the vector for which each element is equal to the same constant. Therefore, the three laws compose to state that, in the absence of any other constraints, all of the node potentials will be equal. In the next section, we discuss the solution of the circuit equations in the presence of energy sources that force a nontrivial solution. In the context of circuit theory, we will consider a *solution* as a complete knowledge of the node potentials or, equivalently, edge voltages or edge currents for a specific circuit topology and a prescribed set of resistance values. Given any one of these sets of values, the others may be generated using the three circuit laws given above.

3.2 Steady-State Solutions

A **voltage source** or a **current source** provides a fixed source of energy to drive the circuit. Without these energy sources, the circuit would yield the trivial solution consisting of a constant potential at each node and no flow of current. Both of these sources supply a fixed voltage or current which is constant with respect to time and are traditionally known as direct current or DC sources. A voltage source forces the voltage between two nodes to be fixed and a current source forces the current through an edge to be fixed.

By convention, a voltage source can be added in series to any resistor in the graph to insert an additional fixed voltage difference between the two nodes of the resistor [360, p. 149]. The value of the voltage added to edge e_k is denoted by b_k, and all such voltage sources can be assembled into a vector $\mathbf{b} \in \mathbb{R}^{|\mathcal{E}|}$. Note that the elements b_k may be positive or negative, and are zero-valued at edges where no voltage source is present. Therefore, when voltage sources are present, the expression for the voltages in the circuit becomes

$$\mathbf{A}\mathbf{x} + \mathbf{b} = \mathbf{p} \tag{3.6}$$

with the edge–node incidence matrix $\mathbf{A}$ defined as above. The expression of Ohm's law, $\mathbf{y} = \mathbf{G}^{-1}\mathbf{p}$, is unchanged. Similarly, a current source with one end attached to ground can be attached to any other node in order to inject current into the node. If we represent the current inserted into node v_i as f_i, and assemble all additional current sources into a vector $\mathbf{f} \in \mathbb{R}^{|\mathcal{V}|}$, then the expression of KCL in the circuit becomes

$$\mathbf{A}^T\mathbf{y} + \mathbf{f} = \mathbf{0}. \tag{3.7}$$

By combining these two laws and including the voltage and current source terms, we return to the single law summarizing the circuit behavior, given by

$$\mathbf{A}^T\mathbf{G}^{-1}\mathbf{A}\mathbf{x} = \mathbf{L}\mathbf{x} = -\mathbf{A}^T\mathbf{G}^{-1}\mathbf{b} - \mathbf{f}. \tag{3.8}$$

We let $\mathbf{q}$ represent the set of source terms on the right hand side of this equation, i.e., $\mathbf{q} = -\mathbf{A}^T\mathbf{G}^{-1}\mathbf{b} - \mathbf{f}$.

Due to the nature of the circuit behavior, the above equations can not be solved uniquely for any given circuit configuration due to the inherent ambiguity in the node potentials—a constant potential can be added to each node without changing the voltages across the edges. In other words, the solution is not affected by adding a constant potential to all nodes. Therefore, a family of admissible solutions exists that are equivalent up to this constant potential. For this reason, it is often convenient to assign a single node v_0 as a reference or ground by eliminating one column of the edge–node incidence matrix $\mathbf{A}$. This **reduced incidence matrix** is often denoted as $\mathbf{A}_0$. The potential x_0 is set to zero and the corresponding node is eliminated from $\mathbf{x}$ and from $\mathbf{f}$. Once an arbitrary reference node is set to ground, then the solution can be calculated relative to the chosen reference.

Example 1 (Solving node potentials in a static circuit) To demonstrate the process of constructing the discrete operators for a circuit and using them to solve for a set of unknown node potentials or edge currents, consider the circuit diagram presented in Fig. 3.2. The edge–node incidence matrix $\mathbf{A}$ for this circuit can be constructed from the diagram after assigning an orientation for each edge. For this graph the incidence matrix is given by

$$\mathbf{A} = \begin{bmatrix} 1 & -1 & 0 & 0 & 0 \\ 1 & 0 & 0 & -1 & 0 \\ 0 & 1 & -1 & 0 & 0 \\ 0 & 0 & 1 & -1 & 0 \\ 0 & 0 & 1 & 0 & -1 \end{bmatrix}$$

based on an assumed orientation. The constitutive matrix consists of the affinity edge weights (conductances). Therefore, $\mathbf{G}^{-1} = \text{diag}([2\ 1\ 1\ 4\ 1]^T)$ and the Laplacian

Fig. 3.2 Circuit diagram for Example 1. The circuit consists of five nodes and five edges, with a fixed current source injecting current into node v_2, a voltage source setting a fixed potential at node v_4, and node v_5 serving as the reference node for ground. Note that the standard symbol for resistance (Ohms) is given by Ω

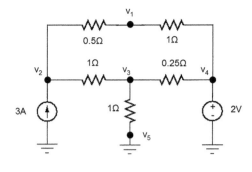

matrix $\mathbf{L} = \mathbf{A}^T \mathbf{G}^{-1} \mathbf{A}$ evaluates to

$$\mathbf{L} = \begin{bmatrix} 3 & -2 & 0 & -1 & 0 \\ -2 & 3 & -1 & 0 & 0 \\ 0 & -1 & 6 & -4 & -1 \\ -1 & 0 & -4 & 5 & 0 \\ 0 & 0 & -1 & 0 & 1 \end{bmatrix}.$$

In this example, the voltage source forces a potential at node v_4, indicating that this node voltage is not free (or unknown), and the voltage source does not coincide with an edge variable. For this reason the voltage source vector $\mathbf{b}$ is expressible in terms of the edge–node incidence matrix operating on the fixed voltages, i.e.,

$$\mathbf{b} = \mathbf{A} \begin{bmatrix} 0 \\ 0 \\ 0 \\ 2 \\ 0 \end{bmatrix} = \begin{bmatrix} 0 \\ -2 \\ 0 \\ -2 \\ 0 \end{bmatrix}.$$

The current source vector $\mathbf{f}$ is straightforward: $\mathbf{f} = [0\ 3\ 0\ 0\ 0]^T$.

With the weighted Laplacian operator $\mathbf{L}$ defined for the circuit, and the vectors $\mathbf{b}$ and $\mathbf{f}$ representing the energy sources, the unknown node potentials $\mathbf{x}$ can be solved with any linear systems solver. However, the full Laplacian matrix is not full rank, and must be reduced to solve the system. Since nodes v_4 and v_5 are fixed, we can solve the reduced system $\mathbf{L}_0 \mathbf{x}_0 = \mathbf{q}_0$, whose entries correspond only to the free nodes in the circuit, v_1, v_2, and v_3 (i.e., the '0' subscript indicates that the entries corresponding to fixed nodes have been removed). Therefore, for this example the node voltages evaluate to

$$\mathbf{x}_0 = \begin{bmatrix} 4.07 & 5.11 & 2.19 \end{bmatrix}^T.$$

3.2.1 Dependent Sources

In the last section we considered voltage (current) sources which supplied a fixed voltage (current) across two branches. However, if the amount of voltage (current) supplied by the source is controlled by the voltage or current of a different branch, then the source is said to be a **dependent source**. In contrast, a fixed source considered above is said to be an **independent source**. Dependent sources can appear in the linear circuit equations as *off-diagonal elements of* $\mathbf{G}^{-1}$.

Before discussing the implications of the matrix $\mathbf{G}^{-1}$ with off-diagonal elements, we first illustrate the concept with an example. Consider the small circuit in Fig. 3.3 in which we employ the matrix

$$\mathbf{G}^{-1} = \begin{array}{c} \\ e_1 \\ e_2 \\ e_3 \\ e_4 \\ e_5 \end{array} \begin{array}{c} \begin{array}{ccccc} e_1 & e_2 & e_3 & e_4 & e_5 \end{array} \\ \left[\begin{array}{ccccc} 1 & 0 & 1 & 0 & 0 \\ 0 & 1 & 0 & 0 & 0 \\ 1 & 0 & 1 & 0 & 0 \\ 0 & 0 & 0 & 1 & 0 \\ 0 & 0 & 0 & 0 & 1 \end{array} \right] \end{array} \qquad (3.9)$$

as the edge weighting matrix. Clearly all of the branch resistances are unit-weighted, but now there is a new relationship between edges 1 and 3. The nature of this relationship becomes apparent if we consider the set of branch voltages

$$\mathbf{p} = \begin{bmatrix} 1 \\ 0 \\ 0 \\ 0 \\ 0 \end{bmatrix}, \qquad (3.10)$$

giving the currents

$$\mathbf{y} = \mathbf{G}^{-1}\mathbf{p} = \begin{bmatrix} 1 \\ 0 \\ 1 \\ 0 \\ 0 \end{bmatrix}. \qquad (3.11)$$

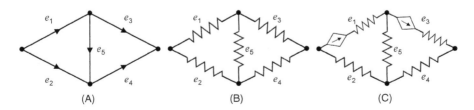

Fig. 3.3 A circuit with dependent current sources

Therefore, even though the voltage occurred over branch 1, a current was induced in branch 3, which allows us to interpret the off-diagonal entries as indicating the presence of *voltage-controlled current sources*, which are often used in circuit theory to model nonlinear elements such as transistors. This example is illustrated in Fig. 3.3.

In Chap. 2, we discussed how the metric tensor definition for a discrete space led to a diagonal matrix for $\mathbf{G}^{-1}$. Therefore, the appearance of off-diagonal elements in $\mathbf{G}^{-1}$ should be regarded only as a convenience that allows us to represent the dependent sources and/or nonlinear circuit elements.

By interpreting a $\mathbf{G}^{-1}$ having off-diagonal elements as representing dependent sources, we can derive an unusual interpretation of the discrete **biharmonic** operator. Recall that the biharmonic operator is defined in conventional mathematics as $\nabla^2 \nabla^2$ and appears frequently in the description of the bending of thick-plate material (as opposed to the membrane deformation described by the standard Laplacian operator). Based on Chap. 2, we may rewrite the discrete biharmonic operator as

$$\mathbf{LL} = \mathbf{A}^\mathsf{T} \mathbf{G}_1^{-1} \mathbf{A} \mathbf{G}_0 \mathbf{G}_0 \mathbf{A}^\mathsf{T} \mathbf{G}_1^{-1} \mathbf{A}. \tag{3.12}$$

By letting

$$\tilde{\mathbf{G}}_1^{-1} = \mathbf{G}_1^{-1} \mathbf{A} \mathbf{G}_0 \mathbf{G}_0 \mathbf{A}^\mathsf{T} \mathbf{G}_1^{-1}, \tag{3.13}$$

then we can interpret the biharmonic operator as a Laplacian operator with a non-diagonal constitutive matrix $\tilde{\mathbf{G}}_1^{-1}$, i.e.,

$$\mathbf{L}^\mathsf{T} \mathbf{L} = \tilde{\mathbf{L}} = \mathbf{A}^\mathsf{T} \tilde{\mathbf{G}}_1^{-1} \mathbf{A}. \tag{3.14}$$

Since $\tilde{\mathbf{G}}_1^{-1}$ has off-diagonal elements, this interpretation allows us to further consider the biharmonic operator as introducing a set of *dependent sources* into the circuit.

As an example, consider the graph in Fig. 3.4 with unity edge and node weights. The biharmonic operator for the graph will consist of

$$\mathbf{L}^\mathsf{T} \mathbf{L} = \begin{array}{c} \\ v_1 \\ v_2 \\ v_3 \\ v_4 \end{array} \begin{array}{cccc} v_1 & v_2 & v_3 & v_4 \end{array} \\ \left[\begin{array}{cccc} 6 & -4 & -4 & 2 \\ -4 & 12 & -4 & -4 \\ -4 & -4 & 12 & -4 \\ 2 & -4 & -4 & 6 \end{array} \right]. \tag{3.15}$$

3.2 Steady-State Solutions

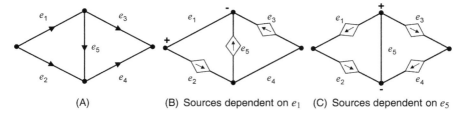

Fig. 3.4 Voltage-controlled current sources used to interpret the discrete biharmonic operator. Panels (**B**) and (**C**) display the current sources which respond to a voltage across e_1 and e_5, respectively, in the direction given by the '+' and '−' signs

This biharmonic operator may be viewed as the Laplacian, $\tilde{\mathbf{L}}$, for $\mathbf{L}^T\mathbf{L} = \tilde{\mathbf{L}} = \mathbf{A}^T\tilde{\mathbf{G}}_1^{-1}\mathbf{A}$, in which

$$\tilde{\mathbf{G}}_1^{-1} = \begin{array}{c} \\ e_1 \\ e_2 \\ e_3 \\ e_4 \\ e_5 \end{array} \begin{array}{c} \begin{array}{ccccc} e_1 & e_2 & e_3 & e_4 & e_5 \end{array} \\ \left[\begin{array}{ccccc} 2 & 1 & -1 & 0 & -1 \\ 1 & 2 & 0 & -1 & 1 \\ -1 & 0 & 2 & 1 & 1 \\ 0 & -1 & 1 & 2 & -1 \\ -1 & 1 & 1 & -1 & 2 \end{array} \right] \end{array}. \qquad (3.16)$$

The current sources which depend on voltages across various branches for this circuit are depicted in Fig. 3.4.

3.2.2 Energy Minimization

The solution for the node potentials given by (3.5) may be considered as the minimum energy solution for a variational problem. Specifically, if we define the total energy for a configuration of voltages **x** as

$$\mathcal{E}[\mathbf{x}] = \mathbf{x}^T\mathbf{L}\mathbf{x}, \qquad (3.17)$$

then, because **L** is both symmetric and positive semi-definite, the steady-state solution

$$\frac{d\mathcal{E}}{d\mathbf{x}} = \mathbf{L}\mathbf{x} = \mathbf{0}, \qquad (3.18)$$

represents a minimum of the energy in (3.17). Note that it is possible to introduce current sources into the above energy by letting $\mathcal{E}[\mathbf{x}] = \mathbf{x}^T\mathbf{L}\mathbf{x} + \mathbf{x}^T\mathbf{f}$. Likewise, it is possible to introduce voltage sources by fixing the potentials **x** and taking the gradient of the energy with respect to **b**. The energy in (3.17) represents electrical *power dissipation* of the circuit and may be rewritten in several equivalent ways.

For example, the power can be phrased as: (a) as the sum of voltages multiplied by currents across all edges,

$$\mathcal{E}[\mathbf{x}] = \mathbf{x}^T \mathbf{L} \mathbf{x} = \mathbf{p}^T \mathbf{y} = \sum_{e_{ij}} p_{ij}\, y_{ij}, \qquad (3.19a)$$

(b) as the sum of squared voltages divided by resistances across all edges,

$$\mathcal{E}[\mathbf{x}] = \mathbf{x}^T \mathbf{L} \mathbf{x} = \mathbf{p}^T \mathbf{G}^{-1} \mathbf{p} = \sum_{e_{ij}} \frac{p_{ij}^2}{r_{ij}}, \qquad (3.19b)$$

or (c) as the sum of squared currents multiplied by resistances across all edges,

$$\mathcal{E}[\mathbf{x}] = \mathbf{x}^T \mathbf{L} \mathbf{x} = \mathbf{y}^T \mathbf{G} \mathbf{y} = \sum_{e_{ij}} y_{ij}^2\, r_{ij}. \qquad (3.19c)$$

These three expressions for power dissipated by each edge are classical. The energy minimization principles expressed above in terms of circuit theory are equivalent to the conventional Dirichlet's principle on a graph that was discussed in Chap. 2.

Using the above expressions for power dissipation, we now show that the energy minimization formulation may also be written in terms of the current variable. Specifically, the currents distribute themselves to optimize

$$\min_{\mathbf{y}} \mathcal{E}[\mathbf{y}] = \min_{\mathbf{y}} \mathbf{y}^T \mathbf{G} \mathbf{y},$$
$$\text{s.t.} \quad \mathbf{A}^T \mathbf{y} = \mathbf{0}. \qquad (3.20)$$

This constrained optimization may be converted to an unconstrained optimization (see Appendix B) by noting that any solution that satisfies the constraints must lie in the nullspace of $\mathbf{A}^T$. If we represent a basis to span the nullspace of $\mathbf{A}^T$ by the matrix $\mathbf{B}$, then we may write $\mathbf{y} = \mathbf{B}\mathbf{z}$ in terms of a new variable $\mathbf{z}$. Adding this change of variable into the optimization of the currents gives us the unconstrained optimization problem

$$\min_{\mathbf{z}} \mathcal{E}[\mathbf{z}] = \min_{\mathbf{z}} \mathbf{z}^T \mathbf{B}^T \mathbf{G} \mathbf{B} \mathbf{z}. \qquad (3.21)$$

Once again, current sources or voltage sources provide constraints that, when included in the energy formulation in terms of $\mathbf{z}$, produce a nontrivial solution. The matrix $\mathbf{B}$ has the convenient form of the *face–edge* incidence matrix and the $\mathbf{z}$ variables are associated with the faces (known as "mesh variables" in the circuit literature). In the context of circuit theory, solution for the circuit variables via optimization of the first energy minimization problem (3.17) in terms of the node variables is known as **node analysis**. Similarly, optimization of the second energy minimization problem of (3.21) in terms of the mesh variables is known as **mesh analysis**. In some circumstances, mesh analysis is much easier or has many fewer variables (e.g., if the circuit consists of a single long cycle). However, in typical circuits, the extra computation required to find cycles and thus generate **B** is not justified by any computational gain in performing mesh analysis instead of node analysis. Consequently, the practical solution of circuit problems is predominated by node analysis.

3.2.2.1 Power Minimization with Nonlinear Resistors

So far we have considered graph edges to represent resistors which act to translate differences in quantities defined at the nodes (electric potentials) into flows through the edges (currents). Linear resistors create flows which are proportional to the potential differences, but we may also consider edges which act as nonlinear resistances as well. We may view the linear resistor circuit as minimizing the *square* of the difference of node variables by writing the circuit power in (3.17) as a summation over edges

$$\mathcal{E}[\mathbf{x}] = \mathbf{x}^T \mathbf{L} \mathbf{x} = \mathbf{x}^T \mathbf{A}^T \mathbf{G}^{-1} \mathbf{A} \mathbf{x} = \sum_{e_{ij}} w_{ij}(x_i - x_j)^2. \tag{3.22}$$

Viewing the circuit power dissipation in this form, we can see that other functions of the difference between nodes may be optimized in the circuit by introducing *nonlinear resistors*. For example, by using resistors with the value $r_{ij} = \frac{1}{w_{ij}}(x_i - x_j)^{2-p}$, the power dissipated by the circuit becomes

$$\mathcal{E}[\mathbf{x}] = \sum_{e_{ij}} w_{ij} |x_i - x_j|^p. \tag{3.23}$$

Therefore, by using a nonlinear resistor we may build a circuit for which the minimum power distribution minimizes the measurement of the potential differences under any p-norm. It has been shown [350] that if $p = 1$ (i.e., using nonlinear resistors with value $r_{ij} = \frac{1}{w_{ij}}|x_i - x_j|$) then the minimum power solution may be found by solving a *maximum-flow/minimum-cut problem* for a particular configuration of sources. Specifically, if all power sources either ground some nodes or fix node potentials to unit voltage above ground, then any partitioning of the nodes obtained by thresholding the electrical potentials at 0.5 represents a minimum cut between the nodes of unit voltage and the grounded nodes. In the second part of this book, many applications will require an optimization of an energy in the form (3.23), which will be called the *Basic Energy Model* in the context of these applications. Therefore, each of the applications using he Basic Energy Model may be interpreted as finding the minimum power dissipation of a circuit with nonlinear resistors (3.23).

3.3 AC Circuits

Both linear and nonlinear resistors map potential differences directly to flows. We now consider edge elements that instead translate between flows and differences via the *change* in these quantities over time. Such a circuit element may be physically realized in the form of the standard circuit elements of a **capacitor** and **inductor**. In this book, we will not consider any of the details of real capacitors and inductors, but rather work with their idealized mathematical behavior.

A capacitor translates the temporal change in potential difference between nodes into a current flow through the edge connecting the nodes. Specifically, the relationship is given by

$$y_{ij} = c_{ij} \frac{\mathrm{d}p_{ij}}{\mathrm{d}t}, \tag{3.24}$$

where the variable c is used to represent **capacitance**, and $\mathrm{d}t$ is the time differential. We may write the capacitance equation (3.24) in integral form for the voltage as

$$p_{ij} = \frac{1}{c_{ij}} \int y_{ij} \, \mathrm{d}t. \tag{3.25}$$

The inductor element has the opposite effect as the capacitor in the sense that the inductor translates the change in current flow through an edge into a potential differences across the edge. Specifically, the relationship is given by

$$p_{ij} = u_{ij} \frac{\mathrm{d}y_{ij}}{\mathrm{d}t}, \tag{3.26}$$

where u_{ij} represents the inductance of the edge e_{ij}.

With these additional elements in a circuit, we may consider the response of the current flow through an edge when the voltage across the edge varies periodically with time (i.e., an **alternating voltage source**). Because the circuits we consider represent linear and time invariant systems, these time varying quantities can be conveniently represented with complex exponentials (or superpositions thereof). If we represent the voltage as the real part of $\mathrm{Re}\{p_{ij}\,\mathrm{e}^{\mathrm{i}\omega t}\}$, then we may substitute the current $\mathrm{Re}\{y_{ij}\,\mathrm{e}^{\mathrm{i}\omega t}\}$ into (3.25) and (3.26) to produce the current flow through an edge containing an inductor, resistor and capacitor in series

$$\left(\mathrm{i}\omega u_{ij} + r_{ij} + \frac{1}{\mathrm{i}\omega c_{ij}}\right) y_{ij}\,\mathrm{e}^{\mathrm{i}\omega t} = p_{ij}\,\mathrm{e}^{\mathrm{i}\omega t}. \tag{3.27}$$

Note that ω represents the oscillation frequency of the sources. Since the factor $\mathrm{e}^{\mathrm{i}\omega t}$ appears on both sides of the equation, it may be ignored to give us a relationship which is *independent of time*

$$\left(\mathrm{i}\omega u_{ij} + r_{ij} + \frac{1}{\mathrm{i}\omega c_{ij}}\right) y_{ij} = p_{ij}. \tag{3.28}$$

The factor applied to the current is a constant which depends on the material properties of the edge (the resistance, inductance and capacitance of the edge). Therefore, we may replace this factor with

$$z_{ij} = \left(\mathrm{i}\omega u_{ij} + r_{ij} + \frac{1}{\mathrm{i}\omega c_{ij}}\right), \tag{3.29}$$

which is known as the **impedance** of the edge and may be considered as a generalized conception of the edge resistance. Writing the time-independent equation for an alternating voltage/current relationship in terms of the impedance gives

$$z_{ij}y_{ij} = p_{ij}, \tag{3.30}$$

which may be viewed as a generalized form of Ohm's Law (3.1) that reflects the "complex resistor" represented by the impedance. Similar to the relationship between resistance and conductance, the reciprocal of the impedance is called the **admittance**, $\psi_{ij} = 1/z_{ij}$.

The introduction of capacitors, inductors and alternating energy sources affects only the relationship of the voltage across an edge to the current through the edge. However, the voltages are still generated as the differences of the electrical potentials, (3.3), and the sum of the currents into a node still equals the currents leaving a node, (3.2). Consequently, if all of the sources have the same frequency of oscillation, ω, then we may write the three laws of circuit theory for an alternating current (AC) circuit as

$$\mathbf{Ax} = \mathbf{p}, \tag{3.31}$$

$$\mathbf{\Psi p} = \mathbf{y}, \tag{3.32}$$

$$\mathbf{A}^T\mathbf{y} = \mathbf{0}, \tag{3.33}$$

where $\mathbf{\Psi}$ represents the diagonal matrix containing the admittance of each edge along the diagonal. As before, these laws combine to give

$$\mathbf{A}^T\mathbf{\Psi A x} = \mathbf{0}, \tag{3.34}$$

which governs the steady-state distribution of currents and potentials in a circuit where $\mathbf{x}$ and $\mathbf{y}$ are understood to be multiplied by $\text{Re}\{e^{i\omega t}\}$.

We may now adopt the viewpoint that the steady state behavior of a circuit with alternating sources is equivalent to the steady state behavior of a circuit with constant sources that has "complex resistors". Given this viewpoint, we may follow the procedure above to produce the response of the circuit to voltage and current sources of various magnitudes (but the same oscillation frequency) to give

$$\mathbf{Lx} = -(\mathbf{A}^T\mathbf{\Psi Ax} + \mathbf{f}). \tag{3.35}$$

Similarly, we may consider this equation as the solution to a minimization of the power dissipated by the circuit given by

$$\mathcal{E}[\mathbf{x}] = \mathbf{x}^T\mathbf{A}^T\mathbf{\Psi A x}. \tag{3.36}$$

Since $\mathbf{\Psi}$ is complex valued, the power dissipation in the above equation is also a complex-valued quantity, which is referred to in the circuit literature as the **complex power**. The magnitude, $|\mathcal{E}[\mathbf{x}]|$, is called the **apparent power**, while the real component $\text{Re}\{\mathcal{E}[\mathbf{x}]\}$ is called the **real power** and the imaginary part $\text{Im}\{\mathcal{E}[\mathbf{x}]\}$ is called the **reactive power**. In a physical circuit, the real power is the only energy dissipated by the system.

3.4 Connections Between Circuit Theory and Other Discrete Domains

Our main purpose for reviewing circuit theory is to provide a physical, realizable model for discrete mechanics that provides a natural analogy for the continuous calculus developed in the last chapter. The utility of seeing circuit theory from the standpoint of linear algebra is that it focuses attention on the *operators* (defined by incidence matrices) that define the circuit laws and the energy minimization problems behind these laws.

Although circuit theory is a natural setting to explore the physical interpretation of the discrete calculus machinery developed in Chap. 2, circuit theory is by no means the only discipline in which these mechanics appear. In fact, just like conventional vector calculus is used to describe the structure of many physical theories, so too is discrete calculus. Specifically, we will use the treatment of circuit theory above to examine the topics of spring networks, random walks on a graph, Markov Random Fields, the use of tree counting in graph theory and linear algebra.

3.4.1 Spring Networks

Spring–mass networks offer another example of a physically realizable system that employs the same equations as we reviewed in circuit theory. A spring–mass network consists of a series of masses connected to each other via springs, where the quantity of each mass may be different and each spring may have a different spring constant. The equations that govern spring networks are used not only to calculate quantities of physical systems but also as computational models for wide ranging applications (e.g., [208, 210, 270]).

The elements of spring–mass networks have been mapped to elements of circuit theory in several different ways. In fact, there are some parts of the literature which map force to current and others which map force to voltage, with accompanying arguments about the utility of one electrical analogy over the other [202, 357]. A standard reference for these analogies is Shearer et al. [341]. The typical focus of attention in these electrical-mechanical analogies is study of the system *dynamics*. Therefore, the analogy is drawn between the mechanical system and an RLC circuit in order to study the system dynamics. These dynamical systems form the basis for studies in the synchronization of oscillators [12, 26], with a strong emphasis on the role of graph topology in the behavior of the system.

In contrast to these dynamical systems, the primary focus of the later chapters of this book are on how energy minimization techniques may be used to find solutions to problems from several different applications. Therefore, since our focus will generally be on *steady-state* problems, we will present a simplified spring-circuit analogy of a steady-state system, following Strang [359]. The equations that govern the steady-state of a spring–mass model differ from the equations that govern a DC electric circuit mainly in the interpretation of the variables. The setting that we consider for a spring–mass model is depicted in Fig. 3.5. Each mass is associated with

3.4 Connections Between Circuit Theory and Other Discrete Domains

Fig. 3.5 A spring–mass system in steady-state

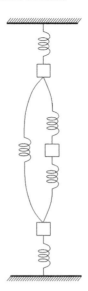

a node and each spring between nodes with an edge. The displacement (location) of each node, v_i, along the line is represented by the variable x_i. Some node, v_0, is fixed at a reference location such that $x_0 = 0$. When two connected nodes are separated, they stretch the spring between them with an **elongation**, p_{ij}. This elongation for each edge may be represented in matrix form via

$$\mathbf{Ax} = \mathbf{p}. \tag{3.37}$$

When a spring is elongated, Hooke's Law states that it responds with a force proportional to the elongation. The constant of proportionality is determined individually for each spring and is known as the **spring constant** of the spring, w_{ij}. Written in matrix form, the constitutive relationship represented by Hooke's Law states that

$$\mathbf{G}^{-1}\mathbf{p} = \mathbf{y}, \tag{3.38}$$

where y_{ij} represents the force exerted by the spring on edge e_{ij} in response the elongation. Each spring incident on a node exerts a force, but the node may also have an external force applied to it, such as gravity. In a steady-state system, these internal and external forces must balance each other, which may be written as

$$\mathbf{A}^\mathsf{T}\mathbf{y} - \mathbf{f} = \mathbf{0}, \tag{3.39}$$

where $\mathbf{f}$ represents the external force (e.g., $f_i = m_i \gamma$ where m_i is the mass of node v_i and γ is the gravitational constant). These three equations combine to allow us to solve for the steady-state solution of the displacements via

$$\mathbf{A}_0^\mathsf{T}\mathbf{G}^{-1}\mathbf{A}_0 = \mathbf{L}_0 \mathbf{x} = \mathbf{f}. \tag{3.40}$$

Table 3.1 List of equivalent quantities between linear electric circuits and spring–mass networks

Symbolic	Electrical circuit	Mass–spring network
x	electric potential	mass displacement
p	voltage (potential difference)	elongation
y	current	responding spring force
w	conductance (1/resistance)	spring constant
v_0	grounded node	reference node
f	currents injected into nodes from ground	external forces on masses
$G^{-1}p = y$	Ohm's Law	Hooke's Law
$A^T y = 0$	Kirchhoff's Current Law	conservation of forces

The matrix L_0 in this context is known as the **stiffness matrix**.

By using a consistent notation, it is straightforward to see in Table 3.1 the corresponding variables in a spring network and an electrical network. Consequently, the (one dimensional) spring network behaves *exactly* like an electrical circuit when both systems are in the steady state.

3.4.2 Random Walks

The connection between random walks on a graph and electrical circuits is unexpected. Although this connection has been recognized for a long time [221, 227, 228], the book by Doyle and Snell [115] provided a clear and introductory exposition that propelled subsequent interest in this connection.

We define a random walk on a graph as an iterative process that tracks a random walker located at a node as it moves from node to node along edges. At each iteration of the random walk, a random walker located at node v_i will transition to one of its neighbor nodes, v_j with probability equal to $q_{ij} = w_{ij}/d_i$, where d_i is the (weighted) degree of node v_i. If we let x_i represent the probability that the random walker is present at node v_i, then we may write the transition rule in matrix form as

$$\mathbf{x}^{[k+1]} = \mathbf{D}^{-1}\mathbf{W}\mathbf{x}^{[k]} \tag{3.41}$$

where $\mathbf{x}^{[k]}$ indicates the k-th iteration step and $\mathbf{W}$ is the adjacency matrix.

Given this definition of a random walk on a graph, we may consider the following problem: What is the probability that a random walker starting at node v_i reaches node v_1 before it reaches node v_0? Let x_i represent the probability that a random walker leaving node v_i reaches node v_1 before it reaches node v_0. By definition of the problem, we know that $x_0 = 0$ and $x_1 = 1$. For the remaining nodes, we can imagine that if we knew the probability that each of the neighbors of node v_i sent a random walker to v_1 before v_0 then we know that x_i is just the sum of these probabilities weighted by the likelihood that the random walker transitions to each

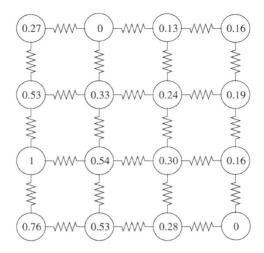

Fig. 3.6 Equivalent circuit to calculate the probability that a random walker leaving each node first arrives at the '1' node before arriving at the '0' nodes. The circuit is a passive resistive network in which the '0' nodes are connected to ground and the '1' node is connected to a unit voltage source with ground. The electrical potential established by this circuit at each node (printed inside the node) equals the probability that a random walker leaving each node arrives at the '1' node before arriving at any '0' node

neighbor. Formally, we can write that

$$x_i = \sum_j q_{ij} x_j = \frac{1}{d_i} \sum_j w_{ij} x_j, \qquad (3.42)$$

or in matrix form

$$\mathbf{x} = \mathbf{D}^{-1}\mathbf{W}\mathbf{x}, \qquad (3.43)$$

$$\mathbf{D}\mathbf{x} = \mathbf{W}\mathbf{x}, \qquad (3.44)$$

$$\mathbf{L}\mathbf{x} = \mathbf{0}. \qquad (3.45)$$

Therefore, this random walk problem amounts to solving $\mathbf{Lx} = \mathbf{0}$ subject to $x_0 = 0$ and $x_1 = 1$. This is exactly the same circuit theory problem that we encountered before when trying to solve for the electrical potentials $\mathbf{x}$ that were induced by grounding v_0 and establishing a unit voltage source between v_1 and ground, i.e., in (3.8). If we replace the unit voltage source by a voltage source of arbitrary magnitude k, then the equivalent random walk problem adjusts by calculating the expected payoff of a game in which a player is paid k if the random walker arrives first at v_1 and is paid zero if the random walker arrives first at v_0. Figure 3.6 illustrates the circuit construction that produces node potentials equal to the probability that a random walker leaving each node is more likely to arrive at node v_1 before arriving at node v_0.

Other random walker problems also have equivalent circuit problems. For example, the **hitting time** of a random walk from node v_a to node v_0, $h(v_a, v_0)$, is defined as the expected number of steps taken by a random walker to travel from v_a to v_0. As before, we can solve for the hitting time by considering the relationship of the hitting time of every node to reach v_0 in relationship to its neighbors. Specifically, if we let x_i represent the hitting time from node v_i to some reference node v_0, then

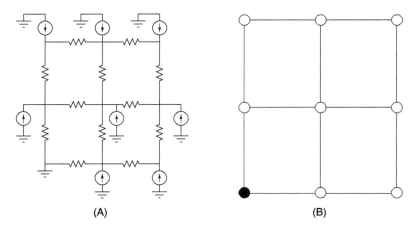

Fig. 3.7 Equivalent circuit to compute the hitting times of a graph. The electrical potential induced at a node in the circuit in (**A**) equals the hitting time for a random walker leaving that node to reach the grounded node. (**B**) The same graph in which the reference (grounded) node is indicated as the filled node. Note that the magnitude of each current source equals the degree of the node

the hitting time at v_i will be equal to the hitting time of its neighbors multiplied by the probability of transitioning to each neighbor, plus one (to account for the extra step taken to a neighbor), i.e.,

$$\mathbf{x} = \mathbf{D}^{-1}\mathbf{W}\mathbf{x} + \mathbf{1}, \tag{3.46}$$

$$\mathbf{D}\mathbf{x} = \mathbf{W}\mathbf{x} + \mathbf{d}, \tag{3.47}$$

$$\mathbf{L}\mathbf{x} = \mathbf{d}, \tag{3.48}$$

where $x_0 = 0$ is fixed. From the standpoint of circuits, the above equation (3.48) may be interpreted as a circuit problem in which node v_0 is grounded, an amount of current equal to each node degree d_i is injected into every node $v_i \neq v_0$ and the resulting potentials x_i equal the hitting time from node v_i to v_0. Figure 3.7 illustrates the circuit construction that produces node potentials equal to the hitting time to pass to node v_0 from every other node [115].

The hitting time $h(v_i, v_j)$ does not necessarily equal $h(v_j, v_i)$, since some nodes in a graph are simply faster to reach via a random walk process than other nodes (e.g., nodes with high degree are easier to reach). Therefore the **commute time** between two nodes is defined as the expected number of steps a random walker will traverse between nodes v_i and v_j and then back again from v_j to v_i, i.e., $c(v_i, v_j) = h(v_i, v_j) + h(v_j, v_i) = c(v_j, v_i)$. It is possible to show that the commute time between two nodes is proportional to the **effective resistance** between the nodes when the graph is viewed as an electrical circuit [72]. The effective resistance between two nodes v_i and v_0, $R_{\text{eff}}(v_i, v_0)$, is defined as the electrical potential induced by injecting one unit of current into node v_i and grounding node v_0. Formally, let the vector $r_k = 1$ if $k = i$ and $r_k = 0$ otherwise. The electrical potentials

3.4 Connections Between Circuit Theory and Other Discrete Domains

established by injecting one unit of current into v_i and grounding node v_0 satisfies

$$\mathbf{L}_0 \bar{\mathbf{x}} = \mathbf{r}. \tag{3.49}$$

The quantity $\bar{x}_i = R_{\text{eff}}(v_i, v_0)$. We may then calculate the hitting time for all nodes to reach v_i from (3.48) as

$$\mathbf{L}\tilde{\mathbf{x}} = \begin{bmatrix} \mathbf{d} \\ \sum_k d_k - d_i \end{bmatrix} \tag{3.50}$$

(where the right hand side is set such that it sums to zero) and the hitting time for all nodes to reach v_0 as

$$\mathbf{L}\hat{\mathbf{x}} = \begin{bmatrix} \sum_k d_k - d_0 \\ \mathbf{d} \end{bmatrix}. \tag{3.51}$$

Then, we may note that

$$\mathbf{L}(\tilde{\mathbf{x}} - \hat{\mathbf{x}}) = \begin{bmatrix} +\sum_k d_k \\ 0 \\ -\sum_k d_k \end{bmatrix}, \tag{3.52}$$

For convenience we may assume that the graph is unweighted (all edges are unity weighted), meaning that

$$\begin{bmatrix} +\sum_k d_k \\ 0 \\ -\sum_k d_k \end{bmatrix} = \begin{bmatrix} +2m \\ 0 \\ -2m \end{bmatrix}, \tag{3.53}$$

where $m = |\mathcal{E}|$. The entry $(\tilde{\mathbf{x}} - \hat{\mathbf{x}})_i$ represents $h(v_j, v_i)$ and the entry $(\tilde{\mathbf{x}} - \hat{\mathbf{x}})_j$ represents $-h(v_i, v_0)$. Consequently, by adding the value $h(v_i, v_0)$ to each entry of $(\tilde{\mathbf{x}} - \hat{\mathbf{x}})$, the entry belonging to v_0 becomes zero and the ith entry represents $h(v_0, v_i) + h(v_i, v_0) = c(v_i, v_0)$. We may therefore consider the v_0 node grounded and restrict our attention of $\mathbf{L}$ and the vector $(\tilde{\mathbf{x}} - \hat{\mathbf{x}})$ to the portion excluding v_0. With this restriction, the difference satisfies

$$\mathbf{L}_0(\tilde{\mathbf{x}} - \hat{\mathbf{x}}) = \begin{bmatrix} 2m \\ 0 \end{bmatrix} = 2m\mathbf{r} = 2m\bar{\mathbf{x}}. \tag{3.54}$$

Therefore, for any pair of nodes in any graph, $c(v_i, v_0) = 2m R_{\text{eff}}(v_i, v_0)$.

In their book, Doyle and Snell also give an interpretation of the current through each edge when one unit of current is injected into node v_i and out of v_j [115], i.e., in the circuit with potentials and currents satisfying (3.49). The current through edge e_{pq} represents the expected number of times that a random walker will pass through edge e_{pq} when the walker is inserted at v_i and exits at v_j. Note that in this interpretation of the current, the current represents the expected number of times that the random walker pass through the edge in the v_p to v_q direction *subtracted from* the expected number of times that the random walker passes through the edge in the v_q to v_p direction. The proof of this statement is somewhat involved, so the interested reader is referred to [115].

We have seen that several problems concerning random walks on a graph have interpretations as problems in circuit theory and that the quantities of voltage and current in a circuit may be interpreted as expectations of certain random walks. This connection has been very helpful for analyzing random walk problems on graphs. One problem in particular for which this connection has been useful is to provide a much simpler proof of Polya's theorem [308], which states that a random walk on an infinite lattice will always return to its starting point when the lattice is either one-dimensional or two-dimensional, but has a finite probability of never returning to its starting point for lattices in dimensions three or greater [115]. In the next section, we leave random walks and consider the Gaussian Markov Random Field (GMRF), a different probabilistic model that leads back to the circuit equations.

3.4.3 Gaussian Markov Random Fields

A Markov Random Field (MRF) describes any collection of random variables defined on nodes $v_i \in \mathcal{V}$ for which the value of each individual variable x_i is conditionally dependent only on the values of its neighbors. This conditional probability can be represented by a simple graph structure recorded by the set $\mathcal{E} \subseteq \mathcal{V} \times \mathcal{V}$. In this structure, each node represents a random variable and each edge represents conditional dependence between the variables. Node (variables) which are independent are not connected by an edge, i.e., if v_i and v_j are independent, then $p(x_i|x_j) = p(x_i)$. In this short section, we will show that the MAP estimate of a certain class of Gaussian MRFs may be interpreted as the electrical potentials of a specific circuit construction.

The classic MRF is the Ising Model for ferromagnetic material in which each variable can assume one of two states, $x_i \in \{-1, +1\}$. In contrast to the two-state model, a *Gaussian* Markov Random Field model assumes that each variable $x_i \in \mathbb{R}$ represents a random variable with a Gaussian distribution, $x_i \sim N(\mu_i, \sigma_i)$, and all individual x_i are correlated with covariance $\boldsymbol{\Sigma}$. The probability distribution for any particular state of the entire MRF (i.e., a particular set of values for each of the random variables) is typically given by the Gibbs distribution

$$P(X = \mathbf{x}) = \frac{1}{Z} \exp\left(-\frac{1}{2}(\mathbf{x} - \boldsymbol{\mu})^\mathsf{T} \boldsymbol{\Sigma}^{-1} (\mathbf{x} - \boldsymbol{\mu})\right), \qquad (3.55)$$

where $\boldsymbol{\Sigma}$ is the covariance matrix, and Z is the *partition function* that serves as a normalization constant, which is set such that the integral of P over $\mathbf{x}$ is unity. The covariance matrix captures the coupling between each pair of nodes, yet represents both direct covariance between neighboring nodes and indirect coupling that results from a propagation of covariance to remote pairs of nodes. The inverse of the covariance matrix, $\boldsymbol{\Sigma}^{-1}$, has several interpretations and is sometimes called the *potential matrix*, the *precision matrix* or the **information matrix** in the literature. The information matrix can provide a more compact representation of the coupling between the random variables x_i and more succinctly captures the neighborhood relations in the GMRF [224].

3.4 Connections Between Circuit Theory and Other Discrete Domains

In many applications [9, 224], the information matrix is assigned to a sum of a weighted graph Laplacian matrix with a non-negative diagonal matrix $\mathbf{T}$, i.e., $\mathbf{\Sigma}^{-1} = \lambda \mathbf{L} + \mathbf{T}$, where λ is the weighting parameter.

Using this choice for the information matrix, the maximum likelihood or ML estimate of the state $\mathbf{x}$ of the Gaussian MRF under the Gibbs distribution (3.55) is given by

$$\widehat{\mathbf{x}}_{\text{ML}} = \underset{\mathbf{x}}{\operatorname{argmax}}\{P\} = \underset{\mathbf{x}}{\operatorname{argmax}}\{(\mathbf{x} - \boldsymbol{\mu})^{\mathsf{T}}(\lambda \mathbf{L} + \mathbf{T})(\mathbf{x} - \boldsymbol{\mu})\}.$$

Consequently, the ML estimate may be generated by solving the linear system

$$(\lambda \mathbf{L} + \mathbf{T})\widehat{\mathbf{x}}_{\text{ML}} = (\lambda \mathbf{L} + \mathbf{T})\boldsymbol{\mu}. \tag{3.56}$$

Similarly, if we adopt an observation model of the Gaussian MRF where the state $\mathbf{x}$ is observed in the presence of noise, the observed state $\mathbf{y}$ is expressed as

$$\mathbf{y} = \mathbf{x} + \mathbf{n}, \tag{3.57}$$

where the noise is zero-mean and Gaussian distributed, $\mathbf{n} \sim \text{N}(0, \mathbf{R}^{-1})$ with covariance matrix $\mathbf{R}^{-1}$. For simplicity here we assume that $\mathbf{n}$ is independent for each variable v_i (i.e., $\mathbf{R}$ is diagonal), and $\mathbf{x}$ is zero mean, $\mathbf{x} \sim \text{N}(0, \mathbf{\Sigma})$. If again the information matrix $\mathbf{\Sigma}^{-1}$ is captured by the weighted graph Laplacian of the MRF graph, $\mathbf{\Sigma}^{-1} = \lambda \mathbf{L} + \mathbf{T}$, then

$$P(\mathbf{y}|\mathbf{x}) = \frac{1}{Z} \exp\left(-\frac{1}{2}(\mathbf{y} - \mathbf{x})^{\mathsf{T}} \mathbf{R}(\mathbf{y} - \mathbf{x})\right). \tag{3.58}$$

Therefore, the maximum *a posteriori* or MAP estimate of

$$P(\mathbf{x}|\mathbf{y}) \propto P(\mathbf{x}) P(\mathbf{y}|\mathbf{x}), \tag{3.59}$$

is given by

$$(\lambda \mathbf{L} + \mathbf{T} - \mathbf{R})\widehat{\mathbf{x}}_{\text{MAP}} = -\mathbf{R}\mathbf{y}. \tag{3.60}$$

Viewed as a circuit, the MAP estimate in (3.60) may be interpreted as the electrical potentials established by a circuit where (a) every pair of neighboring nodes is connected by a resistor with value $1/\lambda$; (b) a resistor connects each node v_i to ground via a resistor with value $1/(T_{ii} - R_{ii})$; and (c) a current source injects $-R_{ii} y_i$ current into each node v_i. A similar circuit interpretation may be given to the ML estimate in (3.56) in which the current sources are given instead by $(\lambda \mathbf{L} + \mathbf{T})\boldsymbol{\mu}$.

We can now examine how this circuit model reflect various conditions in the Markov Random Field through considering three examples. First, if the weighting parameter $\lambda = 0$, then each variable is independent and the MAP estimate is simply $\widehat{x}_i = -R_{ii} y_i / (T_{ii} - R_{ii})$. From a circuit standpoint, this result is explained by the fact that each node is connected to ground by an edge with resistance $R_{ii}/(T_{ii} - R_{ii})$ and that $-R_{ii} y_i$ current is injected into each node. Thus, the resulting voltages are apparent from Ohm's law. As a second example, consider the case of an observed variable at some node v_i such that $x_i = x_{\text{obs}}$. From a circuit perspective, this requirement that $x_i = x_{\text{obs}}$ is equivalent to imposing a voltage source between v_i and ground which establishes the potential x_{obs} at v_i. Consequently, the remaining potentials may be calculated using the steady-state solution for a voltage source (as in (3.8)) where the observed variables form the set of constrained voltages $\mathcal{D}$. As a final example, we can give the circuit interpretation of the probability for any state in the as given by the Gibbs energy in (3.4.3). If $\boldsymbol{\mu} = \mathbf{0}$ then, given any state of the MRF, $\mathbf{x}$, the power dissipated by the circuit equivalent model is given in (3.17) by $\mathcal{E}[\mathbf{x}] = \mathbf{x}^T (\lambda \mathbf{L} + \mathbf{T}) \mathbf{x}$ and therefore the probability of state $\mathbf{x}$ may be interpreted as

$$P(X = \mathbf{x}) = \frac{1}{Z} \exp\left(-\frac{1}{2} \mathcal{G}[\mathbf{x}]\right). \tag{3.61}$$

In other words, the probability of any state in the Gaussian MRF is proportional to the exponential of half the power dissipated by the equivalent circuit. Thus configurations of electrical potentials in the circuit that yield larger power dissipation are less likely to occur under the MRF model. Therefore, the minimum power distribution for the circuit which is produced by nature is the maximum likelihood solution for the equivalent MRF.

In this section we considered a second probabilistic model that could also be interpreted as an electrical circuit. Although circuit interpretations of Gaussian MRFs are not common in the literature, this connection provides another example of an area in which the same variables, operators and equations all participate.

3.4.4 Tree Counting

Almost any quantity in circuit theory may be calculated by counting subtrees of the graph. Counting subtrees is very impractical for computation in most circumstances (although some treatments of computational circuit theory have pursued exactly this approach [75]), but it is sometimes useful to examine a circuit theory problem (or a partial differential equation) in terms of counting subtrees. One utility of the tree viewpoint can be that certain properties of the behavior of a variable under perturbation or noise are simpler to prove by using trees [161]. We believe that the tree interpretations of the circuit equations are particularly valuable because it is difficult to see an analogue for this viewpoint in the continuum equations. In other words, random walks are straightforward to define in the continuum, and electromagnetism provides an intuitive analogy for electrical circuits, yet we are not aware of any equivalent to tree counting in the continuum. Therefore, since this method is available in the discrete domain, we devote some space here to computing solutions to the circuit equations by counting subtrees of the graph. In this section we largely follow the results and notation of Biggs [34], which provides support for all of the results reviewed here.

Define a tree $\mathcal{T}$ as a graph on n nodes and m edges for which there is a single connected component such that $m = n - 1$. Note that some authors refer to this type of tree as a *spanning tree*. One important property of any tree is that a tree consists of no cycles (by contradiction, if a tree did contain a cycle then it would be possible to remove one edge without losing connectivity). A second property of a spanning tree that we will use in this section is that within a spanning tree, there exists a unique path that connects any two nodes. The number of subtrees in a graph is the number of unique subgraphs of the graph that are trees. In this section, any use of the term *tree* should be interpreted as referring to a spanning tree.

By definition of the edge–node incidence matrix, any vector in the right nullspace must take the same value for each pair of nodes connected by an edge. Therefore, if the graph is connected, the right nullspace of the edge–node incidence matrix is spanned by the set of constant-valued functions, i.e., $\mathbf{A1} = \mathbf{0}$. Consequently, the *reduced incidence matrix*, $\mathbf{A}_0$, formed by removing the column corresponding to any node, has a right nullspace of dimension zero. Since the left nullspace of the edge–node incidence matrix has already been discussed as being comprised of the graph cycles, it is clear that if $\mathbf{A}$ represents the incidence matrix of a tree, there can be no such cycles, and thus the left nullspace of the incidence matrix of a tree is also dimension zero. Furthermore, for any tree, the reduced incidence matrix $\mathbf{A}_0$ is both a square matrix and nonsingular. Put differently, any $n \times n$ submatrix of an incidence matrix $\mathbf{A}_0$ is nonsingular if and only if this submatrix represents a tree. The invertibility of this matrix indicates that we may solve for a set of node potentials x_i such that the differences across the edges of a tree fit prescribed values, i.e., a solution $\mathbf{x}$ will always exist for the equation

$$\mathbf{Ax} = \mathbf{p}. \tag{3.62}$$

In summary, if our graph is a tree, then *any* flow in the graph may be represented via potential differences for some set of node potentials due to the guaranteed existence of a solution. Additionally, if we set $x_i = 0$ for some node v_i, then the solution is unique.[1]

The connection between circuit theory and tree counting goes back to the earliest days of circuit theory. Kirchhoff himself proved that the number of trees in a graph, κ, can be calculated as the determinant of the *reduced Laplacian matrix*, $\mathbf{L}_0 = \mathbf{A}_0^T \mathbf{A}_0$ for a reduced incidence matrix $\mathbf{A}_0$ created by removing any node [232], i.e., $\kappa = |\mathbf{A}_0^T \mathbf{A}_0|$. This result is known as the **Matrix-Tree Theorem**. When the Laplacian matrix represents a weighted graph, $\mathbf{L}_0 = \mathbf{A}_0^T \mathbf{G}^{-1} \mathbf{A}_0$, then the determinant $\omega = |\mathbf{A}_0^T \mathbf{G}^{-1} \mathbf{A}_0|$ represents the weighted sum of distinct subgraphs within the graph that are trees, i.e.,

$$\omega = |\mathbf{L}_0| = \sum_{T \in \mathcal{T}} w[T], \tag{3.63}$$

---

[1] In fact, Branin [58] showed that the inverse of the reduced incidence matrix $\mathbf{S} = \mathbf{A}_0^{-1}$ has a semantic interpretation:

$$S(v_i, e_{jk}) = \begin{cases} +1 & \text{if edge } e_{jk} \text{ is traversed in the path from } v_i \text{ to } v_0 \text{ in the positive direction,} \\ -1 & \text{if edge } e_{jk} \text{ is traversed in the path from } v_i \text{ to } v_0 \text{ in the negative direction,} \\ 0 & \text{otherwise.} \end{cases}$$

where the sum is understood to be taken over all such trees $\mathcal{T}$ [36], and we define the weight of each individual tree as the product of the edge weights comprising that tree

$$w[T] \equiv \prod_{e_{ij} \in T} w(e_{ij}). \tag{3.64}$$

We now consider how we may use tree counting to solve problems in circuit theory. The first problem we consider is the solution of the electrical potentials when a voltage source is applied between nodes v_0 and v_1, where v_0 is tied to ground. Consequently, $x_0 = 0$ and $x_1 = 1$. In this case, we may calculate the voltage x_i for any other node via

$$x_i = \frac{1}{\chi} \sum_{F \in \mathcal{F}_{(0|1,i)}} w[F], \tag{3.65}$$

where the sum is over all 2-trees, $\mathcal{F}_{(0|1,i)}$, for which v_1 and v_i are in one component and v_0 is in another component. Recall that a **2-tree** is defined as any graph that is the result of removing one edge from a spanning subtree of $\mathcal{G}$, and that a 2-tree therefore need not be a tree. As before, we let

$$w[F] = \prod_{e \in F} w(e_{ij}), \tag{3.66}$$

and define the scale factor

$$\chi \equiv \sum_{F \in \mathcal{F}_{(0|1)}} w[F], \tag{3.67}$$

where this sum is taken over all 2-trees in the graph in which v_0 is in one component and v_1 is in the other. When the graph is unweighted, $w[\mathcal{F}_{(0|1)}] = w[\mathcal{T}] = 1$, meaning that χ is simply the number of 2-trees separating v_0 and v_1. In this case, (3.65) simply says that the potential at node v_i is given by the percentage of 2-trees separating v_0 and v_1 that groups v_i with v_1. This result may therefore be interpreted as the *expected value* of a random variable that selects 2-trees with uniform probability and gives $x_i = 1$ if v_i is grouped with v_1 and $x_i = 0$ otherwise [267].[2] If the graph is weighted, then process is modified such that the probability of selecting a 2-tree $\mathcal{F}$ is proportional to $w[\mathcal{F}]$.

If we replace our voltage source between v_0 and v_1 with a unit current source such that v_0 is still tied to ground, then the induced potential at node v_i is given

---

[2] It is possible to interpret the formulae concerning 2-trees by considering a second graph $\mathcal{G}^+$ formed from $\mathcal{G}$ by adding an edge between v_0 and v_1. In this case, the 2-trees of $\mathcal{G}$ are the same as the trees of $\mathcal{G}$ that contain the edge spanning v_0 and v_1. Therefore, all of the formulae concerning 2-trees could be written in terms of the trees of $\mathcal{G}^+$ that contain the new edge. This development in terms of the trees of $\mathcal{G}^+$ was the view adopted by Kirchhoff in his original work on this topic [232].

3.4 Connections Between Circuit Theory and Other Discrete Domains

instead by

$$x_i = \frac{1}{\omega} \sum_{F \in \mathcal{F}_{(0|1,i)}} w[F], \qquad (3.68)$$

where, as before, the sum is over all 2-trees, $\mathcal{F}_{(0|1,i)}$, for which v_1 and v_i are in one component and v_0 is in another component. Therefore, the only difference between the potentials induced by a voltage source and a current source are the denominator χ or ω. Consequently, it immediately follows that we may calculate the effective resistance between nodes v_0 and v_1 by

$$R_{\text{eff}}(v_0, v_1) = \frac{\omega}{\chi}. \qquad (3.69)$$

The current through any edge which is induced by a unit current source between v_1 and v_0 also admits an interpretation in terms of trees. Specifically if the unit current is supplied at v_1 and drawn from v_0, then the current y_{ij} through edge e_{ij} is given by

$$y_{ij} = \frac{1}{\omega} \left(\sum_{T^+ \in \mathcal{T}_{\{0,1|i,j\}}^+} w[T^+] - \sum_{T^- \in \mathcal{T}_{\{0,1|i,j\}}^-} w[T^-] \right), \qquad (3.70)$$

where $\mathcal{T}^+$ is a tree in which the path from v_1 to v_0 passes through e_{ij} in the positive direction and $\mathcal{T}^-$ is a tree in which the path from v_1 to v_0 passes through e_{ij} in the negative direction. In the first term the sum is taken over all $\mathcal{T}^+$ trees in the graph in the sum in the second term is taken over all $\mathcal{T}^-$ trees in the graph.

The reader may be concerned that our treatment has addressed only the interpretation of potentials and currents induced by unit voltage sources or current sources. Real circuit problems often involve multiple sources or source magnitudes which are not equal to unity. However, it is important to remember that we are considering *linear* circuits. Therefore, we may decompose any set of sources into a sum of unit sources weighted appropriately by a set of coefficients. Given this decomposition, we could count the trees for each source independently (multiplied by the factor) and sum the results to obtain the solution for the original set of sources. This approach to decomposing sources is common in circuit theory and is known as the **method of superposition**. Consequently, it is sufficient to consider only the solutions for unit voltage and current sources, which form the building blocks for solving more complicated problems.

Example 2 (The Wheatstone bridge) To demonstrate the utility of tree counting for solving for unknowns in an electrical circuit, we will consider the classic *Wheatstone bridge circuit*. The Wheatstone bridge is a circuit that enables the determination of an unknown resistor value r_x. The circuit consists of two known fixed resistors r_1 and r_2 plus a variable resistor or *potentiometer* r_3, together with a fixed voltage source V and a simple current meter or galvanometer. The configuration is shown in Fig. 3.8. To measure r_x, the resistance r_3 of the potentiometer is adjusted

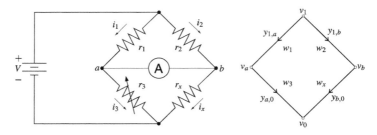

Fig. 3.8 The Wheatstone bridge. The unknown resistance r_x is derived from fixed resistors r_1 and r_2 after the variable resistor r_3 is set so that the current between nodes a and b is measured to be zero by the ammeter. The corresponding weighted, oriented graph is presented *on the right*, with edge weights w_1, w_2, w_3, and w_x on edges $y_{1,a}$, $y_{1,b}$, $y_{a,0}$, and $y_{b,0}$, respectively. Recall that, by convention, the edge weights are conductances and therefore $w_i = 1/r_i$

Fig. 3.9 The four subgraphs of the Wheatstone bridge graph that are trees

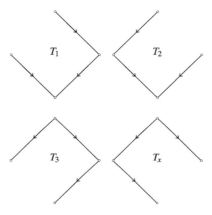

until the current measured between nodes a and b is zero. Nodal analysis shows that the resulting resistance is given by

$$r_x = \frac{r_2}{r_1} r_3. \tag{3.71}$$

To re-derive this result from tree counting, we begin by identifying the graph corresponding to the circuit, which is shown in Fig. 3.8, and the four trees of the graph, provided in Fig. 3.9. The determinant of the weighted reduced Laplacian ω can then be computed, and in this example evaluates to $\omega = w_2 w_3 w_x + w_1 w_3 w_x + w_1 w_2 w_x + w_1 w_2 w_3$.

The next step is to identify the 2-trees $\mathcal{F}_{(0|1)}$ within the graph that separate nodes v_0 and v_1, shown in Fig. 3.10. From these 2-trees the scale factor χ as defined in (3.67) can be calculated and evaluates to $\chi = 2w_1 w_2 + 2w_2 w_3 + 2w_3 w_x + 2w_x w_1$. The effective resistance can then be calculated from these two scale factors, however the value of w_x is still unknown.

The voltage x_a is given in terms of the 2-trees $\mathcal{F}_{(0|1,a)}$ (in which one component contains v_0 and the other component contains both v_a and v_1) identified from the

Fig. 3.10 The eight 2-trees separating v_1 and v_0

three trees containing a connection between nodes v_a and v_1, and similarly for x_b. From (3.65), we see that

$$x_a = \frac{1}{\chi}(2w_1 w_x + 2w_1 w_2)$$

and

$$x_b = \frac{1}{\chi}(2w_2 w_3 + 2w_1 w_2).$$

Therefore when the bridge is balanced by the potentiometer, $x_a = x_b$ and thus

$$w_x = \frac{w_2}{w_1} w_3.$$

Converting the edge weights into resistances we arrive at the final expression for the unknown r_x given by

$$r_x = \frac{r_2}{r_1} r_3.$$

3.4.5 Linear Algebra Applied to Circuit Analysis

In our exposition of linear circuit theory, all of the operators were represented by matrices and the solutions for circuit variables were obtained via linear algebra. Therefore, it should be no surprise that some of the common techniques in linear algebra also have interpretations in terms of circuit theory. Specifically, we will address here the formation of an **equivalent circuit** via a series of circuit transforms.

3.4.5.1 The Delta–Wye and Star–Mesh Transforms

The circuit transform we will consider first is the **delta–wye** transform or, more generally, the **star–mesh** transform. In the context of linear algebra, we will show that one can view an application of these transforms as one step of Gauss–Jordan elimination on the Laplacian matrix.

The delta–wye transform can be viewed as a method for producing a second graph that is missing one node, but for which the electrical connections between all remaining nodes are the same. Formally, we use the delta–wye transform to produce

Fig. 3.11 The original 'Y' graph and the 'Δ' graph obtained by eliminating the central node, v_1, from the 'Y' graph. The delta–wye transform consists of a mapping of resistor values that ensures that the voltages and currents at the terminal nodes v_2, v_3, and v_4 are identical in the two configurations

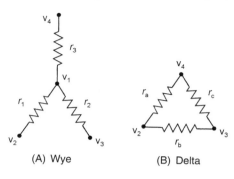

(A) Wye   (B) Delta

a new graph $\mathcal{G}_1$ from an original graph $\mathcal{G}_0$ with the node v_a removed, thus $\mathcal{V}_1 = \mathcal{V}_0 - v_a$, and such that the effective resistance $R_{\text{eff}}(v_i, v_j)$ between all remaining nodes is unchanged after the transformation. Consider the graphs in Fig. 3.11. The delta–wye transform gives us a method for calculating the resistances r_a, r_b and r_c from the original resistors r_1, r_2 and r_3 such that the effective resistances (and thus the voltages and currents at the terminals of the circuit) are preserved. The transform is called "delta–wye" due to the shape of the two graphs (a 'Δ' and 'Y'). It is known from basic circuit theory [294] that the resistances of the 'Δ' circuit can be phrased in terms of resistances of an equivalent 'Y' circuit as

$$r_a = \frac{r_1 r_2 + r_2 r_3 + r_3 r_1}{r_2}, \tag{3.72}$$

$$r_b = \frac{r_1 r_2 + r_2 r_3 + r_3 r_1}{r_3}, \tag{3.73}$$

$$r_c = \frac{r_1 r_2 + r_2 r_3 + r_3 r_1}{r_1}. \tag{3.74}$$

Let us consider the Laplacian matrix for the original 'Y' graph

$$\mathbf{L} = \begin{bmatrix} w_1 + w_2 + w_3 & -w_1 & -w_2 & -w_3 \\ -w_1 & w_1 & 0 & 0 \\ -w_2 & 0 & w_2 & 0 \\ -w_3 & 0 & 0 & w_3 \end{bmatrix}, \tag{3.75}$$

where the central node, v_1, in the 'Y' graph is given by the first row/column. Recall that the edge weights in the Laplacian matrix, w_i, are affinity weights which represent the conductances $w_i = 1/r_i$ (see Chap. 2). If we perform Gauss–Jordan elimination of the central 'Y' node, v_1, in $\mathbf{L}$, then we are left with the matrix

$$\mathbf{L} = \frac{1}{w_1 + w_2 + w_3} \times \begin{bmatrix} w_1 + w_2 + w_3 & 0 & 0 & 0 \\ 0 & w_1 w_2 + w_1 w_3 & -w_1 w_2 & -w_1 w_3 \\ 0 & -w_1 w_2 & w_1 w_2 + w_2 w_3 & -w_2 w_3 \\ 0 & -w_1 w_3 & -w_2 w_3 & w_1 w_3 + w_2 w_3 \end{bmatrix}.$$

3.4 Connections Between Circuit Theory and Other Discrete Domains

The lower-right submatrix of this matrix now corresponds to a graph whose topology matches that of the 'Δ' graph. The resulting conductances in the graph corresponding to this submatrix are given by

$$w_a = \frac{w_1 w_3}{w_1 + w_2 + w_3}, \tag{3.76a}$$

$$w_b = \frac{w_1 w_2}{w_1 + w_2 + w_3}, \tag{3.76b}$$

$$w_c = \frac{w_2 w_3}{w_1 + w_2 + w_3}, \tag{3.76c}$$

which yield the same resistance values given for the delta–wye transform. Consequently, the delta–wye transform is equivalent to removing the node, v_1, in the center of the 'Y' via Gauss–Jordan elimination on the Laplacian matrix.

More generally, a Gauss–Jordan elimination on the Laplacian matrix may be used to eliminate any node and produce a second circuit with the node removed but otherwise equivalent electrical properties. This more general elimination procedure is called a "star–mesh" transform, since the center node of a "star" is replaced by a mesh of resistors that connect all nodes incident on the star. Since each elimination of a node in this manner causes all nodes that were previously connected to the removed node to now be connected to each other, the Gauss–Jordan elimination of a node often adds more edges to the graph than there were originally. These additional edges may cause computational problems since the new graph may require more memory to store and process than the old one. This problem can be addressed by eliminating nodes in a particular order.

3.4.5.2 Minimum-Degree Orderings

Specifically, when using Gaussian elimination to produce "LU" factors (or Cholesky factors) of a sparse Laplacian matrix, the amount of "fill-in" of zeros in the Laplacian matrix caused by the creation of new edges after each elimination may be disastrous for the memory required to store the matrix factors. However, elimination of the nodes in a different order may result in a different level of fill-in. This viewpoint on the elimination of nodes was the primary motivation behind the development of **minimum-degree orderings** that are used to factor sparse matrices [271, 322, 376].

One graph for which the ordering is particularly convenient is any tree. When the graph is a tree, there exists an ordering of the elimination that produces the creation of *zero* new edges and therefore does not create any fill-in for the factored matrix. The principle behind such an ordering is to order the elimination of nodes from the leaf nodes inward, which causes the creation of no new edges in the process. Figure 3.12 gives an example of this elimination procedure on small tree. Algorithm 3.1 accomplishes the ordering in linear time, where the array tree contains, for each node, the index of one neighbor (with no edges overrepresented) and the array degree contains the degree of each node in the tree. This representation is possible

| original | 1st elimination | 2nd elimination | 3rd elimination | Final elimination |

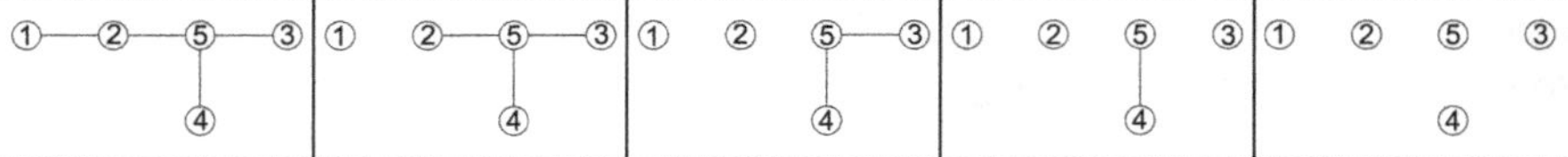

$$\begin{bmatrix} 1 & -1 & 0 & 0 & 0 \\ -1 & 2 & 0 & 0 & -1 \\ 0 & 0 & 1 & 0 & -1 \\ 0 & 0 & 0 & 1 & -1 \\ 0 & -1 & -1 & -1 & 3 \end{bmatrix} \begin{bmatrix} 1 & -1 & 0 & 0 & 0 \\ 0 & 1 & 0 & 0 & -1 \\ 0 & 0 & 1 & 0 & -1 \\ 0 & 0 & 0 & 1 & -1 \\ 0 & -1 & -1 & -1 & 3 \end{bmatrix} \begin{bmatrix} 1 & -1 & 0 & 0 & 0 \\ 0 & 1 & 0 & 0 & -1 \\ 0 & 0 & 1 & 0 & -1 \\ 0 & 0 & 0 & 1 & -1 \\ 0 & 0 & -1 & -1 & 2 \end{bmatrix} \begin{bmatrix} 1 & -1 & 0 & 0 & 0 \\ 0 & 1 & 0 & 0 & -1 \\ 0 & 0 & 1 & 0 & -1 \\ 0 & 0 & 0 & 1 & -1 \\ 0 & 0 & 0 & -1 & 1 \end{bmatrix} \begin{bmatrix} 1 & -1 & 0 & 0 & 0 \\ 0 & 1 & 0 & 0 & -1 \\ 0 & 0 & 1 & 0 & -1 \\ 0 & 0 & 0 & 1 & -1 \\ 0 & 0 & 0 & 0 & 0 \end{bmatrix}$$

Fig. 3.12 Gaussian elimination of the Laplacian matrix of a tree with ordering given by the numbers inside the nodes. Note that the resulting Gaussian elimination has the same sparsity structure as the original matrix when a no-fill ordering is used (e.g., as computed by Algorithm 3.1). Note that the Laplacian matrix is singular—the last elimination produces a row of all zeros. Once the graph has been reduced (grounded), as in (3.77), this is no longer a concern e.g., if node 5 were grounded, the elimination would stop after the third elimination and $x_5 = 0$ would be used to recover the remaining values of the solution. *Top row*: Elimination of the tree—the figures depict the graph represented by the lower triangle of the matrix. *Bottom row*: Laplacian matrix of the tree after each elimination step

---

Algorithm 3.1 Produce a no-fill ordering of a tree

1: void compute_ordering(degree, tree, ground, ordering)
2:   $k \Leftarrow 0$
3:   degree[root] $\Leftarrow 0$ {Fixed so that ground is not eliminated}
4:   ordering[$N - 1$] $\Leftarrow$ ground
5:   **for** each node in the graph **do**
6:     **while** degree[current_node] equals 1 **do**
7:       ordering[k] $\Leftarrow$ current_node
8:       degree[current_node] $\Leftarrow$ degree[current_node]-1
9:       current_node $\Leftarrow$ tree[current_node]
10:      degree[current_node] $\Leftarrow$ degree[current_node]-1
11:      $k \Leftarrow k+1$
12:     **end while**
13:     $k \Leftarrow k+1$
14: **end for**

---

since a tree has $n - 1$ edges (where the root would contain a '0'). Therefore, one could solve the linear system of equations

$$\mathbf{L}_0 \mathbf{x} = \mathbf{f}, \tag{3.77}$$

in linear time when **L** represents a tree. Algorithm 3.2 finds a solution to (3.77) in linear time, given a tree and an elimination ordering. For ease of exposition, we assume that all $w_{ij} = 1$, but the algorithm could be easily modified to handle an arbitrary set of nonnegative weights. Because the full Gaussian elimination has the same sparsity structure as the original matrix when a no-fill ordering is used we need only compute the no-fill ordering, and not the full Gaussian elimination, in order to solve the linear system.

Algorithm 3.2 Given a tree, solve (3.77)

1: solve_system(ordering, f, tree, r, output)
2: {Forward pass}
3: $k \Leftarrow 0$
4: **for** each non-ground node **do**
5:    r[tree[ordering[k]]] $\Leftarrow$ r[tree[ordering[k]]] + r[ordering[k]]/f[ordering[k]]
6:    f[tree[ordering[k]]] $\Leftarrow$ f[tree[ordering[k]]] − 1/f[ordering[k]]
7:    $k \Leftarrow k + 1$
8: **end for**
9:
10: output[ordering[$N-1$]] $\Leftarrow$ r[ordering[$N-1$]]/f[ordering[$N-1$]]
11:
12: {Backward pass}
13: $k \Leftarrow N - 2$ {Last non-ground node}
14: **for** each non-ground node **do**
15:    output[ordering[k]] $\Leftarrow$ output[tree[ordering[k]]] + r[ordering[k]]/f[ordering[k]]
16:    $k \Leftarrow k - 1$
17: **end for**

In this section, we showed that another aspect of circuit theory also appears in other areas of finite mathematics. Specifically, we showed that the methods for circuit transformations and equivalent circuits arise from Gauss–Jordan elimination on the Laplacian matrix. This insight is helpful for understanding the motivation behind matrix ordering algorithms (specifically minimum-degree orderings). When the graph is a tree, we showed that the graph has a straightforward ordering that allows linear systems that represent trees to be solved in linear time. This linear time solution has been exploited before in applications to computer vision (see [160]).

3.5 Conclusion

In his Lecture Notes on Physics, Richard Feynman paused to consider why the same operators and equations recur throughout seemingly different areas of physics [132]. Specifically, Feynman focused on the pervasiveness of the Laplacian operator and questioned whether it was possible that the commonality of the governing equations indicated that all of these physical phenomena were actually the same underlying process on some fundamental level. After examining these different phenomena individually, Feynman concludes that it is highly unlikely that each of these processes are actually the same on a fundamental level and instead offers an alternative explanation for the commonality of the governing equations throughout physics. Feynman's explanation is that the common thread tying together these physical phenomena is that the processes *all occur in the same space* and that this fact alone practically forces a certain relationship between variables.

Similar to Feynman's inquiry, we can seek an explanation of why circuit theory, graph theory, mass-spring networks, random walks and Markov Random Fields (among others) all involve the same equations. In our discrete setting, Feynman's explanation would suggest that the reason for the recurrence of these equations is that all of the phenomena are defined on a set of locations which are connected via some neighborhood structure (i.e., a graph). Therefore, if we are going to describe a relationship between variables defined at the nodes, our basic tool set will have to reflect the *space* upon which these variables are defined. The operators of graph theory—the Laplacian matrix, adjacency matrix and incidence matrix—explicitly represent the space and therefore it should come as no surprise that these operators recur throughout any equations that govern the relationship between variables defined on a graph.

Part II
Applications of Discrete Calculus

Chapter 4
Building a Weighted Complex from Data

Abstract In some applications, both the neighborhood structure of the data and the weights of the cells are naturally and directly defined by the problem at hand (e.g., road networks, social networks, communication networks, chemical graph theory or surface simplification). However, in many other applications the appropriate representation of the data to be analyzed is not provided (e.g., machine learning). Therefore, to use the tools of discrete calculus, a practitioner must determine the topology and weights of the graph or complex *from the data* that is most appropriate for solving the problem. In this chapter, we will discuss different techniques for generating a meaningful weighted complex from an embedding or from the data itself. Our focus will be primarily on generating weighted edges and faces from node and/or edge data, but we additionally demonstrate how these techniques may be applied to weighting higher-order structures.

This chapter marks the beginning of our transition from the theory of discrete calculus (both the mathematical exposition of Chap. 2 and the physical exposition of Chap. 3) to the practice of using these tools in various applications. However, before we can develop content extraction algorithms for processing and analyzing networks (complexes) and the data associated with them, we must first discuss how the network (complex) is constructed. In some applications, the network (complex) construction is given by the problem definition, but in other applications this structure must be imposed on the data.

A recurring characteristic of combinatorial algorithms in image processing, machine learning and computer graphics is that the algorithms are defined on a *weighted* complex in which the weight is used to encode something meaningful about the data or domain. In many circumstances, the domain can be manipulated through the weights to affect the behavior of an algorithm or its outcome, and several examples of this will be discussed in later chapters. However, before we can define a cell weight, we must first discuss how to generate an edge and face set from nodes. Note that in this chapter we will employ the term "cycle" to describe a 2-cell, since a cycle of 1-cells specifies a unique 2-cell.

4.1 Determining Edges and Cycles

In this section, we consider how to produce an edge set from nodes (data points) or a cycle set from edges. This question of how to define an edge set may also be viewed as a question of weighting because setting an edge weight to zero is equivalent to removing the cell, e.g., any graph may be viewed as a fully connected graph with a positive weight to indicate the presence of an edge or a zero-valued weight to indicate the absence of an edge. However, as in Chap. 2, we avoid this issue by assuming that the desired connections in the graph are explicitly specified, e.g., by the incidence matrix, and all cell weights are greater than zero.

Throughout our treatment of discrete calculus we define several algorithms for analyzing and processing both node and edge data. However, the operators of discrete calculus themselves explicitly depend on the choice of edge set (for the gradient and divergence operators) or cycle set (for the curl operator) used to describe the data. Therefore, in order to employ the tools of discrete calculus to analyze and process data, it is essential to define an edge set (to process node data) and cycle set (to process flow data). For many applications the edge set and cycle set are given directly by the problem. For example, a road network already defines an edge set and a two-dimensional surface mesh embedded in three dimensions naturally defines a cycle set. However, in some applications the definition of an edge or cycle set is less clear. For example, in a manifold learning application (see Chap. 7) we often start by having data associated with each node and looking for an appropriate edge set. Even for those applications that define an edge set, the definition of a cycle set may not given by the application. For example, the cycle set may not be defined in an application where the goal is to filter flow data on an abstract graph, or in applications in which we want to use the cycle structure as a feature to measure network characteristics (see Chap. 8). This section is intended to address these situations where we wish to define a meaningful edge set or cycle set.

4.1.1 Defining an Edge Set

An edge set may be viewed as defining a *neighborhood* for each data point. A node (representing a single data point) is considered to be neighbors with every other node for which it shares an edge. For example, the foundational graph theory problem of the Köningsberg bridges assigned two islands to be neighbors if they shared a bridge. Likewise, in chemistry two atoms in a molecule are considered to be neighbors if they share a bond, and in a road network problem two locations are considered to be neighbors if they are directly connected by a road. In these applications, the edge set is a part of the problem formulation and therefore there is no need to ask which objects or data points are neighbors. However, not every application is so clear. For example, if we were studying a problem in land use and collecting data about farms, we would be faced with the problem of deciding which farms are "neighbors" in the sense that two neighboring farms directly affect each other. If we declare that two farms are neighbors only if they are geographically bordering each other, then we ignore the fact that local wind patterns might cause mutual influence on two farms which are geographically distant from each other, or that the structure of the underlying water table might connect two geographically distant farms. Such domain-specific features require information external to the data collected and are generally specific to individual applications. In this section, we address the problem of defining a meaningful edge set for a collection of nodes. The nodes may or may not be embedded geometrically and they may or may not have data associated with them.

In the absence of other information, a natural possibility for defining an edge set is to employ a fully connected graph. A **fully connected graph** contains an edge for every pair of nodes, i.e., the edge set of a fully connected graph is isomorphic to the Cartesian product of the node sets, $\mathcal{E} = \mathcal{V} \times \mathcal{V}$. Although a fully connected graph is in some sense the most agnostic choice for the edge structure, the number of edges, $|\mathcal{E}| = \frac{1}{2}(|\mathcal{V}|^2 - |\mathcal{V}|)$, is prohibitive for a graph of any size. Even a relatively small graph which is fully connected will outstrip the computational power of today's computers,[1] causing us to look at defining a sparse edge set with a size proportional to the size of the node set.

If a fully connected graph is impractical then we must examine other methods of defining a meaningful neighborhood for each node. Fortunately, the neighborhood of a node may be thought to be small relative to the entire set of nodes, leading to a computationally tractable sparse graph. The most natural approach to assigning the neighborhood of a node is from a geometric embedding of the node set. A **geometric embedding** is an association of n spatial coordinates with each point, denoted by the tuple $\tilde{s}_i$ for node v_i, defined as the mapping $\tilde{s} : \mathcal{V} \to \mathbb{R}^n$. In some applications, the coordinates of a node are defined by treating an n-tuple of node data as spatial coordinates. Other applications, such as image processing or geospatial analysis, associate a spatial coordinate with each node which is distinct from the data associated with each node.

Given a geometric embedding, there are several approaches to exploit this metric information to build connections from embedded data. Here we consider three such methods followed by a procedure for adding extra edges to an existing edge set.

4.1.1.1 Edges from an Ambient Metric

Given an embedding and a metric $\mathcal{D}(v_i, v_j)$ measuring the distance between two nodes (e.g., a Euclidean distance based on embedding coordinates), it is common to use the metric to *induce* the edge set $\mathcal{E} = \{e_{ij} \mid \mathcal{D}(v_i, v_j) \leq k\}$, where k is a free parameter. The Euclidean metric $\mathcal{D}(v_i, v_j) = \|\tilde{s}_i - \tilde{s}_j\|_2$ is the usual choice. This

---

[1] A *parallel* machine, with one processor devoted to one node or to a small number of nodes, can potentially be used to address fully connected graphs, since each processor must do work only on the order of $|\mathcal{V}|$. Although such machines are becoming increasingly common, the computers of today are still predominantly unable to process fully connected graphs of any size.

method of using a metric to induce an edge set may cause the resulting graph to contain isolated nodes (e.g., if one of the nodes is distant from all others).

4.1.1.2 Edges by k-Nearest Neighbors

An alternative method for generating the edge set, which overcomes the isolated nodes problem, is k-**nearest neighbors**. This approach for establishing an edge set is defined algorithmically by examining each node and adding an edge between the node and each of the k nodes with a smaller distance than the rest of the node set. The k-nearest neighbor algorithm can be used to generate a *directed* graph (i.e., v_i may be one of the nearest neighbors of v_j, but v_j is not one of the nearest neighbors of v_i), however we generally assume that all graphs are undirected unless otherwise noted (see Chap. 2). Additionally, any edges which are added twice from this procedure (as a result of two nodes each being "nearest neighbors" with each other) are conventionally reduced to a single edge. The k-nearest neighbor method for establishing an edge set gives a guarantee on the degree structure of the resulting graph, that $d_i \geq k$. Consequently, if $k > 0$, the resulting graph will have no isolated nodes.

4.1.1.3 Edges from a Delaunay Triangulation

The methods for generating an edge set described above, i.e., the induced metric method and the k-nearest neighbors method, do not guarantee that the resulting graph is connected. Additionally, these methods do not guarantee that the resulting graph is planar, even when the nodes are embedded in two dimensions. If connectedness or planarity are important for the application (or the optimization procedure), then a third option for establishing an edge set from nodes embedded in two dimensions is the Delaunay triangulation of the data [343, 344]. Recall that the Delaunay triangulation is the unique triangulation of a point set in the plane that is the dual graph to the Voronoï tessellation of the point set. A Delaunay "triangulation" may also be applied to data embedded in higher dimensions to ensure that the resulting structure is a cell complex [22].

4.1.1.4 Adding Edges via the Watts–Strogatz Model

Even when the data is geometrically embedded, it is sometimes profitable to include nonlocal (long-range) edges into the analysis. One example of such long range connections comes from the analysis of "small-world" networks [396, 397]. A small-world network may be defined as a locally connected network with a small diameter. The diameter of a graph is the length of the longest optimal path connecting any two nodes in the graph. Watts and Strogatz [396] showed that a small-world network may be generated from any locally connected graph via the introduction of a very small number of random edges. In other words, a locally connected graph with a large diameter may be converted into a graph with a small diameter simply by adding a small number of edges that connect two randomly chosen nodes. A graph with a small diameter can be helpful in applications because the diameter of a graph may be thought of as a measure of how quickly information spreads throughout the graph. In fact, the convergence rate of conjugate gradients when solving a linear system of equations with the graph Laplacian matrix is known to depend on the graph diameter [174, 175]. Therefore, it has been suggested [167] that a small number of random edges may be added to any locally connected graph (e.g., a lattice representing an image) in order to increase the convergence rate of iterative optimization algorithms.

4.1.2 Defining a Cycle Set

A graph (i.e., a 1-complex) may be fully described by its edge–node incidence matrix **A**, which also provides a gradient operator (as a special case of the exterior derivative) that can be used for computing gradients of node data. This gradient operator is widely used to define a range of data analysis/processing algorithms. However in applications where we seek to process or analyze edge data, such as flow data, the corresponding exterior derivative operator is represented by the face–edge incidence matrix, **B**, which implements the curl operator. In order to define the incidence matrix **B** to provide this operation, it is necessary to generate a *cycle set*, which is a set of faces identified within a given graph. For a general graph, there are several possible cycle sets, each with distinct properties and advantages for particular problems.

In this section we discuss the problem of defining a set of cycles which can then be used to define incidence relations between edges and cycles. The specification of a cycle set may also be viewed as defining neighborhoods of edges, i.e., two edges are neighbors if they share a cycle—just as two edges may be considered neighbors if they share the same node. Defining these neighborhood relationships will allow us to perform operations such as filtering and clustering on edge data in later chapters.

Methods for defining a cycle set have been less studied than the methods for generating an edge set. The reason for this lesser amount of study is that the processing of flow data is generally less common than the processing of node data. However, a theme of this work is that all of the machinery which has been developed for analyzing node data can also be applied to the analysis of flow data by using the dimensionally-appropriate discrete differential operators. Additionally, when discussing the measurement of network properties in Chap. 8, we suggest that the measurement of the cycle structure of a network may reveal new characteristics of network structure. Consequently, we predict an increasing future interest in the definition of cycle sets. In this section we consider the definition of a cycle set from the edges geometrically (i.e., from an embedding) and algebraically. We begin with geometric methods for defining a cycle set (i.e., from an embedding), before moving to algebraic methods for defining the cycles.

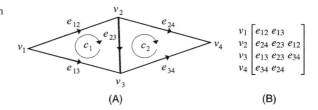

Fig. 4.1 An example rotation system. The faces in the embedding shown in (**A**) are represented combinatorially by the rotation table given in (**B**)

4.1.2.1 Defining Cycles Geometrically: Cycles from an Embedding

When a bridgeless graph[2] is embedded onto a surface such that no edge crossings are generated, then a cycle set (representing the faces) in this embedding may be identified algorithmically by using a device known as a *rotation system*. Given an embedding, we will employ the term *geometric cycles* to describe the cycles which are obtained from the rotation system derived from the embedding. A **rotation** of a node $\rho(v_i)$ is defined as a cyclic permutation of the (oriented) edges incident on the node v_i. A list of rotations defined for each node is called a **rotation system**, which may be stored in table format as a **rotation table**. A rotation for each node may be induced from an embedding onto an orientable surface by ordering the edges consistently in a clockwise or counterclockwise fashion. The algorithm employs a *rotate operation* $\pi(v_i, e_{jk})$ which outputs the next edge after e_{jk} in $\rho(v_i)$ (circulating to the beginning of the list if e_{jk} is the last), and a *trace operation* $\tau(v_i, e_{jk})$ which outputs node v_k if $i = j$ and $\tau(v_i, e_{jk})$ outputs v_j if $i = k$, i.e., the node at the opposite end of the edge. Note that the algorithm requires that the edge orientation is specified by the edge index, i.e., e_{jk} represents the edge that begins at node v_j and terminates at node v_k, as usual.

Figure 4.1 shows a rotation system for an example graph. In this example, the rotation of node v_2 is $\rho(v_2) = \{e_{24}, e_{23}, e_{12}\}$. Therefore, the rotate operation $\pi(v_2, e_{23}) = e_{12}$ and the rotate operation $\pi(v_2, e_{12}) = e_{24}$. Also for this node, the trace operation $\tau(v_2, e_{24}) = v_4$ and the trace operation $\tau(v_2, e_{12}) = v_1$.

Given a rotation system corresponding to an embedding on an orientable two-dimensional surface, the following algorithm may be used to produce the geometric cycles of the embedded graph [177], which is equivalent to identifying the 2-cells of the graph. Each cycle consists of a set of edges oriented in the direction of the edge traversal in the cycle.

1. Initialize every entry of the rotation table to *unused* (note that each edge has two entries).
2. While any unused entry remains:
    a. Start a new (empty) cycle.
    b. Choose unused entry, e_{ij}, in the rotation table, mark as used and add the oriented e_{ij} to the cycle.

---

[2]In graph theory, a **bridge** is an edge whose deletion would result in increasing the number of connected components in the graph. Thus an edge is a bridge if (and only if) it is not contained in a cycle.

c. Repeat:
   i. Set the current entry to the oriented edge $\pi(\tau(v_i, e_{jk}), e_{jk})$.
   ii. Add the current entry (oriented edge) to the cycle and mark as used.
d. Until the current entry equals the first entry in the cycle.
e. Remove the last edge from the cycle and close the cycle.

The cycle set may be used to produce the face–edge incidence matrix **B** in which every row, $\mathbf{B}_i$, corresponds to a cycle and $\mathbf{B}_i(e_{jk}) = \pm 1$ if cycle i contains edge e_{jk}, with the sign determined by agreement between the native edge orientation and the orientation induced on the edge from the cycle, as described in Chap. 2.

A rotation system provides a purely combinatorial representation of an embedding. More strongly, it is known that every rotation system on a graph defines (up to equivalence of embeddings) a unique oriented graph embedding, and that every locally oriented graph embedding defines a rotation system [176, 355]. This statement is sometimes referred to as the **Heffter–Edmonds Principle** [420]. The Heffter–Edmonds Principle may be used to define an algorithm for computing the closed surface of minimum genus onto which a graph may be embedded without edge crossings, known as the **graph genus**. This Heffter–Edmonds algorithm computes the graph genus by simply enumerating all possible rotation systems for a graph and then using the algorithm above to compute the faces for each rotation system [176]. If the *maximum* number of cycles found for a graph using this method is $|\mathcal{F}|$, then the formula relating the Euler characteristic $\chi = |\mathcal{F}| - |\mathcal{E}| + |\mathcal{V}|$ to the genus (see Chap. 8) allows us to compute the genus g of the corresponding closed embedding surface with *minimal* genus for the graph as

$$g = \frac{1}{2}(2 - |\mathcal{F}| + |\mathcal{E}| - |\mathcal{V}|), \tag{4.1}$$

which defines the unique graph genus for the graph. Consequently, the number of geometric cycles of the graph will depend on the genus of the surface upon which the graph was embedded (see Chap. 8 for more discussion of genus). Note that this Heffter–Edmonds algorithm for finding the graph genus is not computationally feasible due to the enormous number of rotation systems (e.g., a $(d+1)$-regular graph has $(d!)^{|\mathcal{V}|}$ rotation systems).

4.1.2.2 Defining Cycles Algebraically

When we use a geometric embedding to define a cycle set, the Heffter–Edmonds principle above states that the number of cycles obtained in the resulting cycle set is determined by the genus of the particular surface onto which the graph is embedded. However, the maximum number of cycles obtainable by the geometric method is limited by the graph genus.

In contrast, we know from Chap. 2 that the dimension of the right nullspace of **A** is equal to $|\mathcal{E}| - |\mathcal{V}| + 1$. Since $\mathbf{AB} = 0$, then the *maximum* number of linearly independent columns of **B**, and therefore the number of linearly independent cycles, is equal to $|\mathcal{F}|_{indep} = |\mathcal{E}| - |\mathcal{V}| + 1$. Consequently, by the relation between the Euler Characteristic and graph genus recalled in (4.1), the geometric cycles for a graph with graph genus greater than zero[3] (i.e., a nonplanar graph) for which $|\mathcal{F}| < |\mathcal{F}|_{indep}$ *do not span the nullspace of* **A**. This statement is known as *MacLane's Planarity Criterion* [268].

Algebraic methods for defining a cycle set take the approach of finding a cycle set which forms a *basis* for the nullspace of **A** and therefore always generates a cycle set for which $|\mathcal{F}| = |\mathcal{E}| - |\mathcal{V}| + 1$. Therefore, algebraic methods for defining a cycle set differ from the geometric methods due to the fact that the number of cycles in the set generated by an algebraic method depend only on the number of edges and nodes, rather than on the connectivity structure which determines the number of cycles obtainable by a geometric method.

There are several methods for producing a cycle basis from an arbitrary graph. The classical method for producing a cycle basis is based on a tree/cotree decomposition[4] of the graph [147]. This method begins by identifying an arbitrary spanning tree subgraph, i.e., $\mathcal{V}_{tree} = \mathcal{V}$ and $\mathcal{E}_{tree} \subseteq \mathcal{E}$ such that $\mathcal{G}_{tree} = (\mathcal{V}_{tree}, \mathcal{E}_{tree})$ is connected and $|\mathcal{E}_{tree}| = |\mathcal{V}_{tree}| - 1$. Given any edge in the cotree, $e_{ij} \in \mathcal{E}_{cotree} = \mathcal{E} - \mathcal{E}_{tree}$, there is a unique path within the tree $\mathcal{G}_{tree}$ from v_j to v_i. Denote this oriented path as

$$\mathcal{P}_{ji} = \{e_{k\ell} \mid \text{if the path includes edge } e_{k\ell} \text{ in the direction from } v_k \text{ to } v_\ell\}. \quad (4.2)$$

Therefore, the oriented cycle corresponding to e_{ij} consists of e_{ij} and the edges in $\mathcal{P}_{ji}$. Each cycle may be represented algebraically by a row of the face–edge incidence matrix **B** in which each edge included in the cycle assumes a value of ± 1 (depending on agreement of the cycle orientation with the native orientation, as in Chap. 2) and each edge not included in the cycle assumes a zero. This method may be used to construct a cycle basis by letting every edge in $\mathcal{E}_{cotree}$ correspond to one row of **B**. Each row of matrix **B** formed in this way is independent because each row contains a nonzero which is unique to that row (corresponding to the cotree edge). Additionally, the cycle set formed in this way is a basis because $|\mathcal{E}_{cotree}| = |\mathcal{E}| - |\mathcal{E}_{tree}| = |\mathcal{E}| - |\mathcal{V}| + 1$, which is equal to the rank of the nullspace of the edge–node incidence matrix **A**. Figure 4.2 shows an example of a tree/cotree basis for a small graph. Although the original graph in this example is planar, note that the cycle set obtained from this spanning tree does *not* correspond to the geometric cycle set (cell faces).

---

[3]Note that when the graph genus equals zero then, by the relationship between the Euler Characteristic and genus, $|\mathcal{F}| = |\mathcal{E}| - |\mathcal{V}| + 2$. In contrast, the number of linearly independent cycles (spanning the nullspace of **A**) is given by $|\mathcal{F}| = |\mathcal{E}| - |\mathcal{V}| + 1$. Therefore, a planar graph implicitly has an *extra* face that is not linearly independent. This is the "outside face" described in Chap. 2.

[4]Recall that, for a connected graph with $\mathcal{V}$ and $\mathcal{E}$, a tree subgraph is identified by the connected subgraph comprised of $\mathcal{V}$ and $\mathcal{E}_{tree} \subseteq \mathcal{E}$, where $|\mathcal{E}_{tree}| = |\mathcal{V}| - 1$. The **cotree** subgraph corresponding to a tree subgraph is the (possibly disconnected) subgraph consisting of $\mathcal{V}$ and the set of the *remaining edges* $\mathcal{E}_{cotree} = \mathcal{E} - \mathcal{E}_{tree}$. Thus the union of the edge set from any given tree with the edge set from its cotree is equivalent to the original graph.

4.1 Determining Edges and Cycles

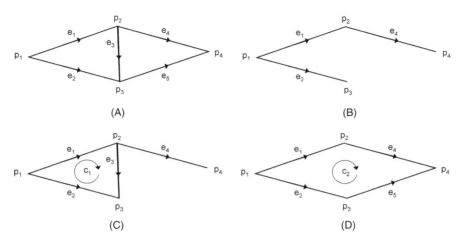

Fig. 4.2 Example of a tree/cotree basis on a small graph. (**A**) The original graph $\mathcal{G}$. (**B**) An arbitrary spanning tree of the graph $\mathcal{G}$. (**C**) The first cycle, corresponding to edge e_3 in the cotree. (**D**) The second cycle, corresponding to edge e_5 in the cotree

The tree/cotree basis is straightforward to generate and efficient to produce. However, this choice of basis may not be ideal for many applications because of several limitations: (i) the tree/cotree basis for a given graph depends on the choice of tree/cotree decomposition and is therefore not unique; (ii) the tree/cotree cycle basis often contains many non-zeros (i.e., in **B**), which results in inefficient storage and often requires longer computation times for any practical usage of the basis; (iii) edge weights are ignored in producing the tree/cotree cycle basis; and (iv) the tree/cotree cycle basis is rarely equivalent to the geometric cycle set of an embedding, even when the graph is planar.

An alternative to the tree/cotree method for defining a cycle basis is the **minimum cycle basis**. The minimum cycle basis finds a cycle basis that optimizes the sum of the weights of the edges comprising each cycle. Specifically, given a cycle basis represented by the face–edge incidence matrix **B** and diagonal edge weighting matrix **G**, define the cycle basis weight as

$$\text{cycle basis weight}(\mathbf{B}, \mathbf{G}) = \mathbf{1}^T |\mathbf{B}|\mathbf{G}\mathbf{1}, \qquad (4.3)$$

where $|\mathbf{B}|$ indicates the absolute value of each entry in **B**. A minimum cycle basis for a graph is any cycle basis (represented by $\hat{\mathbf{B}}$) which minimizes the cycle basis weight (4.3) over the set of all possible cycle bases for the graph, i.e.,

$$\hat{\mathbf{B}} = \underset{\mathbf{B}}{\text{argmin}}\{\text{cycle basis weight}(\mathbf{B}, \mathbf{G})\}. \qquad (4.4)$$

The minimum cycle basis has appeared in applications in electrical engineering [78], structural engineering [66] and surface reconstruction [157, 374].

The minimum cycle basis explicitly finds a basis which minimizes the number of non-zeros in **B** (when all edge weights are equal). Although the minimum cycle basis is not guaranteed to be unique, the minimum cycle basis often is unique in practical situations. Additionally, the geometric cycle basis generated by an embedded planar graph—with the largest cycle, the "outside face", omitted—comprises a minimum cycle basis for the graph (assuming equal edge weights). A minimum cycle basis may be found by an algorithm with a polynomial-time dependence on the number of edges in the graph. The first polynomial time algorithm was developed by Horton [206] who adopted the simple approach:

1. Find a set of cycles which contains a minimum cycle basis as a subset.
2. Sort the cycles by weight.
3. Add cycles to the cycle set starting from the cycles of minimum weight, while rejecting cycles which are not independent (using Gaussian elimination to determine independence).
4. Terminate when the cycle set contains $|\mathcal{F}| = |\mathcal{E}| - |\mathcal{V}| + 1$ cycles.

Horton [206] showed that a cycle set obtained by taking each node v_i and edge e_{jk} can be used to generate a set of cycles containing a minimum cycle basis. This set is generated by computing, for each node/edge pair, the cycle composed of the two shortest paths from the ends of edge e_{jk} (i.e., v_j and v_k) to the node v_i, in which the cycle is rejected if these two paths contain any common nodes (except v_i). Although Horton's algorithm for producing a minimum cycle basis is solvable in polynomial time, it is still computationally expensive. Recently, there has been substantial interest in the problem of finding a more efficient algorithm for computing a minimum cycle basis [30, 100, 155, 226, 278]. Although these new algorithms have a lower computational complexity than Horton's algorithm, they still require a substantial amount of computation.

We have discussed how to generate a cycle set from a geometric embedding or a cycle set which is an algebraic basis for the nullspace of **A**. However, the problem of generating *all* cycles for a graph has also been studied and there exist algorithms for producing this set [261, 273, 316, 375] (see Ref. [33] for a review). These algorithms are often computationally infeasible for graphs of any size, due to the overwhelmingly large number of possible cycles.

4.1.2.3 Cycle Sets and Duality

In Chap. 2 we defined the dual of a 2-complex (i.e., graph with an identified cycle set) by replacing each cycle in the primal graph with a node in the dual graph, which was possible when every edge was incident on exactly two cycles that induced opposite orientations on the edge. A cycle set which traverses each edge exactly twice in opposite directions is called an oriented **cycle double cover**. The oriented cycle double cover conjecture [340] states that every bridgeless graph has an oriented cycle double cover [215]. Since this conjecture is presently open, there is no known method for generating an oriented cycle double cover for any nonplanar graph. However, some heuristics for finding a cycle double cover of an arbitrary graph have been proposed [262]. The geometric cycles for a planar graph embedded on the surface of a sphere define a double cycle cover, and as a result there is a straightforward method for finding the dual of a planar graph.

4.2 Deriving Edge Weights

Given an edge set, obtained either natively by the application or via one of the methods above, we may consider assigning weights to the edges in a meaningful way to assist in the analysis of the network or the data associated with the network. Once again, some applications will define edge weights as part of the problem definition, in which case the methods presented in this section are not relevant. As before, we are primarily concerned with methods for defining edge weights, but will also discuss methods for defining cycle weights and weights of higher-order p-cells.

Edge weights can have several interpretations depending on the application. In physical systems, the edge weights reflect material properties such as branch resistance or a spring constant (see Chap. 3). In geometric applications, edge weights are used to represent the *distance* between nodes, while in many data processing applications the edge weights represent an *affinity* measure between data points. Consequently, in the terminology introduced in Chap. 2, the weights derived from geometry will be considered as *distance weights* and the weights derived from data will be considered as *affinity weights*. Problems in physical system modeling generally include the value of the edge weights as part of the problem definition. In this section, we discuss some of the methods used to define edge weights in geometric and data processing applications.

4.2.1 Edge Weights to Reflect Geometry

Many graphs are embedded geometrically, meaning that each node v_i embedded in n dimensions is associated with an n-tuple of coordinates, $\tilde{s}_i$. Examples of these graphs include road networks, distributed sensors, three-dimensional meshes in computer graphics and pixel data in an image. Weighting these edges to reflect the embedding is very simple: the edge weights are given by the distances between connected nodes measured using the node coordinates,

$$w_{ij} = \|\tilde{s}_i - \tilde{s}_j\|, \tag{4.5}$$

where $\tilde{s}_i$ indicates the node coordinates and $\|\cdot\|$ indicates the Euclidean norm. This metric weighting is the natural choice for computing such quantities as the length of a road connecting two cities in a road network. Weights obtained in this manner are clearly examples of *distance weights* (described in Chap. 2), since they directly represent the distance between nodes.

Weights have also been used to represent geometric quantities on an embedded lattice. Specifically, Boykov and Kolmogorov [54] proposed a method for weighting edges such that the sum of the edge weights spanning two sets of nodes would approximate the length of the *boundary* of the nodes embedded in space as measured via some metric (typically the Euclidean metric). The purpose of this edge weighting was to be able to use combinatorial optimization techniques for minimizing cuts between nodes in order to find a minimal Euclidean boundary between two sets of points in space. Similarly, Schoenemann et al. [61, 334] have defined a set of weights between *pairs* of edges such that the weighted length of an edge cycle approximates the *mean curvature* of the polygon represented by the cycle.

4.2.2 Edge Weights to Penalize Data Outliers

In content extraction applications, such as filtering or clustering, we often define edge weights for the purpose of including extra information to improve the results of a content extraction algorithm. In general, these edge weights are used to reflect the *similarity* or affinity between the data at two nodes. The strategy is effectively to set the edge weights such that they represent similarity when the edge spans two nodes belonging to the same class/cluster and to set the edge weights to represent dissimilarity when the edge spans two nodes belonging to different classes/clusters (i.e., outliers). We consider several methods for establishing edge weights to serve this purpose by looking at (i) similarities derived from metric distances and (ii) similarities derived from robust functions. The weights described in this section are all interpreted as *affinity* weights, as discussed in Chap. 2. The geometric coordinates of the nodes, discussed above, may be substituted for node data in this section to obtain edge weightings for an embedded graph that are robust to outliers (e.g., for mesh filtering applications).

4.2.2.1 Univariate Data

In many applications, edge weights are used to define distances between nodes (inversely, the strength of connection between two nodes) which are incorporated into the data analysis (e.g., filtering, clustering). In the absence of an external source for distance information, the edge weights must be defined from the given data in some way that is suitable to the application. The end result is an operator that is adapted to the data—a classic example, anisotropic diffusion, is described below. We now review several common approaches in the literature for setting edge weights from univariate data and then proceed to consider the case of multivariate data below.

In order to develop the weighting functions used in graph-based algorithms, we begin by considering weighted data processing in a general context beyond that of discrete complexes and graphs. For example, in the calculation of a weighted least-squares regression or of a weighted mean, the purpose of the weights is generally to de-emphasize certain data points or to prevent the undue influence of outliers. Specifically, consider that the mean of the values [0, 1, 2, 3, 4, 1000] is 145, which is

4.2 Deriving Edge Weights

dominated by the outlier 1000. However, by *weighting* the calculation of the mean, i.e.,

$$\text{weighted mean} = \frac{0w_0 + 1w_1 + 2w_2 + 3w_3 + 4w_4 + 1000w_5}{w_0 + w_1 + w_2 + w_3 + w_4 + w_5}, \quad (4.6)$$

with weights $w_0 = w_1 = w_2 = w_3 = w_4 = 1$ and $w_5 = 0$, the weighted mean is equal to 2. By using the above weights, we effectively removed the '1000' sample from the calculation of the mean. The same effect of removing an outlier can be achieved by replacing the calculation of the mean by a calculation of the median. To continue the above example, the median of these numbers would conventionally be given as equal to 2.5. Although the mean and median calculations appear to be very different, we can view the median as a *weighted* mean.[5] Specifically, consider the mean of a set of real numbers $\mathcal{S}$ with N elements $s_i \in \mathcal{S}$ as the solution to the problem

$$x = \underset{x}{\operatorname{argmin}} \left\{ \sum_i (x - s_i)^2 \right\}, \quad (4.7)$$

with the solution given by the classic formula for the sample mean

$$x = \frac{\sum_i s_i}{N}. \quad (4.8)$$

A *weighted* mean is the solution to

$$x = \underset{x}{\operatorname{argmin}} \left\{ \sum_i w_i (x - s_i)^2 \right\}, \quad (4.9)$$

where the minimum is given by

$$x = \frac{\sum_i w_i s_i}{\sum_i w_i}. \quad (4.10)$$

Similarly, the median value is the solution to the problem

$$x = \underset{x}{\operatorname{argmin}} \left\{ \sum_i |x - s_i| \right\}, \quad (4.11)$$

which may be rewritten as

$$x = \underset{x}{\operatorname{argmin}} \left\{ \sum_i \frac{1}{|x - s_i|} (x - s_i)^2 \right\}, \quad (4.12)$$

---

[5]This concept appears again in Chap. 5 and Appendix B.

which can be viewed as the weighted mean in (4.9) with weights equal to $w_i = 1/|x - s_i|$ (assuming that $|x - s_i| \neq 0$). Consequently, if we had a method for approximating these weights, then the computation of the weighted mean would provide for an outlier-resistant approximation to the median.

A major utility of graph algorithms for data processing is that the neighborhood structure is exploited to provide additional information about data—in the sense that each node is generally assumed to be similar to its neighbors. For example, in a filtering operation, we typically assume that the data at each node is similar to the data at neighboring nodes, and in a clustering operation, we generally assume that each node is more likely to belong to the same cluster as its neighbors. This prior assumption reflects a presumed smoothness in our data model, as discussed in Chap. 5. Therefore, we may consider that a node value (e.g., data, cluster membership) is likely to be at the "average" of its neighbors, but we want to design weights such that this average is robust to outliers. Therefore, we can view the purpose of the edge weights as the reduction (or removal) of the influence of neighboring nodes for which the values represent outliers. Consequently, the design of a weighting function for data processing applications should be to set small weight values on edges between nodes that are dissimilar or when one node is likely to be an outlier, i.e., between nodes that are likely to belong to different classes or clusters.

Although the median value is an "average" value for a set of values which is more robust to outliers than the mean value, other so-called "robust" error functions were designed to provide an estimate for an "average" which is even more robust than the median. Robust error functions appeared early in the Markov Random Field-based image filtering literature [144, 146, 259], as will be presented in Chap. 5. A *robust* function is a function that is insensitive to outliers, i.e., that does not allow outliers to have undue influence. Intuitively, these robust functions give linear or quadratic penalties near zero but "level off" after a certain value to produce a nearly constant penalty. A robust error function therefore seems like a good candidate for setting edge weights for data analysis, since the robust error function is designed to ignore outliers.

Historically, in the field of image filtering, robust error functions were initially used to *measure* and penalize gradients of the *processed* image rather than to weight edge variables based on the initial data. The motivation for using a robust error function is that smoothness penalties are not unbounded, which would result in over-smoothness of the filtered data (see Chap. 5 for more details on filtering with discontinuities). Figure 4.3 gives an example of a robust error function in which it is possible to observe that the penalty is roughly quadratic near a zero gradient but levels off to a nearly constant penalty for large gradients.

After the introduction of anisotropic diffusion as an image processing technique by Perona and Malik [306], it became common to use a class of edge weighting functions iteratively to update an intermediate solution. In subsequent work, these edge weighting functions were measured from the gradients of the *input* data and used as edge weights for image segmentation [345], clustering [169] and filtering [158, 419]. In the elegant work of Black et al. [38], it was shown that the weighting functions in common usage could be interpreted as the derivatives of common robust estimators from statistics and further suggested new weighting functions which

Fig. 4.3 An example of a robust error function $\rho_\alpha(z)$, the Welsch function, for different values of α. The function z^2 is plotted as a reference

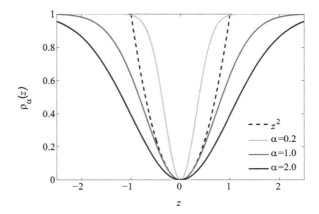

occasionally appear in the literature. In summary, the property of robust error functions to ignore outliers may be exploited.

Following Black et al. [38], we may consider data filtering as the partial minimization of

$$\mathcal{E}[x] = \rho_\alpha(\nabla x) \quad \text{in the continuous setting, or} \quad (4.13)$$

$$\mathcal{E}[\mathbf{x}] = \rho_\alpha(\mathbf{Ax}) \quad \text{in the discrete setting,} \quad (4.14)$$

in which $\mathbf{x}$ is our objective, filtered data and $\rho_\alpha(\cdot)$ is a robust error function with parameter α. By a "partial minimization" we mean that we take $\mathbf{x}$ initially to equal the input data and solve a few iterations of gradient descent on (4.14). When $\rho_\alpha(\mathbf{Ax})$ is a quadratic (as in many of the robust estimators that we will consider shortly), then this iteration is given by

$$\mathbf{x}^{[k+1]} = \mathbf{x}^{[k]} - \Delta t\, \mathbf{A}^\mathsf{T} g_\alpha(\mathbf{Ax}) \mathbf{A} \mathbf{x}^{[k]}, \quad (4.15)$$

and we see that the role of the edge weights are assumed by

$$g_\alpha(z) = \frac{\rho'_\alpha(z)}{z}, \quad (4.16)$$

where $\rho'_\alpha(z) = d\rho_\alpha(z)/dz$. The derivative $\rho'_\alpha(z)$ is sometimes known as the *influence function*. Instead of a gradient descent on (4.14), we can also interpret (4.15) as a forward Euler scheme for solving

$$\frac{d\mathbf{x}}{dt} = -\mathbf{L}\mathbf{x} = -\mathbf{A}^\mathsf{T}\mathbf{G}^{-1}\mathbf{A}\mathbf{x} = -\mathbf{A}^\mathsf{T} g_\alpha(\mathbf{Ax})\mathbf{A}\mathbf{x}, \quad (4.17)$$

with initial condition $\mathbf{x}^{[0]} = \mathbf{s}$, where $\mathbf{s}$ is the data. This interpretation as a forward Euler scheme is how Black et al. [38] derive the anisotropic diffusion filtering method from robust estimator functions of the gradient $\rho_\alpha(\mathbf{Ax})$. Using this derivation we see that setting edge weights based on the g_α expression in (4.16) of a ro-

Table 4.1 Common M-estimators. Edge weights in many data processing applications employ the $g_\alpha(z)$ function to set edge weights with $\mathbf{z} = \mathbf{As}$, where $\mathbf{s}$ represents the node data. For example, if s_i represents image intensity at node v_i, then $w_{ij} = g_\alpha(s_i - s_j)$

Estimator	$\rho_\alpha(z)$	$\rho'_\alpha(z)$	$g_\alpha(z) = \frac{\rho'_\alpha(z)}{z}$
Welsch	$\frac{\alpha^2}{2}\left[1 - \exp(-\frac{z^2}{\alpha})\right]$	$\alpha z \exp(-\frac{z^2}{\alpha})$	$\alpha \exp(-\frac{z^2}{\alpha})$
Cauchy	$\frac{\alpha^2}{2} \log\left[1 + \left(\frac{z}{\alpha}\right)^2\right]$	$\alpha \frac{z}{1+(\frac{z}{\alpha})^2}$	$\alpha \frac{1}{1+(\frac{z}{\alpha})^2}$
Tukey	$\begin{cases} \frac{z^2}{\alpha^2} - \frac{z^4}{\alpha^4} + \frac{z^6}{3\alpha^6} & \text{if } \|z\| \leq \alpha \\ \frac{1}{3} & \text{if } \|z\| > \alpha \end{cases}$	$\begin{cases} z[1 - (\frac{z}{\alpha^2})^2] & \text{if } \|z\| \leq \alpha \\ 0 & \text{if } \|z\| > \alpha \end{cases}$	$\begin{cases} [1 - (\frac{z}{\alpha^2})^2] & \text{if } \|z\| \leq \alpha \\ 0 & \text{if } \|z\| > \alpha \end{cases}$
Huber	$\begin{cases} \frac{z^2}{2} & \text{if } \|z\| \leq \alpha \\ \alpha(\|z\| - \frac{\alpha}{2}) & \text{if } \|z\| > \alpha \end{cases}$	$\begin{cases} z & \text{if } \|z\| \leq \alpha \\ \alpha \operatorname{sign}\{z\} & \text{if } \|z\| > \alpha \end{cases}$	$\begin{cases} 1 & \text{if } \|z\| \leq \alpha \\ \alpha \frac{\operatorname{sign}\{z\}}{z} & \text{if } \|z\| > \alpha \end{cases}$
Fair	$\alpha^2 \left[\frac{\|z\|}{\alpha} - \log\left(1 + \frac{\|z\|}{\alpha}\right)\right]$	$\frac{z}{1 + \frac{\|z\|}{\alpha}}$	$\frac{1}{1 + \frac{\|z\|}{\alpha}}$
p-norm, $p < 1$	$\frac{\|z\|^p}{\alpha p}$	$\operatorname{sign}\{z\} \frac{\|z\|^{p-1}}{\alpha}$	$\frac{\|z\|^{p-2}}{\alpha}$

bust error function will cause filtering and other operations to reduce the influence of spatial outliers when processing data.

We may now examine several of the robust error functions for which the derivatives are commonly used as weighting functions. Table 4.1 lists common robust functions, known as *M-estimators* in the statistical literature (e.g., [422]). Other robust functions exist and even more could be invented. The most commonly used functions for setting edge weights from data are the Welsch or the Cauchy functions, although the Tukey biweight function also appears occasionally. Some evidence exists to suggest that the Cauchy function performs best for the applications of image segmentation [164] and that the Tukey function performs best for image filtering [38], but very few systematic studies have been undertaken to determine when or how to choose one of these functions.[6]

The Gaussian weighting function $g(z) = \exp(-z^2/\alpha)$, corresponding to the robust Welsch function, is also common because it may be derived from other principles. For example, the Gaussian weighting function is derived by Belkin and Niyogi [23] because the Gaussian function is the Green's function for the (continuous) Laplacian. The Gaussian weighting function may also be motivated as tracking "perceptual differences" based on Weber's Law and the exponential response curve of the human eye to stimulus changes [113]. This function may also be derived from an assumption that the node data is sampled from a manifold, in which case the Gaussian kernel is a common interpolation function.

---

[6] Based on the available evidence at the time of writing, the authors would suggest using a Cauchy function for most applications, although the Welsch function and Tukey functions also seem to perform well. The remaining robust functions have yet to prove as generically useful.

4.2.2.2 Computing Weights from Multivariate Data

The previous discussion on setting weights to penalize outliers required the computation of a comparison between data values. This comparison was computed by calculating a difference between the data variables. This discussion implicitly assumed that the underlying data was univariate (i.e., each node was associated with a single data variable). If our data is instead multivariate (i.e., each node is associated with a tuple containing multiple data values), then the comparison involved in outlier detection is no longer a simple difference computation.

Weighting Based on Norms of Differences

The standard approach to comparing multivariate data is to use the norm of the difference between two multivariate data points. This norm can be used in place of the difference for any of the robust error weighting functions described in the previous section. For example, if $\tilde{s}_i$ and $\tilde{s}_j$ represent the multivariate data at two nodes, then an extension of the Welsch weighting function would be

$$w_{ij} = \exp\left(-\frac{\|\tilde{s}_i - \tilde{s}_j\|^2}{\alpha}\right). \tag{4.18}$$

Note that we continue to distinguish between *vector data* and *multivariate data*, since 1-cochains or edge variables are typically used to represent vector fields whereas "multivariate data" refers to multiple, independent 0-cochains (e.g., multi-channel, multi-modal or multi-spectral data) at each node. This distinction between "vector data" and "multivariate date" is often ignored in the image processing literature where, for example, RGB color data is often referred to as "vector data", while we would consider it here to be "multivariate data". Weighting vector data (flow variables) will be discussed in Sect. 4.3.

Although any of the weighting functions described for univariate data may be used with multivariate data by replacing the data difference with the norm of the difference (in the same manner as in (4.18)), we have additional latitude with multivariate data in the sense that we may choose an additional norm for measuring the difference $\tilde{s}_i - \tilde{s}_j$. Along with the usual p-metrics (i.e., Manhattan distance, Euclidean distance, Chebyshev distance), there are other choices that are specific to multivariate data. As before, the norm used for measuring $z = \|\tilde{s}_i - \tilde{s}_j\|$ may also be specified via a $K \times K$ inner product matrix $\mathbf{H}$ such that

$$z = \|\tilde{s}_i - \tilde{s}_j\| = \sqrt{(\tilde{s}_i - \tilde{s}_j)^\mathsf{T} \mathbf{H} (\tilde{s}_i - \tilde{s}_j)}. \tag{4.19}$$

When $\mathbf{H} = \mathbf{I}$, then this norm is just the Euclidean norm. Therefore, in the remainder of this section we will discuss different choices of $\mathbf{H}$ with the assumption that the edge weighting function for multivariate data is in one of the forms in Sect. 4.2.2.1, e.g.,

$$w_{ij} = \exp\left(-\frac{\|z\|^2}{\alpha}\right) = \exp\left(-\frac{\|\tilde{s}_i - \tilde{s}_j\|^2}{\alpha}\right)$$
$$= \exp\left(-\frac{(\tilde{s}_i - \tilde{s}_j)^\mathsf{T}\mathbf{H}(\tilde{s}_i - \tilde{s}_j)}{\alpha}\right). \tag{4.20}$$

Therefore there are $|\mathcal{V}|K$ data values for the entire graph, where K is the number of scalar data fields associated with each node.

For applications such as clustering or image segmentation in which the goal is to separate the *foreground* cluster/object from the *background* cluster/scene, an effective approach for measuring distances between multichannel data is the **Mahalanobis distance**. The Mahalanobis distance is useful when examples are available of the multichannel data sampled from the foreground. In this situation, the Mahalanobis distance allows us to measure distances relative to the within-cluster scatter measured from the training examples. Specifically, if we have a set of nodes $\mathcal{S}$ containing foreground samples, then we may set the inner product matrix $\mathbf{H} = \text{cov}(\mathbf{P}_\mathcal{S})^{-1}$, where $\text{cov}(\mathbf{P}_\mathcal{S})$ indicates the covariance matrix of the data in set $\mathcal{S}$. The calculation of the covariance matrix and its inverse are computationally efficient; the construction of the covariance matrix is computable in linear time with respect to $|\mathcal{S}|$ and the inverse is computed only on the (generally small) $K \times K$ covariance matrix.

A different method for measuring distances between multichannel data assumes that the data is embedded on a lower-dimensional manifold and, by determining the manifold, computes the difference of the data relative to the lower-dimensional manifold (see Chap. 7 for details on manifold learning). One example of using manifold learning to choose a metric is the Locality Preserving Projections (LPP) method of He and Niyogi [193]. This method uses the Laplacian Eigenmaps projection [23] to effectively project the higher-dimensional data onto a lower dimensional manifold using the eigenvectors of the Laplacian matrix (see Chap. 7 for more details). Specifically, if $\mathbf{L}$ is the (weighted or unweighted) Laplacian matrix of the graph, then we can set $\mathbf{H}$ to be the generalized eigenvectors for the problem

$$\mathbf{PLP}^\mathsf{T}\mathbf{x} = \lambda \mathbf{PDP}^\mathsf{T}\mathbf{x}, \tag{4.21}$$

where $\mathbf{D}$ is the diagonal matrix of node degrees. If we let $\mathbf{R}$ represent the set of generalized eigenvectors satisfying (4.21), then the weighting matrix is given by $\mathbf{H} = \mathbf{R}_k \mathbf{R}_k^\mathsf{T}$ where k is the set of eigenvectors corresponding to the k smallest generalized eigenvalues. An attractive property of this method is that this generalized eigenvector problem is only of size $K \times K$, where K is the number of data channels. The LPP method was applied in [165] to the problem of clustering (and alpha-matting) color images.

Weighting Based on Neighborhood

Another method for setting edge weights has recently become popular in the image processing community. This approach establishes the edge weight as a function of the difference in *data neighborhood* between two nodes [15, 63, 123]. Since the neighborhoods of two nodes are directly compared, this type of weighting is most suitable for a shift-invariant graph (see Chap. 5), since each node will have the same neighborhood structure. Generally, these neighborhood comparisons are made between every pair of nodes, resulting in a fully connected graph (or at least a graph with long range edges). The principle of this approach is that data patterns often re-occur over large spatial areas and that the use of non-local neighborhoods between instances of the pattern can lead to increased performance of filtering tasks. The "non-local means" method of [63] uses a shift-invariant graph with an ordered neighborhood and assigns a weight between two nodes via

$$w_{ij} = g(\|\mathbf{x}_{\text{nbhd}(i)} - \mathbf{x}_{\text{nbhd}(j)}\|), \tag{4.22}$$

where $g(\cdot)$ is any robust weighting function discussed above and $\mathbf{x}_{\text{nbhd}(i)}$ represents the collection of data values, arranged into a vector, from the nodes in the local neighborhood of v_i. For example, if we employ $g(\cdot)$ as the Welsch function, then

$$w_{ij} = \exp\left(-\frac{\|\mathbf{x}_{\text{nbhd}(i)} - \mathbf{x}_{\text{nbhd}(j)}\|^2}{\alpha}\right). \tag{4.23}$$

Since a weight is established between every pair of nodes, the final graph is effectively fully connected. Although this approach results in a fully connected graph, which is unusable in many practical situations (due to memory requirements, etc.), one can use a drop tolerance for removing any edges with a weight less than some predefined threshold θ. A small value of θ retains a nearly fully connected graph and a large value of θ removes all but the strongest connections.

4.2.3 Edge Weights to Cause Repulsion

So far we have considered weights that are employed to act as soft barriers between sets of nodes. These soft barriers were motivated from the perspective of reducing or removing the undue influence of outliers. This type of weighting is useful in filtering applications in which our intention is not to filter over a discontinuity, or segmentation/clustering applications in which our goal is to use these partial indications of a discontinuity to help produce a final labeling of the nodes. In these cases, the edge weights are typically nonnegative.

However, some studies have additionally considered the use of edge weights with negative values in filtering and clustering applications [194, 257, 413, 414]. In these cases, the negative edge weights represent the strength of *repulsions*. In the context of a filtering application, these negative edge weights act to force nodes on either side of the edge to assume different values. For example, in a clustering/segmentation application, the negative weight tries to force nodes on either side of the edge to assume different labels. The optimization of models including negative edge weights can be more difficult that the usual case of nonnegative weights, since the energies to be optimized may be non-submodular [244] or non-positive definite. However, local optima and incomplete solutions (e.g., quadratic pseudo-boolean optimization) are possible. In some specific circumstances (e.g., planar graphs), an optimal solution may be guaranteed. See Appendix B for more details on optimization in the presence of negative weights.

4.2.4 Edge Weights to Represent Joint Statistics

Instead of using weights to ignore outliers or represent repulsion, a different utility for edge weights has been to represent the expected correlation of the data between nodes, in which the correlation may be either positive or negative. An application which benefits from the use of weights to represent correlation is pattern restoration. The pattern restoration problem is to input a set of training patterns in which spatial correlations are learned between the data and then to use this learning to restore corrupted data drawn from the same distribution. These approaches to generating weights have not appeared very often in the literature and the methods used for generating weights in these circumstances have been narrowly defined for the application and the optimization method. Therefore, we refer the reader to [93, 243] for some examples of this approach.

4.2.5 Deducing Edge Weights from Observations

On certain occasions, one wishes to know the inhomogeneities of the material properties within a physical system, but has access only to the *action* of the material on various pairs of inputs and outputs. For example, in circuit theory the material inhomogeneities are the resistances of the circuit that, given a set of currents, act to generate a set of voltages and consequently result in power dissipation throughout the circuit. These inhomogeneities are represented by a *constitutive relation* and thus we will refer to the problem of determining the constitutive under these circumstances as the **Constitutive Determination Problem** (CDP). In general, the CDP is not solvable with access to a single test/response measurement, since different constitutive laws may produce the same input/output. Examples of the CDP problem are the determination of the resistor values for an electrical circuit or impedance tomography (e.g., [240]).

The CDP problem that we are concerned with is formulated by a search for edge weights, **w**, for a graph with a known edge set. In this circumstance we know the topology of the graph but not its edge weights, thus the metric component of the graph Laplacian matrix $\mathbf{L}$ is the unknown, and we have access to K observations of the system test–response. Denote the collection of tests as $\mathcal{T} = \{\mathbf{t}_1, \mathbf{t}_2, \ldots, \mathbf{t}_K\}$ and the corresponding responses as $\mathcal{R} = \{\mathbf{r}_1, \mathbf{r}_2, \ldots, \mathbf{r}_K\}$. We may consider that the tests are "designed" and the responses are "measured", via $\mathbf{L}\mathbf{r}_i = \mathbf{t}_i$. An example problem is this: Given a circuit with unknown conductances (corresponding to weights in our case), we may give test functions (i.e., injecting/draining a distribution of current in the nodes, in which the amount of current injected and drained is balanced) and measure the circuit response (i.e., the electrical potentials). Our goal is to determine the conductance values. The conductance values will not, in general, be unique for a single test–response pair. Therefore, our first question is to determine when we will have a unique solution.

We begin by recalling that the Laplacian matrix may be decomposed into the factors

$$\mathbf{L} = \mathbf{A}^\mathsf{T}\mathbf{G}^{-1}\mathbf{A}. \quad (4.24)$$

Given this decomposition of the Laplacian matrix, we may write

$$\mathbf{L}\mathbf{r}_i = \mathbf{A}^\mathsf{T}\mathbf{G}^{-1}\mathbf{A}\mathbf{r}_i = \mathbf{A}^\mathsf{T}\mathbf{P}\mathbf{w} \quad (4.25)$$

where $\mathbf{w}$ is the vector of edge weights. If $\mathbf{p} = \mathbf{A}\mathbf{r}$ represents the response gradients, then we may form a diagonal matrix $\mathbf{P} = \text{diag}(\mathbf{p})$ such that the diagonal entries of $\mathbf{P}$ are the elements of the vector $\mathbf{p}$, i.e., $P_{jj} = p_j$. Given K test–response observations, the solution of (4.25) may then be written

$$[\mathbf{P}_1\mathbf{A}, \mathbf{P}_2\mathbf{A}, \ldots, \mathbf{P}_K\mathbf{A}]^\mathsf{T}\mathbf{w} = [\mathbf{t}_1, \mathbf{t}_2, \ldots, \mathbf{t}_K]^\mathsf{T}, \quad (4.26)$$

for the vector $\mathbf{w}$ and the tests–responses given by $\mathbf{t}_i$ and $\mathbf{P}_i$. There is a problem, however, with this problem formulation: the number of equations is Kn, for $n = |\mathcal{V}|$, and the number of unknowns is $m = |\mathcal{E}|$. For a connected graph with n nodes, the minimum number of edges is $m = n - 1$ but, in general, $m > Kn$. In other words, for a small K, we may not be able to solve (4.26) exactly.

We now consider the question of determining the value of K which is necessary to solve the system exactly. In order to solve the CDP exactly we require that K is chosen so that $Kn \geq m$ and that the K test-response pairs are independent in the algebraic sense relative to (4.26). For a general r-regular graph with n nodes, $m = \frac{1}{2}rn$. Therefore, the number of independent test–response pairs necessary to find an exact solution of the CDP is $K \geq \frac{1}{2}r$.

4.2.5.1 The Underdetermined Case

If there are an insufficient number of equations for the number of unknowns, our system (4.26) is *underdetermined*. In this case, we need to add additional constraints, or regularize the problem, in order to produce a unique solution. A necessary constraint for physical systems is that $w_i \geq 0$ for all edges. However, this constraint is still insufficient to yield a unique solution.

Although several additional constraints could be incorporated in order to constrain the problem to find a unique answer (depending on domain knowledge of the solution) a constraint that yields a particularly straightforward method of calculation is to look for the solution with the smallest sum of weights. With these constraints, we have the following problem

$$\min_{\mathbf{w}} \mathbf{1}^T \mathbf{w},$$
$$[\mathbf{P}_1 \mathbf{A}, \mathbf{P}_2 \mathbf{A}, \ldots, \mathbf{P}_K \mathbf{A}]^T \mathbf{w} = [\mathbf{t}_1, \mathbf{t}_2, \ldots, \mathbf{t}_K]^T, \quad (4.27)$$
$$\mathbf{w} \geq \mathbf{0}.$$

The notation "$\mathbf{1}^T$" above is used to indicate the vector of all ones to provide an unweighted combination of the components in $\mathbf{w}$. This problem is a linear programming problem (see Appendix B) that may be solved with standard, efficient techniques.

4.2.5.2 The Overdetermined/Inconsistent Case

A CDP may be overdetermined if K is sufficiently large that $Kn > m$. If the test-response pairs are precise, (4.26) may be solved simply by taking any M independent columns and solving the resulting linear system. However, in some cases, the test-response measurements may be noisy, or may suffer from numerical precision limitations. In these cases, there will not be a solution satisfying each equation of (4.26).

The inconsistent case is the most difficult because the test-response pairs may not be treated as constraints due to their inconsistency. This problem is compounded by the fact that a standard least-squares approach is inappropriate since the weights are required to be nonnegative. Therefore, we are led to a quadratic programming problem (see Appendix B).

A least squares formulation of (4.26) is

$$\mathbf{S}^T \mathbf{S} \mathbf{w} = \mathbf{S}^T \mathbf{t}, \quad (4.28)$$

where $\mathbf{S} = [\mathbf{P}_1 \mathbf{A}, \mathbf{P}_2 \mathbf{A}, \ldots, \mathbf{P}_K \mathbf{A}]^T$ and $\mathbf{t} = [\mathbf{t}_1, \mathbf{t}_2, \ldots, \mathbf{t}_K]^T$. Introducing the positivity constraint yields the quadratic programming problem

$$\min_{\mathbf{w}} (\mathbf{S}^T \mathbf{S} \mathbf{w} - \mathbf{S}^T \mathbf{t})^T (\mathbf{S}^T \mathbf{S} \mathbf{w} - \mathbf{S}^T \mathbf{t}) = \mathbf{w}^T \mathbf{S}^T \mathbf{S} \mathbf{S}^T \mathbf{S} \mathbf{w} - \mathbf{t}^T \mathbf{S} \mathbf{S}^T \mathbf{S} \mathbf{w},$$
$$\mathbf{w} \geq \mathbf{0}. \quad (4.29)$$

Before concluding this section, we note that in many practical cases of interest it is not possible to probe and/or observe all of the measured responses. For example, in the impedance tomography application, it is only possible to probe and measure responses on the surface of a subject's head. A treatment of these more difficult cases is beyond the scope of the present work, although the framework above can serve as a starting point for these investigations.

4.3 Obtaining Higher-Order Weights to Penalize Outliers

In the previous section, we discussed how to generate *edge* (i.e., 1-cell) weights from a *node* or *scalar field* (i.e., 0-cochain) data. Specifically, we examined how edge weighting functions to penalize the influence of outliers could be used in filtering or clustering applications. Additionally, we have seen in Chaps. 2 and 3 that these edge weights could be interpreted physically as material properties, representing, e.g., diffusion constants, electrical resistances or spring constants. However, in Chap. 5 we will see that our techniques for filtering scalar fields or node data (i.e., 0-cochains) naturally extend to the filtering of flow fields or vector data (i.e., 1-cochains). Therefore, it is natural to examine how we can meaningfully weight flow data for filtering or clustering applications. The purpose of this section is to address this issue and show that these weights also have natural physical interpretations. We will begin by addressing the weighting of higher-order structures for purposes of filtering/clustering flow fields (1-cochains), but these techniques will be generalized at the end of this section to the weighting of data associated with any dimensional structure (p-cochains). As before, we distinguish the higher-dimensional cells from the *higher-order cliques* which are employed occasionally in some applications [212, 239, 245]. See Chap. 2 for more on the relationship between higher-order cliques and higher-order cells.

When using edge weights to penalize the influence of outliers, our weighting functions were of the form $g(\nabla x)$ or $g(\mathbf{A}x)$ in which g was a function that was maximal at zero, $g(-\mathbf{A}x) = g(\mathbf{A}x)$, and $g(\mathbf{A}x)$ decayed with increasing values of the magnitude of $\mathbf{A}x$. From the standpoint of penalizing outlier data at neighboring nodes, the gradient operator (or the coboundary operator on node data, represented by the incidence matrix $\mathbf{A}$) serves to both encode the neighbors of a node and to compute a comparison of the data at a node with its neighbors. Consequently, generalizing our interpretation of weighting as outlier-identification to higher-order cells requires us to identify the operators which both encode the neighborhood of each cell and the comparison of data across neighbors. In the context of filtering, the assumption of this weighting strategy is that large magnitude gradients are likely to be *signal* and low magnitude gradients are likely to be *noise*, consistent with a model of high-frequency noise with discontinuities (see Chap. 5 for more details). The high-frequencies of a flow field are the components of the flow field having a projection onto the high frequency eigenvectors of the edge Laplacian $\mathbf{L}_1 = \mathbf{B}^\mathsf{T}\mathbf{B} + \mathbf{A}\mathbf{A}^\mathsf{T}$. Consequently, the high-frequency components of the flow field may be decomposed into those components of the flow field with a large curl and those components having a large divergence. Recall from Chap. 2 the definition of the weighted edge Laplacian matrix

$$\mathbf{L}_1 = \mathbf{A}\mathbf{G}_0\mathbf{A}^\mathsf{T}\mathbf{G}_1^{-1} + \mathbf{G}_1\mathbf{B}^\mathsf{T}\mathbf{G}_2^{-1}\mathbf{B}. \tag{4.30}$$

While the interpretation of edge weights as a measure of either the similarity or difference between two nodes is natural and arises in many applications, the interpretation of higher-order weights can often be less straightforward as there are fewer examples from physics to provide intuition through examples. To better understand the role of these higher-order weights for the purposes of suppressing the effects of outlier data on filtering and clustering operations, we will consider in depth the special case of deriving weight values from data to avoid outliers when filtering "flow data" (1-cochains) with the weighted edge Laplacian, which can be understood in terms of fluid dynamics. In Chap. 2 we noted the physical interpretation of the cycle weights $\mathbf{G}_2$ and node weights $\mathbf{G}_0$ as *viscosity*, which impedes fluid flow around cycles or through nodes. More specifically, we may view the cycle weights as corresponding to *shear viscosity* (the "first coefficient of viscosity") that acts as a diffusion coefficient on the vorticity. We may also view the node weights as corresponding to *bulk viscosity* (the "compression viscosity" or "second coefficient of viscosity"), which is neglected for incompressible fluids (i.e., for divergence-free fluids).[7]

In the context of processing node data considered in the previous section, edge weighting is used to reduce the influence of outlier data connected to a node via its incident edges. In a filtering or clustering operation (see Chaps. 5 and 6), two flows may influence each other directly if the two flows are on edges (i) that are part of the same cycle or (ii) that connect to the same node. In either case, the two flows are said to be *co-incident* (to the cycle or to the node, respectively). If the two co-incident flows are very different, e.g., if one flow is very strong and the other is very weak, then this difference will induce a strong *circulation* on the cycle with which they are co-incident, or a strong *divergence* on the node with which they are co-incident. Consequently, any cycle with an outlier circulation (e.g., a curl at a cycle that is much larger than other curls in the vicinity) or a node with an outlier divergence (e.g., a divergence at a node that is much larger than other divergences in the vicinity) may be weighted such as to reduce the influence of the cycle or node where the outlier data is found.

This weighting to reduce the influence of flows in a particular cycle or node for flow processing is analogous to the case considered above where edge weighting was used to reduce the influence of an outlier node data value connected via an edge. For a physical example, consider two liquids flowing in parallel at a different speed, as depicted in Fig. 4.4. This difference in speed between the liquids induces a curl on the interface. Consequently, if we wanted to filter noisy measurements of the direction/magnitude of these flows, then the filtering operation should not smooth across the liquid interface. The magnitude of the curl on this interface serves as an indicator that cycles which span the interface should be weighted to reduce the influence of flows from different liquids across the interface.

We may follow the approach taken for weighting node data to derive weights from flow data by simply applying the same weighting functions as before to the $\mathbf{By}$ and $\mathbf{A}^T\mathbf{y}$ components, which represent the curl and divergence operators, respectively. For example, given a flow field $\mathbf{y}$, and cycle c_i, we may employ the Welsch

---

[7]When $\mathbf{G}_2^{-1}$ and $\mathbf{G}_0$ are matrices, these viscosities are linear and therefore the flow field may be viewed as a *Newtonian fluid*. The more complicated analysis of non-Newtonian fluids would describe the behavior of the flows when $\mathbf{G}_2^{-1}$ and $\mathbf{G}_0$ are replaced by nonlinear functions.

4.3 Obtaining Higher-Order Weights to Penalize Outliers

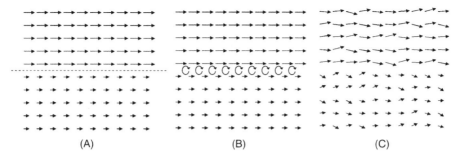

Fig. 4.4 The use of curl to detect flow boundaries. The curl may be used to establish cycle weights (shear viscosity) for processing the flow data in the same way that gradients in node data are used to establish edge weights for processing node data. (**A**) Two flows with a different velocity. The *dashed line* indicates the separation between the flows. (**B**) The different velocities of the two flows induce a strong circulation on the interface. (**C**) The same flows with noise. The interface circulations allow us to detect the flow boundaries, permitting a separate filtering for each flow

function to derive the cycle weight

$$w_i = g_\alpha(\mathbf{B}_i \mathbf{y}) = \exp\left(-\frac{(\mathbf{B}_i \mathbf{y})^2}{\alpha}\right). \qquad (4.31)$$

Note that the expression $\mathbf{B}_i$ represents the row of $\mathbf{B}$ corresponding to cycle c_i. As with the scalar case, $\mathbf{G}_2^{-1}$ is then the diagonal matrix comprised of these cycle weights w_i along the diagonal. Note that since (4.31) represents an expression of an *affinity* weight, it is associated with the *dual* metric tensor, $\mathbf{G}_2^{-1}$. Similarly, the weight penalizing divergence at node v_j may be given as

$$f_j = g_\alpha(\mathbf{A}_j^\mathsf{T} \mathbf{y}) = \exp\left(-\frac{(\mathbf{A}_j^\mathsf{T} \mathbf{y})^2}{\alpha}\right), \qquad (4.32)$$

where $\mathbf{A}_j^\mathsf{T}$ represents the row of $\mathbf{A}^\mathsf{T}$ corresponding to node v_j.

4.3.1 Weights Beyond Flows

The procedure for generating outlier-suppressant weights from flow and scalar data may be extended to produce higher-order weights from data associated with any p-cell by considering the weighted higher-order Laplacian operating on p-cochains. Once again, the p-Laplacian is the primary focus of interest because "low-frequency" signals on the p-cells are measured relative to the corresponding higher-order Laplacian (see Chap. 5). The higher-order p-cochain Laplacian operator consists of two terms, acting on the $(p+1)$- and $(p-1)$-cells. Specifically, recall the general definition of the Laplace–de Rham operator from Chap. 2,

$$\mathbf{L}_p = \mathbf{N}_p \mathbf{N}_p^* + \mathbf{N}_{p+1}^* \mathbf{N}_{p+1} \qquad (4.33)$$

where $\mathbf{N}_p$ is the p-coboundary operator and $\mathbf{N}_p^*$ is its adjoint. By substituting in the definition of the adjoint coboundary operators in terms of the discrete Hodge star operator, for an arbitrary value of p we arrive at the general, weighted Laplacian for p-cochains as

$$\mathbf{L}_p = \mathbf{N}_p \mathbf{G}_{p-1} \mathbf{N}_p^\mathsf{T} \mathbf{G}_p^{-1} + \mathbf{G}_p \mathbf{N}_{p+1}^\mathsf{T} \mathbf{G}_{p+1}^{-1} \mathbf{N}_{p+1}, \qquad (4.34)$$

where $\mathbf{N}_p$ is the incidence matrix mapping $(p-1)$-cochains to p-cochains and $\mathbf{G}_p$ represents the weight matrix on p-cells.

Continuing with higher-order weights, we may follow the approach used in deriving weights from flow data to produce higher order weights for filtering or clustering applications. For example, these weights may be derived from p-cochain data via a Welsch function. Specifically, the weight for $(p+1)$-cell c_i derived from p-cochain data $\mathbf{z}$, is given by

$$w_i = g_\alpha(\mathbf{N}_{p,i}\mathbf{z}) = \exp\left(-\frac{(\mathbf{N}_{p,i}\mathbf{z})^2}{\alpha}\right), \qquad (4.35)$$

and the weight for the $(p-1)$-cell c_j is given by

$$w_j = g_\alpha(\mathbf{N}_{p-1,j}^\mathsf{T}\mathbf{z}) = \exp\left(-\frac{(\mathbf{N}_{p-1,j}^\mathsf{T}\mathbf{z})^2}{\alpha}\right). \qquad (4.36)$$

As before, $\mathbf{G}_{p+1}^{-1}$ is the diagonal matrix comprised of the $(p+1)$-cell weights along its diagonal and $\mathbf{G}_{p-1}^{-1}$ is the diagonal matrix comprised of the $(p-1)$-cell weights along its diagonal.

4.4 Metrics Defined on a Complex

We end this chapter by considering different distance functions which may be derived from a weighted complex. In Chap. 2 we considered the p-cell weights, $\mathbf{G}_p$, as playing the analogous role of the metric tensor in conventional calculus. The metric tensor represented by $\mathbf{G}_p$ provides an inner product between p-cells. This definition allowed us in Sect. 2.3.4.4 to define a *volume* (e.g., length, area) for any p-chain τ_p as

$$\text{volume}(\tau_p) = \mathbf{1}^\mathsf{T} \mathbf{G}_p \tau_p = \mathbf{w}_p^\mathsf{T} \tau_p, \qquad (4.37)$$

which is interpreted as the integral of the volume cochain over a chain. This notion of the volume of a p-chain is closely related to the total distance along a chain, which we shall now explore.

Recall that a **metric** is formally defined by a **distance operator** $\mathcal{D}(a,b)$ on a pair of points a and b on some manifold $\mathcal{M}^n$ that satisfies the following properties:

1. $\mathcal{D}(a,b) = 0$ if and only if $a = b$, *(discernibility)*
2. $\mathcal{D}(a,b) > 0$ if $a \neq b$, *(non-negativity)*

4.4 Metrics Defined on a Complex

3. $\mathcal{D}(a,b) = \mathcal{D}(b,a)$, (*symmetry*)
4. $\mathcal{D}(a,c) \leq \mathcal{D}(a,b) + \mathcal{D}(b,c)$. (*triangle inequality*)

As discussed in Sect. 2.2, the inner product on p-chains **a** and **b**, in terms of the primal metric tensor $\mathbf{G}_p$, given by

$$\langle \mathbf{a}, \mathbf{b} \rangle_{\mathbf{G}_p} = \mathbf{a}^\mathsf{T} \mathbf{G}_p \mathbf{b}, \quad (4.38)$$

can be used to define a norm

$$\|\mathbf{a}\| = \mathbf{a}^\mathsf{T} \mathbf{G}_p \mathbf{a} \quad (4.39)$$

and, subsequently, a suitable distance operator on chains

$$\mathrm{dist}_{\mathbf{G}_p}(\mathbf{a}, \mathbf{b}) = \langle \mathbf{a} - \mathbf{b}, \mathbf{a} - \mathbf{b} \rangle_{\mathbf{G}_p} = (\mathbf{a} - \mathbf{b})^\mathsf{T} \mathbf{G}_p (\mathbf{a} - \mathbf{b}). \quad (4.40)$$

Thus the inner product given by the metric tensor can be extended to define a *natural distance operator*, which imposes the structure of a pre-Hilbert space on the vector space of p-chains. Similarly, we may also define an equivalent metric between p-cochains **x** and **y** using the dual metric tensor $\mathbf{G}_p^{-1}$ as

$$\mathrm{dist}_{\mathbf{G}_p^{-1}}(\mathbf{x}, \mathbf{y}) = \langle \mathbf{x} - \mathbf{y}, \mathbf{x} - \mathbf{y} \rangle_{\mathbf{G}_p^{-1}} = (\mathbf{x} - \mathbf{y})^\mathsf{T} \mathbf{G}_p^{-1} (\mathbf{x} - \mathbf{y}). \quad (4.41)$$

Note that, when **u** and **v** are p-chains whose coefficients consist only of binary-valued elements (i.e., each coefficient belongs to the set $\mathbb{B} = \{0, 1\}$), then the distance between them is equivalent to the *volume of the intersection of the two chains subtracted from the union of the two chains*, i.e.,

$$\mathrm{dist}_{\mathbf{G}_p}(\mathbf{u}, \mathbf{v}) = \mathrm{volume}[(\mathbf{u} \cup \mathbf{v}) - (\mathbf{u} \cap \mathbf{v})], \quad \text{if } \mathbf{u}, \mathbf{v} \in \mathbb{B}^{n_p}. \quad (4.42)$$

Given this definition of a distance measure in terms of an inner product, we may interpret other common distance measures, which provide a valid metric, as stemming from a distance operator with another positive definite matrix substituted in for the metric tensor $\mathbf{G}_p$. For example, a distance measure can be defined by employing the weighted Laplacian matrix as the inner product matrix. Recall that the weighted Laplacian matrix, which operates on 0-cochains, is defined as the positive semi-definite matrix $\mathbf{L} = \mathbf{G}_0 \mathbf{A}^\mathsf{T} \mathbf{G}_1^{-1} \mathbf{A}$. If we assume that the weights on nodes are uniform such that $\mathbf{G}_0$ is the identity matrix, then **L** becomes simply the symmetric matrix $\mathbf{A}^\mathsf{T} \mathbf{G}_1^{-1} \mathbf{A}$. In this case, we can identify 0-cochains with 0-chains since the metric tensor on 0-chains is the Euclidean metric (see Sect. 2.2.1.3), therefore the weighted Laplacian matrix can equivalently operate directly on 0-cochains or 0-chains.

The matrix **L** is only positive *semi*-definite, and therefore is not a suitable inner product matrix because it would violate the discernibility criterion for a metric given above. However, in the special case in which the distance is measured *between a pair* *of nodes*[8] rather than the general case of an arbitrary pair of 0-chains, the matrix **L** provides a valid metric. If we represent two individual nodes in the complex, v_i and v_j, by (non-negative) binary-valued indicator vectors **i** and **j** in $\mathbb{B}^{|\mathcal{V}|}$, where the only non-zero entry in **i** corresponds to node v_i and the only non-zero entry in **j** corresponds to node v_j, then new metrics can be constructed in terms of the weighted Laplacian matrix, including

$$\text{dist}_\mathbf{L}(\mathbf{i}, \mathbf{j}) = \langle \mathbf{i} - \mathbf{j}, \mathbf{i} - \mathbf{j}\rangle_\mathbf{L} = (\mathbf{i} - \mathbf{j})^T \mathbf{L}(\mathbf{i} - \mathbf{j}) \tag{4.43}$$

and also

$$\text{dist}_{\mathbf{L}^\dagger}(\mathbf{i}, \mathbf{j}) = \langle \mathbf{i} - \mathbf{j}, \mathbf{i} - \mathbf{j}\rangle_{\mathbf{L}^\dagger} = (\mathbf{i} - \mathbf{j})^T \mathbf{L}^\dagger(\mathbf{i} - \mathbf{j}), \tag{4.44}$$

where $\mathbf{L}^\dagger$ represents the pseudoinverse of **L**. Note that this defines a distance between an arbitrary pair of nodes, which is given only by the weights prescribed to edges via $\mathbf{G}_1^{-1}$ since the node weights are uniform—due to the fact that the metric tensor $\mathbf{G}_0$ is Euclidean in this case. The matrices **L** and $\mathbf{L}^\dagger$ therefore provide definitions of new metrics for the cell complex to measure distances between pairs of nodes, which are derived from the same edge weights that provided the natural distance operator defined in (4.40).

This definition of distance between a pair of nodes in terms of the weighted Laplacian matrix has an important interpretation. Again, if **i** represents node v_i and **j** represents node v_j, then the distance between nodes v_i and v_j with respect to the pseudoinverse of the weighted Laplacian matrix **L** is given by

$$\mathcal{D}_{\mathbf{L}^\dagger}(v_i, v_j) = \text{dist}_{\mathbf{L}^\dagger}(\mathbf{i}, \mathbf{j}) = R_{\text{eff}}(v_i, v_j), \tag{4.45}$$

where R_{eff} denotes the *effective resistance* between nodes v_i and v_j (see Chap. 3 for more about effective resistance). The effective resistance was previously discussed as a metric by Klein et al. [234, 235] in which it was named the *resistance distance* to reflect the metric quality of the effective resistance.

This construction allows us to immediately extend this definition of resistance distance to higher-order resistance distances between any two p-cells. If we represent two individual p-cells, c_i and c_j, by (non-negative) binary-valued indicator vectors **i** and **j** in $\mathbb{B}^{n_p}$,

$$\mathcal{D}_{\mathbf{L}_p^\dagger}(c_i, c_j) = \text{dist}_{\mathbf{L}_p^\dagger}(\mathbf{i}, \mathbf{j}) = \langle \mathbf{i} - \mathbf{j}, \mathbf{i} - \mathbf{j}\rangle_{\mathbf{L}_p^\dagger} = (\mathbf{i} - \mathbf{j})^T \mathbf{L}_p^\dagger(\mathbf{i} - \mathbf{j}). \tag{4.46}$$

Note that the higher-order Laplacian $\mathbf{L}_p$ may have full rank, in which case $\mathbf{L}_p^\dagger = \mathbf{L}_p^{-1}$.

While the resistance distance arises in many applications, it does not conform to the intuitive notion of the geometric distance between two points. The common definition of distance can, however, be expressed in this same framework, but with

---

[8] Desbrun et al. [102] distinguish between a **local metric** which is defined only for pairs of nodes which share an edge and a **global metric** which is defined for any pair of nodes in a connected graph. In this terminology, our treatment will focus exclusively on global metrics.

an important caveat. If we continue to adopt the view of **i** and **j** as 0-chains which indicate a single node, then we may define a new distance between nodes as

$$\mathcal{D}_{SP'}(v_i, v_j) = \min_{\tau} \mathbf{1}^T \mathbf{G}_1 \tau = \min_{\tau} \mathbf{w}^T \tau \qquad (4.47)$$
$$\text{s.t.} \quad \mathbf{A}^T \tau = \mathbf{i} - \mathbf{j},$$

where τ represents a 1-chain. The minimization finds the 1-chain with the smallest distance subject to the constraint that it connect nodes v_i and v_j. We use τ to represent this 1-chain because the solution to this problem is an oriented path from v_i to v_j. In other words, $\mathcal{D}_{SP'}(v_i, v_j)$ represents the weighted length of the shortest path from v_i to v_j. However, written in this way, the "distance" does not provide a metric because the coefficients comprising τ may take either positive or negative values to represent the orientation of the path relative to the reference orientation of the edges. However, there exists a modification of the problem to avoid this issue by considering a modified version of (4.47),

$$\mathcal{D}_{SP}(v_i, v_j) = \min_{\pi} \mathbf{1}^T \begin{bmatrix} \mathbf{G} & 0 \\ 0 & \mathbf{G} \end{bmatrix} \pi = \min_{\pi} \begin{bmatrix} \mathbf{w} \\ \mathbf{w} \end{bmatrix}^T \pi$$
$$\text{s.t.} \quad \begin{bmatrix} \mathbf{A} \\ -\mathbf{A} \end{bmatrix}^T \pi = \mathbf{i} - \mathbf{j}, \qquad (4.48)$$
$$\pi \geq 0.$$

where now π represents the 1-chain. Intuitively, this modification can be viewed as replacing each edge in the cell complex with two edges having opposite orientation. Therefore, π can be constrained to be strictly positive (i.e., by choosing the edge with positive orientation at each position along the chain from v_i to v_j) and is capable of representing any possible path connecting the two nodes. Note that, due to the "over-representation" of edges, each edge along the path is represented twice in π. Since, as it is defined, the distance $\mathcal{D}_{SP}$ is strictly positive, the distance now represents a metric (the triangle inequality is satisfied due to the distance being defined by a minimum). The value of $\mathcal{D}_{SP}(v_i, v_j)$ may be computed easily by finding the shortest path from v_i to v_j (e.g., with Dijkstra's algorithm [87]). This shortest-path metric is used often in graph theory and will be revisited again in Chap. 8.

4.5 Conclusion

The structure and weights of the cell complex (graph) are given by the problem definition in some applications, such as the analysis of road networks. However, when we apply discrete calculus techniques to a wider class of applications that have no natural structure and weights, we can often employ an embedding to define both the structure and the weights. When an embedding is not available, the data itself may be used to define structure and weights. Depending on the intended analysis operation, we showed how the weights may be used to encode the geometry of the embedding, to avoid the influence of outliers, to represent repulsion, to represent statistical relationships, or to be consistent with a series of observations. We saw that these techniques for defining weights from data also extend directly to the generation of weights on higher-order cells. Finally, the structure and weights were used to further define several metrics between two chains/cochains and between two cells (e.g., nodes).

More attention was spent in this chapter on producing a weighting for filtering/clustering/segmentation applications in which the weights are used as affinities that affect the spread of filtered data or clustering labels. These filtering and clustering applications comprise the content of the next two chapters.

Chapter 5
Filtering on Graphs

Abstract Measured data often includes noise. A data point measured in isolation offers little opportunity to tease signal apart from noise. However, this separation of noise from the signal becomes more possible when multiple data points are acquired which have a relationship with each other. A spatial relationship, such as the edge set of a graph, permits the use of the collective data acquisition to make better decisions about the true data underlying each measurement. This process whereby the spatial relationships of the data are used to provide better estimates of the noiseless data is called a *filtering* or a *denoising* process. In this chapter, we outline the assumptions used to justify spatial filtering, describe the equivalent of Fourier analysis on a general graph and discuss how different parameter settings of a small number of variational approaches to filtering lead to a large number of commonly used filters. Although our focus in this chapter is on the filtering of node data (0-cochains), we also discuss how these techniques may be applied to the filtering of edge data (i.e., flows, or 1-cochains) and to the filtering of data associated with higher-dimensional cells.

Data filtering is a common procedure in any kind of data processing and analysis application. In this chapter, we assume that our object of interest is a graph with data $\tilde{s}_i$ assigned to each node v_i and a meaningful neighborhood definition given by the edge set. The standard assumption is that noise has been added to all of the data. In the absence of a data model, there is very little that one can do to extract the signal from the noisy observation. However, data associated with a discrete cell complex has meaningful neighborhood relationships and we may generally assume that the noise has a high spatial frequency. Therefore, the goal of most filtering operations is to remove the high frequency noise, while being careful to preserve the high frequency signal (often modeled as spatial discontinuities).

We begin this chapter by addressing the filtering topic in a traditional context in which the general goal is to produce a filter that removes some frequency range in the data (e.g., a lowpass filter). The focus here will be on Fourier-based techniques in the context of data associated with an arbitrary graph. Following this exposition of

Fourier techniques on a graph, we address the lowpass filtering of data as an energy optimization problem. These optimization-based filtering operations will then be modified to preserve rapid data changes (i.e., discontinuities). Specifically, we may model the filtering process as smoothing noise *within* a region (node set), but not *between* regions. We then proceed to show how to filter via gradient manipulation and discuss nonlocal filtering. These filtering techniques are then generalized beyond the processing of node data to filtering procedures that remove noise in edge data (e.g., flows, traffic) or data associated with higher-dimensional cells. Finally, the chapter ends by showing applications of these filtering techniques.

5.1 Fourier and Spectral Filtering on a Graph

The traditional approach to filtering data sampled on an equally spaced grid in arbitrary dimensions is to apply a digital filter intended to suppress certain frequencies without disrupting others. Digital filtering approaches of this nature comprise an enormous literature which we do not intend to review (a standard text on this subject is Oppenheim and Schafer [297]). Our first goal in this section will be to examine *when* we can apply standard Fourier methods to data defined on a graph.

The Fourier transform was originally developed by Fourier to produce solutions to the diffusion (heat) equation

$$\frac{\partial u}{\partial t} = \nabla^2 u, \qquad (5.1)$$

where u is a real-valued function defined on $\mathbb{R}^N$ and $\nabla^2 = \nabla \cdot \nabla$ is the diffusion operator. Specifically, in $\mathbb{R}^N$, the standard Fourier basis constitute the *eigenfunctions* of the Laplacian operator. We would therefore expect that the columns of the Discrete Fourier Transform (DFT) matrix would likewise represent the *eigenvectors* of the Laplacian matrix from discrete calculus, appearing in the diffusion equation (see Chap. 2)

$$\frac{\partial \mathbf{x}}{\partial t} = -\mathbf{L}\mathbf{x}. \qquad (5.2)$$

In this section, we will employ s and k as index variables rather than the conventional i and j to avoid confusion with $i = j = \sqrt{-1}$. The DFT matrix, $\mathbf{Q}$, has rows s and columns k defined as

$$Q_{sk} = e^{-\frac{2\pi i}{N}(s-1)(k-1)}, \qquad (5.3)$$

where N is the number of nodes in the graph. The first issue we address is to determine when $\mathbf{Q}$ will comprise the eigenvectors of the Laplacian matrix, analogous to the continuous case. We can proceed to show that $\mathbf{Q}$ will form the eigenvectors of the Laplacian matrix, if and only if the Laplacian matrix is *circulant*. Recall that a **circulant matrix** is defined as a matrix in which each row is a circular shift of the

5.1 Fourier and Spectral Filtering on a Graph

previous row, i.e., $\mathbf{H}(i, j) = \mathbf{H}(i-1, j-1)$, for $i > 1$, $j > 1$, $\mathbf{H}(i, 0) = \mathbf{H}(i-1, N)$. The general form of a circulant matrix can be seen more easily to follow

$$\mathbf{H} = \begin{bmatrix} h_1 & h_2 & h_3 & \cdots & h_N \\ h_N & h_1 & h_2 & \cdots & h_{N-1} \\ \vdots & \vdots & \vdots & \ddots & \vdots \\ h_2 & h_3 & h_4 & \cdots & h_1 \end{bmatrix}. \qquad (5.4)$$

Note that a single vector, $\mathbf{h}$, is sufficient to generate a circulant matrix if we set the first column of $\mathbf{H}$ to this vector, i.e., $\mathbf{H}(1, j) = \mathbf{h}_j$. An important aspect of circulant matrices is that a circulant matrix embodies a circular convolution operation represented in matrix form.[1] Given this view of circulant matrices, we can now proceed to connect the DFT to a circulant matrix via the convolution theorem.

Before we investigate when a graph structure gives rise to a circulant Laplacian, we must first prove that the DFT vectors are eigenvectors of any circulant matrix.

Theorem 5.1 *The columns of the DFT matrix, $\mathbf{Q}$, are eigenvectors of any circulant matrix, $\mathbf{H}$.*

This theorem can be proved by first considering a lemma. Define a *shift matrix* as the circulant matrix, $\mathbf{S}$, generated by the row vector $\mathbf{v} = [0, 1, 0, 0, \ldots, 0]^T$. In other words, $\mathbf{S}$ is the identity matrix with each row undergoing a circular shift one entry below the diagonal. While the transformation effected by $\mathbf{S}$ is a delay, its transpose $\mathbf{S}^T$ represents a shift forward and is a matrix with non-zero entries above the diagonal.

Lemma 5.1 *Vector $\mathbf{q}_k$, the kth column of the DFT matrix $\mathbf{Q}$, is an eigenvector of $\mathbf{S}$ with eigenvalue $\lambda = e^{\frac{2\pi i}{N}(k-1)}$.*

Proof The effect of the shift matrix applied to the column vector $\mathbf{q}_k$ will be to produce a new vector, $\tilde{\mathbf{q}}_k = \mathbf{S}\mathbf{q}_k$ such that $\tilde{\mathbf{q}}_k[s] = \mathbf{q}_k[s-1]$ for $s < N$ and $\tilde{\mathbf{q}}_k[N] = \mathbf{q}_k[1]$. However, since the columns of the DFT matrix represent equal partitions of the unit circle [297], this shift can be accomplished by multiplying each entry by $\lambda = e^{\frac{2\pi i}{N}(k-1)}$. Therefore,

$$\tilde{\mathbf{q}}_k[s] = \mathbf{q}_k[s-1] = e^{-\frac{2\pi i}{N}(s-2)(k-1)} = e^{-\frac{2\pi i}{N}(s-1)(k-1)} e^{\frac{2\pi i}{N}(k-1)} = \mathbf{q}_k[s] e^{\frac{2\pi i}{N}(k-1)},$$

giving the lemma. □

Now we are prepared to give the proof for Theorem 5.1.

---

[1] Note that, while circulant matrices represent *circular* convolution, **Toeplitz matrices**, which comprise a distinct class of matrices, represent *linear* convolution. A thorough treatment of these matrices is available in [172].

Proof Let us restrict our attention to proving that the kth column of the DFT matrix $\mathbf{Q}$, $\mathbf{q}_k$, is an eigenvector of any circulant matrix $\mathbf{H}$. Note that row s of $\mathbf{H}$ is generated by the first row of $\mathbf{H}$, $\mathbf{g}$ (which is a flipped and rotated version of $\mathbf{h}$), multiplied $s-1$ consecutive times by the shift matrix, which can be represented as $\mathbf{S}^{s-1}$, i.e., $\mathbf{H}_s = \mathbf{S}^{s-1}\mathbf{g}$. Furthermore, assign the inner product of the generator $\mathbf{g}$ of $\mathbf{H}$ with the kth column of the DFT matrix to the scalar α, i.e., $\mathbf{g}^\mathsf{T}\mathbf{q}_k \equiv \alpha$. Now,

$$\mathbf{H}_s^\mathsf{T} \mathbf{q}_k = (\mathbf{S}^{s-1}\mathbf{g})^\mathsf{T} \mathbf{q}_k = \frac{\alpha}{\lambda^{s-1}} = \alpha\,e^{-\frac{2\pi i}{N}(s-1)(k-1)} = \alpha \mathbf{q}_k[s].$$

Therefore, $\mathbf{H}\mathbf{q}_k = \alpha \mathbf{q}_k$. □

Corollary 5.1 *The eigenvalues of $\mathbf{H}$ are the Fourier transform of the generating vector $\mathbf{h}$, i.e., $\lambda = \mathbf{Q}\mathbf{h}$.*

Corollary 5.2 (The *Convolution Theorem*) *Since convolution is a finite, linear operation, it may be represented by a matrix. Furthermore, since convolution applies the same kernel to every location, the matrix representing convolution is circulant. Therefore, the convolution of any two signals $\mathbf{h}$ and $\mathbf{x}$ may be computed by the Fourier transform, i.e., $\tilde{\mathbf{x}} = \mathbf{H}\mathbf{x} = \frac{1}{N}\mathbf{Q}\mathbf{\Lambda}\mathbf{Q}^\mathsf{T}\mathbf{x}$, where the diagonal matrix of eigenvalues $\mathbf{\Lambda}$ is equal to $\mathrm{diag}(\lambda)$.*

Note that the factor of $\frac{1}{N}$ is included to satisfy Parseval's (or Plancherel's) Theorem which effectively states that $\frac{1}{\sqrt{N}}\mathbf{Q}$ is a unitary operator, i.e., that $\frac{1}{N}\mathbf{Q}\mathbf{Q}^\mathsf{T} = \mathbf{I}$.

Contrary to the usual interpretation of the Fourier transform as a decomposition into signal frequencies, we view the Fourier transform here as being tightly coupled to the concept of *shift invariance* of the graph.

> Specifically, a graph is called **shift invariant** if there exists a permutation of the node ordering such that the Laplacian matrix representing the graph is circulant. Therefore, any shift-invariant graph has the DFT basis as the eigenvectors of its Laplacian matrix.

We may now examine the graphs that are shift-invariant. One requirement for shift-invariance of a graph is that the graph must be *regular*. Recall that a regular graph is any graph such that every node has the same number of neighbors (degree). Common examples of shift-invariant graphs are infinite lattices, cycles and fully-connected graphs, as depicted in Fig. 5.1. On all of these shift-invariant graphs, we can solve the diffusion equation (5.2) by decomposition of the initial state onto the DFT basis and evolution of each component independently in time. Although regularity is a necessary condition for a graph to be shift-invariant, not all regular graphs are also shift-invariant graphs.

Treatments of the Fourier transform for specific lattices may now be viewed as special cases of shift-invariant lattices. For example, DuBois [116] treats the Fourier

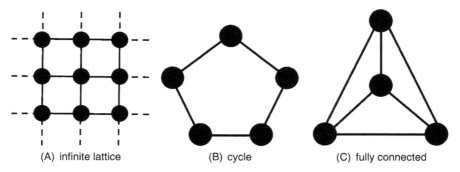

(A) infinite lattice   (B) cycle   (C) fully connected

Fig. 5.1 Examples of shift-invariant lattices. A shift-invariant lattice is represented by a *circulant* Laplacian (and adjacency) matrix, which permits the decomposition of signals defined on the nodes of these graphs onto the Fourier basis. (**A**) Infinite (or wrapping) lattice. An infinite lattice is the standard assumption that justifies use of the Fourier transform, even on finite lattices. (**B**) A cycle graph. (**C**) A fully connected graph

transform on a "quincunx" lattice. A second example of a special Fourier transform treatment for a specific lattice is given by the literature on Fourier descriptors, which gives applications of Fourier transforms on cycle graphs [88, 256, 417]. In the signal processing literature, the signal is typically a one-dimensional signal that varies over time. In this context, a shift-invariant system is considered *time-invariant* and the corresponding theory of such signal analysis is called **LTI system theory**, where LTI stands for Linear and Time-Invariant.

Standard signal processing techniques based on the Fourier transform may be applied to functions defined on a shift-invariant graph. A somewhat surprising consequence of this analysis is that Fourier-based filtering *has nothing to do with the graph topology* as long as the graph is shift-invariant, i.e., the signal at every node is treated equally. However, despite this remarkable fact, the evolution of processes closely associated with the Fourier basis, such as the diffusion equation, will depend on the topology of the graph, since the eigenvalues of shift-invariant graphs change with graph topology.

5.1.1 Graphs that Are Not Shift-Invariant

Although the Fourier transform can be used to analyze signals on a wide range of graphs that arise in practice, not all useful graph structures are shift-invariant. Furthermore, a graph that is seemingly regular can lose shift-invariance for two common reasons: (i) The graph has a dynamically changing or adaptive connectivity; (ii) Although the topology is shift-invariant, the weighting is not. The first situation occurs in surface processing (computer graphics) and space-variant vision in which the graph represents a nonuniform domain. The second situation occurs commonly in image processing, in which the lattice has an inhomogeneous weighting to reflect changes in image properties (e.g., object boundaries).

In principle, filtering is also straightforward to implement on shift-variant graphs that do not possess the regularity necessary for Fourier-based filtering. As before, we can simply calculate the eigenvectors, $\mathbf{Q}$, for the Laplacian matrix $\mathbf{L}$ of any graph. The only drawback with a shift-variant graph is that these eigenvectors are not the convenient columns of the DFT matrix (5.3), nor will the eigenvectors be an orthonormal set of complex exponentials. Therefore, it is necessary to calculate the eigenvectors explicitly for each graph, and the familiar Fast Fourier Transform is not available for efficiently projecting a signal into frequency space.

It is straightforward to use the eigenvectors of the Laplacian matrix of a shift-variant graph for filtering in a manner analogous to standard Fourier-based filtering. Specifically, **spectral filtering** could be implemented by:

1. Computing the eigenvectors of the Laplacian matrix, $\mathbf{Q}$;
2. Projecting the data onto the eigenvector space, $\tilde{\mathbf{x}} = \mathbf{Q}\mathbf{x}$;
3. Modifying the frequency components, which are the values of $\tilde{\mathbf{x}}$, to generate $\hat{\mathbf{x}}$ such that the desired filter is achieved; then
4. Backprojecting $\hat{\mathbf{x}}$ via $\mathbf{Q}^{-1}\hat{\mathbf{x}}$ to obtain the filtered signal with the modified frequency components.

Although this procedure is straightforward for applying spectral filtering on an arbitrary graph, it is computationally expensive (and memory intensive) to calculate and apply the full set of eigenvectors of a shift-variant Laplacian matrix. Fortunately, good alternatives exist to explicit calculation of the eigenvectors of a shift-variant graph. The filtering approach adopted by Taubin [371, 372] begins by observing that applying a diffusion equation with a forward Euler method will have the effect of reducing high frequencies. By phrasing the filtering as a difference

$$\mathbf{x}^{[k+1]} = \mathbf{x}^{[k]} - \lambda \mathbf{L}\mathbf{x}^{[k]} = \mathbf{x}^{[k]} - \lambda \mathbf{Q}^T \mathbf{\Lambda} \mathbf{Q}\mathbf{x}^{[k]}, \qquad (5.5)$$

where $\mathbf{x}^{[k]}$ represents the outcome of the kth iteration, the high frequencies will be subtracted out during the update step, but the low frequencies will largely remain at the same magnitude. The parameter λ controls the speed of the filtering during the iterations, but care must be taken to avoid instability by setting λ small enough to avoid violating the CFL conditions.[2] The iteration represented by (5.5) is commonly called **Laplacian smoothing** in finite element analysis and computer graphics. However, Taubin observed that there was a "shrinking bias" in the context of mesh filtering, as a result of the fact that the low frequencies will eventually also dissipate, leaving only the constant eigenvector (i.e., after enough iterations, all of the nodes would have the same values). A similar effect is observed in image processing in which the diffusion process of (5.5) eventually yields a constant gray value of the image intensities. To counteract this shrinking bias, Taubin preserved the low frequencies while damping the high frequencies by alternating diffusion steps of the form in (5.5) with iterations of the form

$$\mathbf{x}^{[k+1]} = \mathbf{x}^{[k]} + \mu \mathbf{L}\mathbf{x}^{[k]}, \qquad (5.6)$$

---

[2]Note that 'λ' is often used to represent an eigenvalue (e.g., Lemma 5.1). We follow Taubin's notation for his λ–μ algorithm by using 'λ' as a parameter when discussing Taubin's algorithm.

5.1 Fourier and Spectral Filtering on a Graph

where the parameter $\mu > \lambda$, which has the effect of replacing the relatively well-preserved low frequencies that were lost during the smoothing step (5.5) without replacing the high frequencies that were almost completely removed. When $\mu \approx \lambda$, a sharp, "wall filter", is obtained for the low frequencies, while setting $\mu \gg \lambda$ has the effect of a slow filter roll-off.

An immediate concern with Taubin's algorithm as described is that the appropriate values of the λ and μ parameters vary with the graph size, since the magnitude of the eigenvalues of the Laplacian will change. Taubin's resolution of this problem is to employ the symmetric, normalized Laplacian from Chap. 2, $\tilde{\mathbf{L}} = \mathbf{D}^{-\frac{1}{2}}\mathbf{L}\mathbf{D}^{-\frac{1}{2}}$ (where $\mathbf{D}$ is the diagonal matrix of node degrees). This normalized Laplacian is identical to a scaled or reweighted $\mathbf{L}$ for a shift-invariant graph, but has the additional property that the eigenvalues lie in the range [0, 2], which was shown in [81]. Consequently, the values of the parameters λ and μ can remain relatively fixed in order to achieve a particular filtering characteristic with respect to the symmetric normalized Laplacian. Taubin's "λ–μ" filter is very useful in practice due to its simplicity, efficiency and intuitive behavior. However, this algorithm does behave differently from the full spectral filtering approach (i.e., projecting directly onto the eigenvectors of the Laplacian matrix), since Taubin's approach only implicitly employs the spectrum of eigenvalues of the graph. For example, in the case of a shift-invariant graph, the full spectral filtering implementation of a lowpass "wall" filter would be to (i) project the signal onto the eigenvectors (i.e., take a forward DFT), then (ii) set the high frequency components to zero, and finally (iii) project the signal onto the transpose of the eigenvectors (i.e., take an inverse DFT). This procedure would be the same regardless of the distribution of the eigenvalues of the graph, e.g., for any of the shift-invariant graphs such as those in Fig. 5.1. However, the diffusion-based approach would proceed at a much faster rate for, e.g., the fully-connected graph than the ring graph in Fig. 5.1, resulting in a need to adjust the values of λ and μ to achieve with Taubin's λ–μ method the same effect as the full spectral-based method.

Taubin's λ–μ algorithm is defined by the following steps.

1. Choose values for λ and μ parameters.
2. Construct the symmetric normalized Laplacian matrix, $\tilde{\mathbf{L}} = \mathbf{D}^{-\frac{1}{2}}\mathbf{L}\mathbf{D}^{-\frac{1}{2}}$.
3. Given an initial $\mathbf{x}^{[0]}$, solve K iterations, alternating between the diffusion step and the unshrinking step, i.e.,

$$\mathbf{x}^{[2k+1]} = \mathbf{x}^{[2k]} - \lambda \tilde{\mathbf{L}} \mathbf{x}^{[2k]}, \tag{5.7}$$

$$\mathbf{x}^{[2k+2]} = \mathbf{x}^{[2k+1]} + \mu \tilde{\mathbf{L}} \mathbf{x}^{[2k+1]}. \tag{5.8}$$

Taubin's method computes only a lowpass filter directly. However, by subtracting the lowpass filtered signal from the original signal, it is possible to generate a highpass filtered signal. By making further combinations of highpass and lowpass filters, Taubin's method may be used to generate bandpass filters of signals on a graph.

5.1.2 The Origins of High Frequency Noise

Now that we have given a precise meaning to "high-frequency" signals on an arbitrary graph, we would like to pause to examine the origins of the common assumption that noise is predominantly contained in high frequencies. The high-frequency character of noise can be justified from two different viewpoints: as an acquisition model and as a model of independent, identically distributed (i.i.d.) noise.

To see the acquisition model viewpoint, consider a linear acquisition matrix, **H**, a true signal **x**, an additive noise vector **ν**, an ideal observed signal **f** = **Hx** and a noisy observed signal **y**, such that

$$\mathbf{y} = \mathbf{H}(\mathbf{x} + \mathbf{\nu}) = \mathbf{f} + \mathbf{H}\mathbf{\nu}. \tag{5.9}$$

Phrased in this way, our goal in filtering is to recover either the ideal observed signal **f** or the true signal **x**. However, for either goal, it is helpful to remove high-frequency noise that corrupts the observation.

We know from the last section that if our graph is shift-invariant and the acquisition matrix **H** is shift-invariant, then we can analyze the effects of **H** via the DFT and, in particular, the eigenvalues of the acquisition matrix are comprised of the DFT of any single row of **H**. In general, if our graph is shift-invariant then we can consider the acquisition matrix to be shift-invariant since this shift-invariance simply means that the signal acquisition device operates the same everywhere. A typical acquisition model is that each node performs a weighted sum of the underlying data in a small region about the point. However, since the DFT of a small summation kernel such as a box kernel or a Gaussian kernel concentrates more power in the low frequency range (where a greater extent of the spatial window increasingly concentrates power in the low frequencies), then the high-frequency content of the true signal **x** will be suppressed and we may assume that residual high-frequency content in the observation **y** is contributed by the noise vector **ν**. Similarly, if we try to recover the true signal **x** via inversion of the acquisition matrix **H** (i.e., deconvolution), then the high-frequencies of the observation **f** + **Hν** will be amplified and it would be better to filter out the high frequencies in the observed data before inversion.

The other viewpoint for motivating the removal of high frequencies in the observed data is that if the noise term **ν** is i.i.d. across the samples then the expected value for the difference of the values of **ν** over an edge will be zero. Consequently, a reasonable model is to assume that large gradients are noise and attempt to remove them.

Despite the reasonable assumptions underpinning the motivation for removal of high frequency noise, most real data do contain some high frequency content that we do *not* want removed. A common procedure for keeping the high frequency content in the signal is to assume that the high frequency content generally takes the form of discontinuities in which the data jumps from one smoothly varying region to another. In the context of image processing, this assumption leads to an image model where individual objects have a smoothly varying intensity but that neighboring objects may have an arbitrarily large jump in intensity. Of course, this assumption is not true for all kinds of data (e.g., a textured object within an image), but it works well for many kinds of real data and refocuses the filtering operation on the detection of discontinuity locations, while applying lowpass filtering everywhere else.

Despite the usefulness of the above lowpass model for data filtering, it is important to recognize that some noise contains significant power in the low frequencies, which is equivalent to the presence of spatial correlations (called "pink noise"). Additionally, filtering can also be applied to problems in which the purpose of the filtering operation is to correct observation or acquisition *error*, which may have any frequency content. Finally, not all data domains are shift-invariant. However, for a shift-variant data domain, the above assumptions may also imply a high-frequency noise model. Although the signal defined on the shift-variant graph may contain high-frequency noise, the eigenvectors of the graph Laplacian corresponding to high-frequency may behave differently from the standard high-frequency eigenvectors employed in Fourier analysis. However, even in the case of a shift-variant data domain, the techniques in this chapter will allow for the removal of high-frequency components of the observed data, even if we allow discontinuities across some edges.[3]

5.2 Energy Minimization Methods for Filtering

Many filtering methods may be viewed as procedures to minimize an energy. In this section, we review several energy minimization models and show how these models lead to commonly used filtering algorithms. We begin this section by describing the basic energy minimization models before proceeding to describe methods for filtering in the presence of implicit or explicit discontinuities.

5.2.1 The Basic Energy Minimization Model

In the previous section, a traditional view of denoising was taken in which the goal was to remove the high frequencies (noise) while preserving the low frequencies (signal). A different approach to accomplishing the same goal is to view the desired denoised signal, **x**, as a minimum of the energy[4]

$$\mathcal{E}_{\text{BEM}}[x] = \int x \Delta x \, dt = \int \|\nabla x\|_2^2 \, dt, \tag{5.10a}$$

---

[3]Note that in image processing the term *edge* is used to mean discontinuity (e.g., "edge detection"). However, since the context of this entire book is the analysis/processing of graphs (complexes) and data defined on graphs, we reserve the word *edge* to refer strictly to a 1-cell (i.e., we use *edge* in the sense of graph theory).

[4]The term *energy* is used throughout the book to represent an objective function which is optimized to produce a useful application-specific solution. In this case, the solution represents the filtered (denoised) signal. Although the term *energy* is not generally intended to have a physical relationship to energy, note that the energy described in (5.10b) is actually the power dissipation for an electric circuit (when **x** represents the electrical potentials at every node), as given in Chap. 3.

$$\mathcal{E}_{\text{BEM}}[\mathbf{x}] = \mathbf{x}^\mathsf{T}\mathbf{L}\mathbf{x} = (\mathbf{A}\mathbf{x})^\mathsf{T}(\mathbf{A}\mathbf{x}). \tag{5.10b}$$

This energy model returns a high energy for a signal **x** dominated by high-frequency components and returns a low energy for a signal **x** dominated by low-frequency components. Therefore, finding a signal that minimizes this energy will produce a low-frequency signal. However, by connecting the frequency components of **x** with the norm of the gradient, we may generalize this model to measure the gradients with any p-norm.

$$\mathcal{E}_{\text{BEM}}[x] = \int \|\nabla x\|_p^p \, dt, \tag{5.11a}$$

$$\mathcal{E}_{\text{BEM}}[\mathbf{x}] = \mathbf{1}^\mathsf{T}|\mathbf{A}\mathbf{x}|^p = \sum_{e_{ij} \in \mathcal{E}} |x_i - x_j|^p, \tag{5.11b}$$

where the parameter p controls the norm of the energy functional and the summation in (5.11b) is over every edge in the graph. Note that the conditions for a norm are violated when $p < 1$ (specifically the triangle inequality), therefore we will employ the term p-*norm* to refer to

$$\|\mathbf{x}\|_p = \left(\sum_i |\mathbf{x}_i|^p\right)^{\frac{1}{p}}, \tag{5.12}$$

even though we allow $0 < p < 1$. We refer to this energy as the **Basic Energy Model**. The Basic Energy Model pervades this entire book. In future chapters, we see that many common algorithms can be viewed as instances of the Basic Energy Model with different values of p and different interpretations of the variable **x**. In this chapter, the Basic Energy Model leads to mean, median, mode and minimax filters as well as Laplacian smoothing and anisotropic diffusion. In Chap. 6, the Basic Energy Model leads to several clustering methods in the literature, including max-flow/min-cut, random walks, geodesic clustering, watersheds, spectral clustering, normalized cuts and the isoperimetric partitioning algorithm. Going further, in Chap. 7 we see how the same Basic Energy Model leads to the Laplacian Eigenmaps manifold learning technique. We believe that the unification of so many content extraction (data processing) methods into one functional is a major contribution of this work.

The trivial minimum of the Basic Energy Model given by a constant **x** is a useless filtering of a noisy signal. However, a noisy signal may have its energy reduced iteratively without reducing the energy to the undesirable global minimum. There are two standard methods for iteratively minimizing the Basic Energy Model—iterative minimization of the model for each node individually and gradient descent.

5.2.1.1 Iterative Minimization

We begin by describing how to iteratively minimize the Basic Energy Model for each node. If we were given a solution, $\mathbf{x}^{[k]}$, at iteration k, we can fix the solution

5.2 Energy Minimization Methods for Filtering

everywhere except for one node v_i and consider finding the update which would minimize $\mathcal{E}_{\text{BEM}}[x_i]$ for that node. By definition of (5.11b), the minimum is given by

$$x_i = \operatorname{argmin}\left\{\sum_{e_{ij}\in\mathcal{E}} |x_i - x_j|^p\right\}. \tag{5.13}$$

When $p = 2$, x_i is assigned to have the *mean* value of its neighbors, while $p = 1$ causes x_i to be assigned the *median* value of its neighbors. Likewise, $p = 0$ assigns x_i to be the *mode* value of its neighbors, and as $p \to \infty$, x_i approaches the *minimax* value of its neighbors. Therefore, if the initial estimate $\mathbf{x}^{[0]} = \mathbf{s}$, *the mean, median, mode and minimax filters can all be viewed as operations designed to incrementally minimize* (5.11b) under different p-norms. Note that many implementations of these filters also include a self-connected edge at each node such that the node's value is included in the mode, mean, median or minimax calculation.

> Minimization of the energy functional given in the Basic Energy Model of (5.11b) for p-norms given by $p = 0$, $p = 1$, $p = 2$, and $p \to \infty$, results in the commonplace mode, median, mean, and minimax filters, respectively. Consequently, these filters can all be generalized to arbitrary graphs with any neighborhood structure.

Instead of minimizing the Basic Energy Model by optimizing the solution for each node independently, we can also find the solution via gradient descent (see Appendix B). Since gradient descent is typically used to optimize the Basic Energy Model only for the cases of $p = 2$ or $p = 0$, our discussion will be limited to these cases. When $p = 2$, the Basic Energy Model takes the form

$$\mathcal{E}_{\text{BEM}}[\mathbf{x}] = \mathbf{1}^{\mathsf{T}}|\mathbf{A}\mathbf{x}|^2 = \mathbf{x}^{\mathsf{T}}\mathbf{A}^{\mathsf{T}}\mathbf{A}\mathbf{x} = \mathbf{x}^{\mathsf{T}}\mathbf{L}\mathbf{x}, \tag{5.14}$$

with gradient

$$\frac{\partial \mathcal{E}_{\text{BEM}}[\mathbf{x}]}{\partial \mathbf{x}} = \mathbf{L}\mathbf{x}. \tag{5.15}$$

Therefore, the iterative update to perform gradient descent of an initial noisy signal according to the Basic Energy Model when $p = 2$ would be

$$\mathbf{x}^{[k+1]} = \mathbf{x}^{[k]} - \lambda \mathbf{L}\mathbf{x}^{[k]}. \tag{5.16}$$

We saw this filtering algorithm before in (5.5) when it was called Laplacian smoothing. In this context, the *time derivative* of $\mathbf{x}$ was set equal to the negative gradient of the Basic Energy Model, i.e., $\partial \mathbf{x}/\partial t = -\partial \mathcal{E}/\partial \mathbf{x}$. By establishing this equality, a forward Euler solution of the diffusion equation

$$\frac{\partial \mathbf{x}}{\partial t} = -\mathbf{L}\mathbf{x}, \tag{5.17}$$

performs gradient descent on the Basic Energy Model where the time step of the forward Euler operation is represented by λ.

The classic nonlinear anisotropic diffusion model formulated for image processing by Perona and Malik [306] may be viewed as a descent algorithm that minimizes the Basic Energy Model (5.11b) when $p = 0$. In this section, we follow Black et al. [38] to describe this relationship.

Consider the Basic Energy Model for $p = 0$, i.e.,

$$\mathcal{E}_{\text{BEM}}[\mathbf{x}] = \mathbf{1}^T |\mathbf{A}\mathbf{x}|^0 = \sum_{e_{ij}} |x_i - x_j|^0. \tag{5.18}$$

Recall that the value of the ℓ_0-norm $|\cdot|^0$ is zero when the argument is zero and is unity otherwise. Although Black et al. [38] showed that the nonlinear anisotropic diffusion work of Perona and Malik unintentionally approached this energy indirectly, similar filtering models were directly considered at about the same time (see [50, 144, 146]). These models were motivated by the desire to represent implicit discontinuities in the reconstruction. Unfortunately, the $p = 0$ energy in (5.18) is both non-convex and non-differentiable. The non-differentiable aspect of the energy may be overcome by replacing $|\cdot|^0$ with an *M-estimator*, which is an error function used in statistics for which the penalty of large errors increases very slowly for higher error values. M-estimators were used to derive several different methods for generating weights in Chap. 4. For purposes of exposition here, we may approximate the ℓ_0-norm $|\cdot|^0$ via the Welsch function from Chap. 4

$$\rho_\alpha(z) \propto 1 - \exp\left(-\frac{z^2}{\alpha}\right), \tag{5.19}$$

where α is a parameter that controls the approximation to $|\cdot|^0$, such that the approximation fidelity improves as $\alpha \to \infty$. Figure 4.3 in Chap. 4 gives an example of the Welsch function.

Using the Welsch function as an approximation to $|\cdot|^0$ gives us an approximation for the Basic Energy Model energy for the case $p = 0$ as

$$\mathcal{E}_{\text{BEM}}[\mathbf{x}] = \mathbf{1}^T \rho(\mathbf{A}\mathbf{x}), \tag{5.20}$$

where the error function $\rho(\cdot)$ is assumed to operate on the individual components of any vector or matrix input. We may take an initial solution for $\mathbf{x}$, $\mathbf{x}^{[0]} = \mathbf{g}$ and apply gradient descent to reduce the energy (see Appendix B). The gradient of (5.20) is given by

$$\frac{\partial \mathcal{E}_{\text{BEM}}[\mathbf{x}]}{\partial \mathbf{x}} = 2\alpha \mathbf{A}^T \operatorname{diag}(\bar{\rho}(\mathbf{A}\mathbf{x}))\mathbf{A}\mathbf{x}, \tag{5.21}$$

where

$$\bar{\rho}(z) = \exp\left(-\frac{z^2}{\alpha}\right). \tag{5.22}$$

5.2 Energy Minimization Methods for Filtering

Applying the gradient descent operation to iteratively update **x** gives the iteration

$$\mathbf{x}^{[k+1]} = \mathbf{x}^{[k]} - \lambda \mathbf{A}^\top \operatorname{diag}(\bar{\rho}(\mathbf{A}\mathbf{x}^{[k]}))\mathbf{A}\mathbf{x}^{[k]} = \mathbf{x}^{[k]} - \lambda \mathbf{L}^{[k]}\mathbf{x}^{[k]}, \tag{5.23}$$

where the Laplacian operator $\mathbf{L}^{[k]}$ is updated at each iteration to reflect the changing edge weights given by $\mathbf{G}^{-1} = \operatorname{diag}(\bar{\rho}(\mathbf{A}\mathbf{x}^{[k]}))$. This update rule is exactly the iteration given by Perona and Malik to describe anisotropic diffusion, in which the weights were given by $\bar{\rho}(z)$. In this way, the nonlinear anisotropic diffusion of Perona and Malik is naturally interpreted as an *approximation* to the unweighted Basic Energy Model when $p = 0$, while gradient descent on the Basic Energy Model for $p = 2$ corresponds to a *linear* diffusion smoothing process.

5.2.2 Extended Basic Energy Model

The Basic Energy Model defined in (5.11b) has the distinct problem that the energy is trivially minimized by a trivial solution in which $x_i = $ constant. Therefore, we may extend the Basic Energy Model by viewing the Basic Energy Model as a *smoothness term* to which we may add a *data attachment term* that causes the original data to push back against the smoothness, phrasing the solution as a balance between the model and the measured data. This allows for a non-trivial solution and provides a mechanism for incorporating prior information into the solution in the form of regularization. The generalized formulation may be given as

$$\mathcal{E}_{\text{EBEM}}[x] = \int \|\nabla x\|_p^p \, dt + \lambda \int |x - s|^p \, dt, \tag{5.24a}$$

$$\mathcal{E}_{\text{EBEM}}[\mathbf{x}] = \mathbf{1}^\top |\mathbf{A}\mathbf{x}|^p + \lambda \mathbf{1}^\top |\mathbf{x} - \mathbf{s}|^p$$

$$= \sum_{e_{ij}} |x_i - x_j|^p + \lambda \sum_{v_i} |x_i - s_i|^p, \tag{5.24b}$$

where **s** represents the observed (noisy) data and the regularization parameter λ acts to trade off between fidelity of the denoised signal to the original data and smoothness of the denoised signal. Although this new energy model appears substantially different from the Basic Energy Model, we may view the new energy given in (5.24b) as a variant of the Basic Energy Model in which every node $v_i \in \mathcal{V}$ with value x_i has a "phantom" neighbor $u_i \in \mathcal{P}$ with value s_i such that the phantom node u_i is connected only to v_i by an edge weighted with value λ. These phantom nodes are sometimes called **dongle nodes** or **t-links** ("terminal" links in [56]). Therefore, this new energy model may be viewed as an example of the Basic Energy Model with additional nodes which are fixed to Dirichlet boundary conditions (see Chap. 2 for definition and Appendix B for optimization). Since the new energy model may be viewed as equivalent to the Basic Energy Model with an extended node set (where the new nodes are fixed as boundary conditions), we refer to the energy described in (5.24b) as the *extended* **Basic Energy Model**.

The primary advantages of the Extended Basic Energy Model are that the energy is minimized by a nontrivial solution, and that the adjustable influence of the regularization term allows explicit control over the contribution of the filtering. In the case of $p = 2$, an intuitive interpretation exists for the Extended Basic Energy Model. We previously saw that when $p = 2$, gradient descent on the Basic Energy Model could be interpreted as a diffusion process on the noisy measured signal which is solved with a forward Euler method. However, if we solve the diffusion process with a *backward* Euler method, then the diffusion time is $1/\lambda$ [104]. To see this connection, consider again the linear diffusion equation

$$\frac{\partial \mathbf{x}}{\partial t} = -\mathbf{L}\mathbf{x}, \tag{5.25}$$

solved using a backward Euler method [312]

$$\frac{\mathbf{x}^{[k+1]} - \mathbf{x}^{[k]}}{\Delta t} = -\mathbf{L}\mathbf{x}^{[k+1]}, \tag{5.26}$$

$$\mathbf{x}^{[k+1]} = \mathbf{x}^{[k]} - \Delta t\, \mathbf{L}\mathbf{x}^{[k+1]}, \tag{5.27}$$

$$(\Delta t\, \mathbf{L} + \mathbf{I})\mathbf{x}^{[k+1]} = \mathbf{x}^{[k]}, \tag{5.28}$$

$$\left(\mathbf{L} + \frac{1}{\Delta t}\mathbf{I}\right)\mathbf{x}^{[k+1]} = \frac{1}{\Delta t}\mathbf{x}^{[k]}. \tag{5.29}$$

This backward Euler method is equivalent to a solution of the Extended Basic Energy Model when $p = 2$, since the minimum of $\mathcal{E}_{\text{EBEM}}[\mathbf{x}]$ is given by

$$(\mathbf{L} + \lambda \mathbf{I})\mathbf{x} = \lambda \mathbf{s}. \tag{5.30}$$

Consequently, the solution for the backward Euler solution for the diffusion problem in (5.29) is equivalent to (5.30) when $\mathbf{s} = \mathbf{x}^{[0]}$ and $\lambda = 1/\Delta t$. Using this connection allows us to intuitively view the λ parameter in the Extended Basic Energy Model as the reciprocal number of iterations employed for solving the Basic Energy Model.

The form of the Extended Basic Energy Model, which consists of a smoothness term and a data term, is also adopted by several other energy minimization filtering models. In the next section we consider another of these models which makes a modification to the smoothness term.

5.2.3 The Total Variation Model

In the domain of image processing, the underlying graph is a lattice embedded into the Euclidean plane. When $p = 2$, the solution to the Basic Energy Model behaves as if the lattice were a representation of the continuous plane. However, this behavior is not observed when $p \neq 2$. Therefore, when $p \neq 2$, the filtering of images produces undesirable gridding artifacts (sometimes called "metrication artifacts").

5.2 Energy Minimization Methods for Filtering

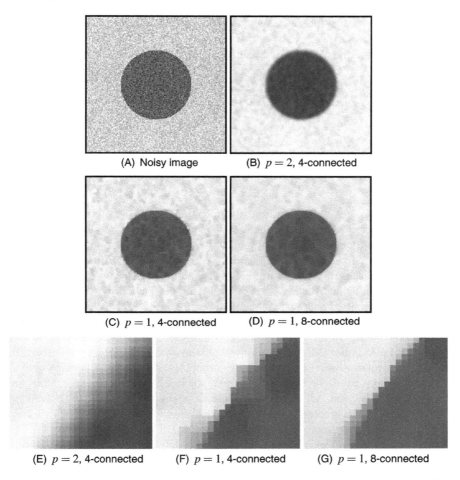

Fig. 5.2 Gridding artifacts observed in the solution to the Basic Energy Model when $p = 1$ but not $p = 2$. Note that the introduction of more edges (an 8-connected lattice) substantially reduces these gridding artifacts. The *bottom row* shows closeup views of the circle boundary to illustrate the variation in gridding artifacts

These gridding artifacts may be reduced by extending the number of edges in the lattice [54]. However, by modifying the smoothness term of the Basic Energy Model such that the smoothness is measured by quantities at *nodes* rather than quantities on *edges*, it was shown that these gridding artifacts may be reduced [386]. This is the approach of the **Total Variation Model**. Figure 5.2 shows the gridding artifacts produced by filtering with the Basic Energy Model when $p = 1$ (a median filter) for a 4-connected and 8-connected lattice. The 8-connected lattice provides substantially reduced gridding artifacts at the cost of additional edges. Although the solution to the $p = 2$ model does not exhibit gridding artifacts, the boundary localization is blurred more for the $p = 2$ solution than the $p = 1$ solution. Preserving both boundary localization and a reduction of gridding artifacts is obtained by the Total Variation Model.

The Total Variation Model based on nodal smoothness is described by the energy functional

$$\mathcal{E}_{\text{TV}}[x] = \int \|\nabla x\|_2^p \, dt + \lambda \int |x - s|^p \, dt, \tag{5.31a}$$

$$\mathcal{E}_{\text{TV}}[\mathbf{x}] = \mathbf{1}^\mathsf{T}(|\mathbf{A}^\mathsf{T}||\mathbf{A}\mathbf{x}|^2)^{\frac{p}{2}} + \lambda \mathbf{1}^\mathsf{T}|\mathbf{x} - \mathbf{s}|^p$$

$$= \sum_{v_i} \left(\sum_{j \forall e_{ij}} |x_i - x_j|^2 \right)^{\frac{p}{2}} + \lambda \sum_{v_i} |x_i - s_i|^p. \tag{5.31b}$$

Within the image processing literature, this model is usually credited to [326] and is sometimes referred to as the "Rudin, Osher, Fatemi" model. In contrast to the formulation of the Extended Basic Energy Model given in (5.24b), the minimization of the Total Variation Model searches for the minimum vector p-norm measured at each node. Generally, the Total Variation Model given in (5.31b) is more difficult to solve than the Extended Basic Energy Model since the 1-cochains produced by the gradient operator are then transfered back to 0-cochains. Despite this enhanced difficulty, the total variation model (5.31b) is convex when $p \geq 1$ and is thus solvable by any descent algorithm [123]. When $p = 2$, the Extended Basic Energy Model and the Total Variation Model are equivalent, and when $p = 1$ this model is a *total variation* minimization from which this model takes its name. Filtering via total variation minimization has been well studied (see [69, 326]) and fast algorithms are known for this case [68, 98]. The total variation model was formulated initially in the continuum and on a graph much later [300] and then further extended to weighted graphs and arbitrary p-norms [49, 123].

A problem with all of these energy minimization algorithms is the assumption that *all* high frequencies represent noise. Solutions of these models with lower values of p typically preserve high-frequency content. We may therefore alter our filtering approach to preserve high-frequency signal content by modeling signals as consisting of smooth regions which are separated by a small number of sharp discontinuities. These discontinuities are often thought of as region boundaries. The discontinuities may be incorporated into the energy minimization models by *weighting* changes in the filtered signal differently. Spatial gradients in the filtered signal which occur over discontinuities are penalized less than spatial gradients inside regions. Since these weights are used only as approximations to the discontinuity locations, we call this approach *filtering with implicit discontinuities*.

5.3 Filtering with Implicit Discontinuities

The assumption that low frequency content represents signal while high frequency content represents noise is often not valid in real data, particularly in image data.

5.3 Filtering with Implicit Discontinuities

Specifically, it is often more accurate to assume that we want to smooth more over similar data at neighboring nodes while smoothing less over dissimilar data at neighboring nodes. The model considered in this section stops short of looking for explicit object boundaries to avoid smoothing over (covered next in Sect. 5.4), but rather performs smoothing on a *weighted* graph (using affinity weights). Generally, the edge weights are derived from some signal feature that can be used to roughly detect object boundaries (e.g., image intensity, image color, node coordinates, texture coefficients). See Chap. 4 for different options to set weights. For the rest of this section, assume that we have a set of nonnegative, normalized, real-valued weights, w_{ij}, that give some indication of discontinuity locations (i.e., $w_{ij} \to 0$ for edges bridging discontinuities and $w_{ij} \to 1$ for edges bridging nodes likely to take the same filtered value).

There are three general approaches to the weighted filtering case: (i) *Spectral filtering* methods compute the eigenvectors of the weighted Laplacian, project the signal, dampen the high frequencies, and reconstruct. Practically, it is preferable to use Taubin's λ–μ algorithm on the weighted Laplacian (as described in Sect. 5.1). (ii) *Edge-based filtering* methods find the solution that tries to fit the data while penalizing smoothness measured across edges. (iii) *Node-based filtering* finds the solution that tries to fit the data while penalizing smoothness measured at nodes.

The Basic Energy Model and the Extended Basic Energy Model are modified easily to incorporate weights, i.e.,

$$\mathcal{E}_{\text{BEM}}[\mathbf{x}] = \sum_{e_{ij}} w_{ij}^p |x_i - x_j|^p, \tag{5.32}$$

$$\mathcal{E}_{\text{EBEM}}[\mathbf{x}] = \sum_{e_{ij}} w_{ij}^p |x_i - x_j|^p + \lambda \sum_{v_i} |x_i - s_i|^p. \tag{5.33}$$

The Basic Energy Model (5.33) now corresponds to minimization via a *weighted* mode, median, mean or minimax filter. The Extended Basic Energy Model (5.33) may be minimized for $p \geq 1$ using any of the convex optimization techniques discussed in the previous section.

Two options are available for introducing weights into the total variation model described in (5.31b). As before, we can use edge weights to modify the scalar-valued *components* that comprise the gradient vector field across edges, or we may use node weights to modify the *norms* of the gradient vector field defined at the nodes. Component weighting was described in [49, 123, 423] and may be formulated as

$$\mathcal{E}_{\text{TV}}[\mathbf{x}] = \sum_{v_i} \left(\sum_{j \forall e_{ij}} w_{ij} |x_i - x_j|^2 \right)^{\frac{p}{2}} + \lambda \sum_{v_i} |x_i - s_i|^p. \tag{5.34a}$$

As discussed in [123], the introduction of nonnegative weights still preserves the convexity of the Total Variation Model for $p \geq 1$, and therefore any descent algorithm may be used to perform the optimization.

The scalar-valued vector norms may be weighted similarly to the edge weights based on the estimation of whether or not a node is a boundary node. Chapter 4 provides examples of node weighting and we will proceed by assuming that we have a set of nonnegative, normalized, real-valued weights, w_i, that give some indication of discontinuity locations (i.e., $w_i \to 0$ for nodes on the border of discontinuities and $w_i \to 1$ for "internal" nodes). Given these node weights, we may consider a node-weighted formulation of (5.31b) as

$$\mathcal{E}_{\text{TV}}[\mathbf{x}] = \sum_{v_i} w_i \left(\sum_{j \forall e_{ij}} |x_i - x_j|^2 \right)^{\frac{p}{2}} + \lambda \sum_{v_i} |x_i - s_i|^p. \qquad (5.34\text{b})$$

Once again, the introduction of node weights preserves the convexity of (5.34b) when $p \geq 1$ and therefore the optimization may be performed using any descent algorithm. Due to the historical derivation of this filtering model from continuum mechanics, the previous node-weighted formulation of the Total Variation Model described in (5.34b) has been more common in the literature (e.g., [11, 386]). However, since discontinuities in 0-cochains are defined *between* nodes, we generally recommend the edge-weighting formulation described by (5.34a). The node-based weighting loses both *directional* discontinuity information and *precision* of the boundary location, since boundaries lie *between* groups of nodes.

We have considered the application of the various filtering approaches on a weighted graph in which the weights were used to encode knowledge of discontinuities. However, these discontinuities were viewed as real-valued weights and were not constrained to form any sort of closed *boundary* surrounding different data clusters. In the next section, we will remove the weights and introduce an explicit boundary variable that explicitly encodes the discontinuities over which smoothing is not permitted.

5.4 Filtering with Explicit, but Unknown, Discontinuities

Instead of treating the discontinuities implicitly with weights, we can formulate the discontinuities explicitly as a boundary that separates some nodes from others over which we permit no smoothing. In a sense, these explicit discontinuities could be viewed as a special case of the models in the previous section in which the weights are restricted to be binary-valued. However, most treatments of explicit discontinuity models impose the additional constraint that the discontinuities form a *closed boundary*. It is rare in practice to know where this boundary is, so the boundary must also be estimated in the filtering process. Unfortunately, this additional unknown variable usually destroys the convexity of the implicit discontinuity models that were previously considered, forcing the estimation of local minima.

When the discontinuities form a closed boundary, the standard approach is to view the filtered values inside the region as a window into a foreground function, x, and the values outside the region as a window into a background function, y. Therefore, in an explicit discontinuity model with a closed boundary, one seeks to find the optimum value of both the reconstructed foreground variable and the reconstructed background variable.

5.4 Filtering with Explicit, but Unknown, Discontinuities

The prototypical energy for filtering with explicit discontinuities is the "piecewise smooth" Mumford–Shah model [289] (strongly related to the Geman and Geman model of [145] and the "weak membrane model" of Blake and Zisserman [39]). In this work, we follow the level set literature to consider the piecewise smooth model [289, 384], formulated as

$$\mathcal{E}_{\text{MS}}[x_{\mathcal{R}}, x_{\bar{\mathcal{R}}}; \mathcal{R}] = \int_{\mathcal{R}} \|\nabla x_{\mathcal{R}}\|_2^2 \, dt + \int_{\bar{\mathcal{R}}} \|\nabla x_{\bar{\mathcal{R}}}\|_2^2 \, dt$$

$$+ \lambda \int_{\mathcal{R}} |x_{\mathcal{R}} - s|^2 \, dt + \lambda \int_{\bar{\mathcal{R}}} |x_{\bar{\mathcal{R}}} - s|^2 \, dt + \nu \Gamma(\mathcal{R}), \quad (5.35)$$

where we introduce the new variable $\mathcal{R}$ which is a subset of the domain and the function $\Gamma(\mathcal{R})$ measures the boundary length of the set. In a more generalized context, we may consider $\mathcal{R} \subseteq \mathcal{V}$, represented by a binary-valued 0-chain, $\mathbf{r}$, indicating membership in $\mathcal{R}$, i.e., $r_i = 1$ if $v_i \in \mathcal{R}$ and $r_i = 0$ otherwise. In this more generalized setting, we may define $\Gamma(\mathcal{R})$ as

$$\Gamma(\mathcal{R}) = \sum_{e_{ij}} w_{ij} |r_i - r_j|. \quad (5.36)$$

When $w_{ij} = 1$ everywhere, this definition equates $\Gamma(\mathcal{R})$ with the number of edges spanning $\mathcal{R}$ and $\bar{\mathcal{R}}$ (although any affinity weighting function could also be applied). This model implicitly assumes that the data may be divided into two groups (e.g., "foreground" and "background"), although more complex models have been considered as well [389] which could be easily adapted to a discrete framework. In the context of a 4-connected image lattice embedded into the two-dimensional Euclidean plane in the usual fashion, the definition of $\Gamma(\mathcal{R})$ given in (5.36) measures the boundary of the region defined by $\mathcal{R}$ with an ℓ_1 metric. If a Euclidean (or other) measure were desirable, the weights could be modified to reflect the desired metric, as described in Chap. 4.

Following the treatment in [163], the discrete formulation of (5.35) may be given as

$$\mathcal{E}_{\text{MS}}[\mathbf{x}_{\mathcal{R}}, \mathbf{x}_{\bar{\mathcal{R}}}; \mathbf{r}] = \mathbf{r}^T |\mathbf{A}^T| |\mathbf{A}\mathbf{x}_{\mathcal{R}}|^2 + (\mathbf{1} - \mathbf{r})^T |\mathbf{A}^T| |\mathbf{A}\mathbf{x}_{\bar{\mathcal{R}}}|^2$$

$$+ \lambda \mathbf{r}^T |\mathbf{x}_{\mathcal{R}} - \mathbf{s}|^2 + \lambda (\mathbf{1} - \mathbf{r})^T |\mathbf{x}_{\bar{\mathcal{R}}} - \mathbf{s}|^2 + \Gamma(\mathcal{R}). \quad (5.37)$$

Although this formulation is no longer convex, additional information is gained by finding a solution for the explicit boundary, $\mathcal{R}$, which allows interpretation of a minimum for (5.37) as a segmentation or clustering algorithm. Consequently, this model will be discussed in greater detail in Chap. 6. Note that the Mumford–Shah model defined on a graph was previously used to establish a set of filter coefficients that could be applied iteratively to perform filtering, see [342].

5.5 Filtering by Gradient Manipulation

A different approach to filtering may be achieved by manipulating the *gradients* of the data and reconstructing a least-squares fit of the node data, **x**. Specifically, if we consider the gradient vectors $v = \nabla u$, then we may apply any manipulating function to these vectors, $\eta(v)$ and then look for the new scalar-valued function for which these vectors are the gradients, $\nabla \tilde{u} = \eta(v)$. Of course, not every set of vectors is the gradient of some scalar function (i.e., the vector field may contain a curl component). Therefore, the standard approach would be to find a scalar function with a gradient field that is as close as possible to $\eta(v)$ in the least-squares sense (see, e.g, [131, 393, 410]). Specifically, the goal is to find the scalar field $\tilde{u}$ that minimizes

$$\mathcal{E}[\tilde{u}] = \int \|\nabla \tilde{u} - \eta(v)\|_2^2 \, dt, \tag{5.38}$$

which takes a minimum at the solution to the Poisson equation

$$\nabla^2 \tilde{u} = \nabla \cdot \eta(v). \tag{5.39}$$

Beyond least-squares minimization, other options are available for reconstructing a scalar field u from a vector field with a nonzero curl component (see [2]).

In a discrete setting, the manipulation of the gradient field occurs over the 1-cochain **y**. Specifically, $\mathbf{y} = \mathbf{A}\mathbf{x}$, which is manipulated with $\eta(\mathbf{y})$ and a least-squares solution is produced via the energy functional

$$\mathcal{E}[\tilde{\mathbf{x}}] = (\mathbf{A}\tilde{\mathbf{x}} - \eta(\mathbf{y}))^\mathsf{T}(\mathbf{A}\tilde{\mathbf{x}} - \eta(\mathbf{y})), \tag{5.40}$$

which takes a minimum when

$$\mathbf{L}\tilde{\mathbf{x}} = \mathbf{A}^\mathsf{T}\mathbf{y}. \tag{5.41}$$

Although traditional applications of this model have not considered weighted edges, these edge weights could easily be introduced into this context. For example, the least-squares solution of the problem in (5.41) would simply be modified to solve the Poisson equation

$$\mathbf{L}\tilde{\mathbf{x}} = \mathbf{A}^\mathsf{T}\mathbf{G}^{-1}\mathbf{A}\tilde{\mathbf{x}} = \mathbf{A}^\mathsf{T}\mathbf{G}^{-1}\mathbf{y}. \tag{5.42}$$

5.6 Nonlocal Filtering

Recent studies have suggested using nonlocal neighborhood relationships to perform filtering [15, 63]. Instead of an edge set based upon a local neighborhood and gradient-based weighting, these methods have advocated for employing a fully-connected graph in which each edge weight is dependent upon the statistical relationship or similarity of the data within a local neighborhood of each node. In other words, each node is considered to have a *local* neighborhood and a *distant* neighborhood (which is a superset of the local neighborhood), where the edges of the distant neighborhood derive their edge weights by comparing local neighborhoods. In this way, weights form connections at two different scales. The principle with these nonlocal techniques is that many patterns are repeated throughout a dataset and therefore the restoration of the pattern at one location can benefit from looking at patterns from other locations. Of course, the introduction of a fully-connected graph makes the computation intense and slow. Despite this computational hurdle, the quality of the results is sufficiently impressive that the technique remains an active area of research.

Since this idea was introduced in the field of image processing, a regular data grid is assumed in which it is possible to measure neighbors in a small window around two pixels. Specifically, the weights can be derived as

$$w_{ij} = \sigma(\|\mathbf{x}_{\text{nbhd}(i)} - \mathbf{x}_{\text{nbhd}(j)}\|), \tag{5.43}$$

where $\sigma(\cdot)$ is any affinity weighting function discussed in Chap. 4 and $\mathbf{x}_{\text{nbhd}(i)}$ represents the collection of data values, arranged into a vector, from the nodes in the local neighborhood of v_i. The weights generated in this fashion may be used directly in any of the methods described in previous sections. This distinction between local and distant neighborhoods also carries over easily to arbitrary regular graphs, but not to irregular graphs. One possible avenue for generalization of these methods to irregular graphs is to modify $\sigma(\cdot)$ to input two distributions as arguments and output a distance between those distributions. For example, $\sigma(\cdot)$ could measure the difference in entropy of the data values in the two local neighborhoods, even if the neighborhoods were of different size. Another approach for generalization of this method to irregular graphs is to consider two nodes to be distant neighbors only if they have the same degree.

5.7 Filtering Vectors and Flows

The focus of the previous sections has been the filtering of 0-cochains or functions defined on nodes. In this section we discuss how to apply these same techniques to filtering 1-cochains or functions defined on edges, such as vectors or flows. For the remainder of this section, we will refer to all 1-cochains as flows. We begin our application into flow filtering by making the same assumption as we did with the filtering of scalar fields—that the target of our filtering operation is to suppress high frequencies. This assumption can be motivated by the same arguments as in Sect. 5.1.2 if the 1-cochains are acquired directly (i.e., if our acquisition device measures flows, differences, gradients or other vectors). The filtering of high frequency components in a flow field is identical to the case of filtering a scalar field: *low-pass filtering entails suppressing the components corresponding to high frequency eigenvectors of the Laplacian.* Instead of suppressing the high frequencies of the *scalar* Laplacian, flow filtering suppresses the high frequencies of the *vector* Laplacian discussed in Chap. 2.

Recall that the continuous formulation of the vector Laplacian is

$$\nabla^2 = (\nabla\nabla\cdot) - (\nabla \times \nabla \times), \qquad (5.44)$$

and the (unweighted) edge Laplacian

$$\mathbf{L}_1 = \mathbf{B}^\mathsf{T}\mathbf{B} + \mathbf{A}\mathbf{A}^\mathsf{T}, \qquad (5.45)$$

where **B** is the face–edge incidence matrix. Although many aspects of our smoothing models for scalar fields translate directly, there are also important differences. We can note several aspects of the edge Laplacian in the context of filtering:

1. The node Laplacian for a connected graph has a rank one nullspace corresponding to the constant vector, which means that a diffusion process governed by the scalar Laplacian will always approach a constant. The edge Laplacian has a rank zero nullspace when $|\mathcal{F}| = |\mathcal{E}| - |\mathcal{V}| + 1$ (see Chap. 4), meaning that "diffusion" with the edge Laplacian will always drive the solution to zero.
2. The edge structure may also be shift-invariant for several common types of complex. For example, a wrapping (or infinite) lattice will have a circulant edge Laplacian (assuming that the cycle set consists of all the local cycles). If the edge Laplacian is circulant, then the DFT may be used to efficiently filter flows, as described above for the case of nodal filtering.
3. From the definition of the edge Laplacian, it is clear that a "smooth" flow field is one for which there is a small curl *and* a small divergence. In other words, the ideal smooth flow is one in which all of the flow vectors are pointing in the same direction. Figure 5.3 illustrates the principle of vector smoothing and Fig. 5.4 gives an example on an arbitrary graph. These examples illustrate vector "diffusion" in which the initial high-frequency flow field is smoothed.

Thus filtering higher-order data is a straightforward generalization of the intuitive filtering of nodal data presented above.

> The filtering of a flow field or any edge data is possible using the same formalism for filtering scalar fields or node data by using the eigenvectors of the edge Laplacian.

This direct extension of filtering to higher-order data provides a clear example of the generality of the discrete framework.

5.7.1 Translating Scalar Filtering to Flow Filtering

If the edge Laplacian is circulant, then standard DFT filtering techniques may be applied since the vectors in the DFT matrix will also be the eigenvectors of the edge

5.7 Filtering Vectors and Flows

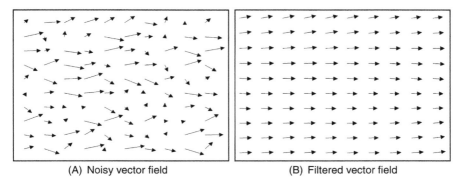

(A) Noisy vector field

(B) Filtered vector field

Fig. 5.3 An example of vector smoothing in the plane. (**A**) The initial, noisy vector field with nonzero curl and divergence throughout the field. (**B**) The vector field smoothed via diffusion (the Basic Energy Model applied to flows (5.49b) with $p = 2$). In the Euclidean plane, this smoothing is equivalent to smoothing the two coordinate components of each vector independently (i.e., treating both as scalar fields). In general, vector smoothing reduces both the curl *and* divergence of the vector field, and adds spatial coherence or correlation to the vector-valued data

Laplacian. However, if the edge Laplacian is not circulant, then we may still apply Taubin's smoothing method to the flow field. Specifically, if we have flow field **y**, then we may alternate "diffusion" and "unshrinking" steps via

$$\mathbf{y}^{[2k+1]} = \mathbf{y}^{[2k]} - \lambda \mathbf{L}_1 \mathbf{y}^{[2k]}, \tag{5.46}$$

$$\mathbf{y}^{[2k+2]} = \mathbf{y}^{[2k+1]} + \mu \mathbf{L}_1 \mathbf{y}^{[2k+1]}. \tag{5.47}$$

In order to utilize the gradient-based filtering techniques that we defined for scalar functions, we need to ask how to define a "gradient" on a vector/flow field. A natural choice of the "gradient" of a flow field might be to view the gradient as the 0-coboundary operator (see Chap. 2) and simply replace it with the 1-coboundary operator, i.e., the curl operator **B**. However, such an approach would view a minimization of the gradient operator as the goal rather than viewing the gradient operator as a proxy for dampening the high-frequency eigenvectors of the Laplacian, as originally derived in (5.10b). For scalar functions, the Laplacian consists of $\mathbf{L} = \mathbf{A}^T \mathbf{A}$ and therefore an iterative reduction of $\mathbf{1}^T |\mathbf{A}\mathbf{x}|^p$ will have the effect of filtering high frequencies in the initial data. However, when we consider the edge Laplacian, then $\mathbf{L}_1 = \mathbf{B}^T\mathbf{B} + \mathbf{A}\mathbf{A}^T$ and we can see that simply replacing gradient with curl, or replacing a minimization of $\mathbf{1}^T|\mathbf{A}\mathbf{x}|^p$ with $\mathbf{1}^T|\mathbf{B}\mathbf{y}|^p$, addresses only the first term of $\mathbf{L}_1$. Therefore, if we intend to extend the gradient-based techniques to flow filtering, we must minimize both $\mathbf{1}^T|\mathbf{B}\mathbf{y}|^p$ *and* $\mathbf{1}^T|\mathbf{A}^T\mathbf{y}|^p$, i.e., in order to dampen high frequencies we must simultaneously minimize both the curl and the divergence of the flow field.

We may now reformulate the gradient-based techniques employed in Sect. 5.2 for scalar functions in the context of flow field filtering by again considering the energy minimization models in this context. The formulation of the Basic Energy

Fig. 5.4 An example of flow filtering in an arbitrary graph. A lowpass flow filtering process attempts to reduce both flow divergence and flow curl. Node v_1 and cycle c_1 have been designated as locations where the initial flow field has high divergence and curl, respectively. (**A**) The initial flow field. Note the strongly nonzero divergence at v_1, equaling $10 + 5 - 2 - 3 = 10$ and the nonzero curl around cycle c_1 equaling $5 + 5 + 5 = 15$. (**B**) The vector field smoothed via diffusion after one iteration (forward Euler on the Basic Energy Model (5.49b) with $p = 2$ and timestep $\Delta t = 0.1 = 1/\lambda$). Even after one iteration, the divergence at v_1 has been reduced to $7.4 + 2.9 - 4.4 - 1.1 = 4.8$ and the curl around cycle c_1 has been reduced to $2 + 1.7 + 2.7 = 6.4$. (**C**) The vector field smoothed via diffusion after five iterations (forward Euler on the Basic Energy Model for flows (5.49b) with $p = 2$ and timestep $\Delta t = 0.5 = 1/\lambda$). The divergence at v_1 has been reduced to $3.94 + 0.99 + 0.19 - 4.74 = 0.38$ and the curl around cycle c_1 has been reduced to $1.42 + 0.53 + 0.19 = 2.14$

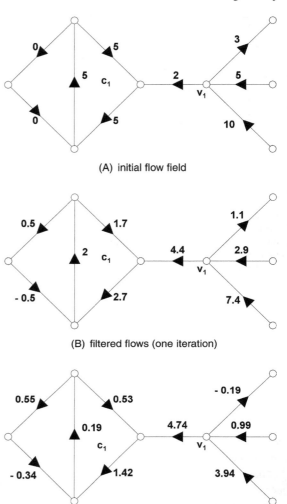

Model for flow filtering therefore gives

$$\mathcal{E}_{\text{BEM}}[y] = \int \|\nabla \cdot \vec{y}\|_p^p \, dt - \int \|\nabla \times \vec{y}\|_p^p \, dt, \tag{5.48a}$$

$$\mathcal{E}_{\text{BEM}}[\mathbf{y}] = \mathbf{1}^\mathsf{T} |\mathbf{B}\mathbf{y}|^p + \mathbf{1}^\mathsf{T} |\mathbf{A}^\mathsf{T} \mathbf{y}|^p. \tag{5.48b}$$

Recall that the sign discrepancy between the continuous and discrete formulations was addressed in Chap. 2. In the scalar case, the minimum of $\mathcal{E}_{\text{BEM}}[\cdot]$ was trivial (constant) and this case is no different (zero). However, just as in the scalar case, a few steps of an iterative minimization algorithm for (5.48b) will serve to quickly dampen the highest frequencies.

In the scalar case we employed a data term to avoid a trivial minimum of our gradient term and the same technique may be used again for flow fields. If we have a noisy observation of our flow field (represented by function **s**), then we may trade off between smoothness and the noisy data via a minimization of

$$\mathcal{E}_{\text{EBEM}}[y] = \int |\nabla \cdot \vec{y}|^p \, dt - \int \|\nabla \times \vec{y}\|_p^p \, dt + \lambda \int |\vec{y} - \vec{s}|^p \, dt, \quad (5.49a)$$

$$\mathcal{E}_{\text{EBEM}}[\mathbf{y}] = \mathbf{1}^T |\mathbf{A}^T \mathbf{y}|^p + \mathbf{1}^T |\mathbf{B}\mathbf{y}|^p + \lambda \mathbf{1}^T |\mathbf{y} - \mathbf{s}|^p. \quad (5.49b)$$

All of the above filtering techniques may be modified to include implicit discontinuities (weights) by using the appropriately weighted operators. Recall from Chap. 2 that the weighted edge Laplacian is given by $\mathbf{L}_1 = \mathbf{A}\mathbf{G}_0\mathbf{A}^T\mathbf{G}_1^{-1} + \mathbf{G}_1\mathbf{B}^T\mathbf{G}_2^{-1}\mathbf{B}$ for node weighting $\mathbf{G}_0$, edge weighting $\mathbf{G}_1$ and face weighting $\mathbf{G}_2$. Letting $\mathbf{G}_1 = \mathbf{I}$ gives

$$\mathcal{E}_{\text{BEM}}[\mathbf{y}] = \mathbf{1}^T \mathbf{G}_0 |\mathbf{A}^T \mathbf{y}|^p + \mathbf{1}^T \mathbf{G}_2^{-1} |\mathbf{B}\mathbf{y}|^p, \quad (5.50)$$

$$\mathcal{E}_{\text{EBEM}}[\mathbf{y}] = \mathbf{1}^T \mathbf{G}_0 |\mathbf{A}^T \mathbf{y}|^p + \mathbf{1}^T \mathbf{G}_2^{-1} |\mathbf{B}\mathbf{y}|^p + \lambda \mathbf{1}^T |\mathbf{y} - \mathbf{s}|^p. \quad (5.51)$$

In the next section we use the generalizations of the filtering models to flow filtering to further extend these filtering techniques to functions defined on cells of any dimension (i.e., to filtering a general p-cochain).

5.8 Filtering Higher-Order Cochains

Now that we have examined the Fourier and variational approaches for filtering scalar (node) and vector (edge) functions, we complete the exposition by briefly considering the filtering of functions defined on higher-dimensional cells (e.g., faces). As before, the filtering of high frequencies depends on a definition of a higher-order Laplacian. As we saw in Chap. 2, the general definition of the higher-order p-Laplacian matrix for arbitrary p-cells is given as

$$\mathbf{L}_p = \mathbf{N}_p \mathbf{N}_p^* + \mathbf{N}_{p+1}^* \mathbf{N}_{p+1}. \quad (5.52)$$

If we consider a specific value of p, then we may still employ DFT-based techniques if the $\mathbf{L}_p$ matrix is circulant. Even if $\mathbf{L}_p$ is not circulant, then we may still apply Taubin's algorithm with the steps

$$\mathbf{z}^{[2k+1]} = \mathbf{z}^{[2k]} - \lambda \mathbf{L}_p \mathbf{z}^{[2k]}, \quad (5.53a)$$

$$\mathbf{z}^{[2k+2]} = \mathbf{z}^{[2k+1]} + \mu \mathbf{L}_p \mathbf{z}^{[2k+1]}, \quad (5.53b)$$

where **z** is the p-cochain variable.

Similarly, when $\mathbf{G}_p = \mathbf{I}$, the variational approaches may be defined by producing a minimum of

$$\mathcal{E}_{\text{BEM}}[\mathbf{z}] = \mathbf{1}^T \mathbf{G}_{(p-1)} |\mathbf{N}_p^T \mathbf{z}|^q + \mathbf{1}^T \mathbf{G}_{(p+1)}^{-1} |\mathbf{N}_{(p+1)} \mathbf{z}|^q, \tag{5.54}$$

$$\mathcal{E}_{\text{EBEM}}[\mathbf{z}] = \mathbf{1}^T \mathbf{G}_{(p-1)} |\mathbf{N}_p^T \mathbf{z}|^q + \mathbf{1}^T \mathbf{G}_{(p+1)}^{-1} |\mathbf{N}_{(p+1)} \mathbf{z}|^q + \lambda \mathbf{1}^T |\mathbf{z} - \mathbf{s}|^q. \tag{5.55}$$

Note that we used q as the exponent parameter to avoid confusion with p used to designate the p-cell. The above equations detail the more general case of variational filtering methods with implicit boundaries (weights). If we desired to use the variational filtering methods without implicit boundaries (unweighted) we may simply set $\mathbf{G}_{(r+1)} = \mathbf{G}_r = \mathbf{G}_{(r-1)} = \mathbf{I}$ in the equations above.

5.9 Applications

The filtering procedures described in this chapter may be applied to any situation in which data measured at discrete locations contains noise to be removed. In this section, we consider several applications of the filtering procedures applied to very different types of data. At the end of these experiments, we summarize our observations to give the reader a guide for when to employ the various filtering techniques. In all of the experiments presented here using the total variation model, we set $p = 1$, since setting $p = 2$ in the total variation model is equivalent to setting $p = 2$ in the Extended Basic Energy Model. Unless otherwise noted, the total variation filtering used an unweighted edge model. The λ parameter was set individually for each algorithm.

5.9.1 Image Processing

5.9.1.1 Regular Graphs and Space-Invariant Processing

A classical application of filtering is image processing. In this problem domain, the pixels are identified with nodes, edges are derived from a local neighborhood (e.g., a 4-connected or 8-connected lattice), and the pixel intensities are the data associated with each node. Therefore, the **x** variable in the filtering routines above is the unknown filtered image intensities that are solved for, while the initial data **s** is identified with the noisy image intensities.

In the first experiment, a synthetic image of a black circle in a white background was corrupted with noise by adding an independent random variable to each pixel with a uniform distribution. Figure 5.5 displays the results of the various filtering procedures discussed above. We may make several observations from these results.

First, the localization of circle boundaries (i.e., the image discontinuities) improves in the Basic Energy Model as the parameter p decreases. Therefore, the

5.9 Applications

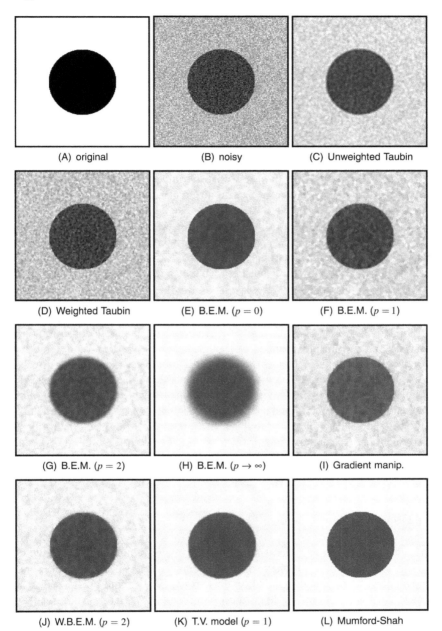

Fig. 5.5 Image filtering of a synthetic image of a black circle on a white background which has been corrupted by additive i.i.d. random noise with a uniform distribution. Note that unweighted filtering with the Basic Energy Model consistently produces worse boundary localization as p increases. However, for any filtering model, weighted filtering is generally better at preserving boundary location. Since the source image exactly matches the Mumford–Shah model, this filtering result is nearly perfect

mean filter localizes boundaries better than the minimax filter, the median filter localizes boundaries better than the mean filter and the (approximated) mode filter localizes boundaries better than the median filter. A different method of providing better localization boundaries is to use a *weighted* filter, as seen by the improvement in the Taubin technique and the improvement in the mean filter attained by using edge weights obtained from the Welsch function in Chap. 4. In the gradient manipulation example, the gradients were manipulated by setting all gradients to zero for which the magnitude was smaller than a fixed threshold. Since this image content allowed such a simple approach, the reconstruction was nearly perfect from this algorithm. Finally, we note that this example perfectly fits the model of the Mumford–Shah functional in the sense that the underlying image consists of two objects (foreground and background) with different intensities. If the boundary between these two objects may be localized well, then this filtering procedure smooths only within these boundaries to achieve a near-perfect filtered result.

The second experiment uses a photograph which is much more complicated than the circle in the previous experiment (see Fig. 5.6). Unlike the previous image, the noiseless image contains significant high spatial frequency content (in the child's hair). In this experiment, zero-mean Gaussian noise was added independently to each pixel to corrupt the original image. As before, we see that as p increases in the Basic Energy Model, the boundaries are progressively blurred in the filtered image. Additionally, the use of a weighted graph increases the boundary localization for the Taubin and mean filtering methods. Although the total variation model continues to produce a good filtering, the Mumford–Shah model does not perform well on this image. The Mumford–Shah model explicitly assumes that there are two regions (possibly consisting of multiple connected components) which are smoothly varying in intensity except at the boundary transitions. Since the example image contains many regions of sharp intensity changes (textures), the Mumford–Shah model is forced to choose (through parameter settings) between many small regions or large overly smoothed regions. The parameters were set in this experiment to produce many small regions. Results of this experiment are displayed in Fig. 5.6.

5.9.1.2 Space-Variant Imaging

Although standard image processing applies to images which are uniformly sampled, there are several situations in which the image data is acquired with nonuniform samples. Some image acquisition devices explicitly acquire data which does not have a Cartesian sampling (e.g., ultrasound medical images). Additionally, almost all known biological vision systems acquire light data nonuniformly in space [209]. Although most biological vision systems employ sampling schemes which are difficult to describe mathematically as a function of space, there has been more success in mathematically describing the sampling of visual space employed by humans and by non-human primates such as the macaque monkey. The macaque is of particular interest because it is considered to have a similar retinal organization to humans and similar visual capabilities [336], and there is vast amounts of data on the

5.9 Applications

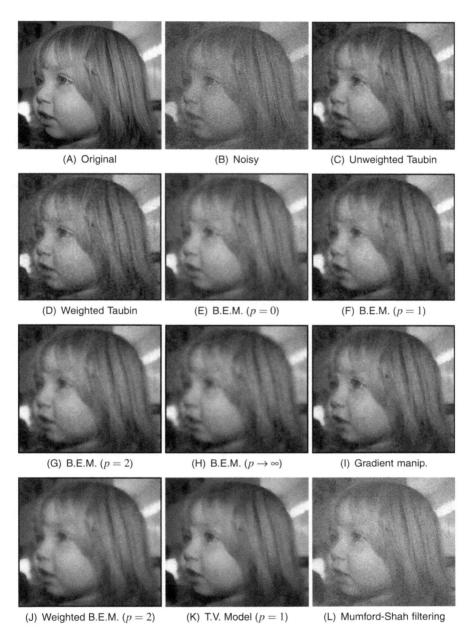

Fig. 5.6 Image filtering of a photograph containing high-frequency texture. This image has been corrupted by i.i.d. random noise with a zero-mean Gaussian distribution. As before, weighted filtering is generally better at preserving boundary location. However, since this image does not match well with the Mumford–Shah model, the filtered image is not nearly as close to the noiseless image as it was in the previous example

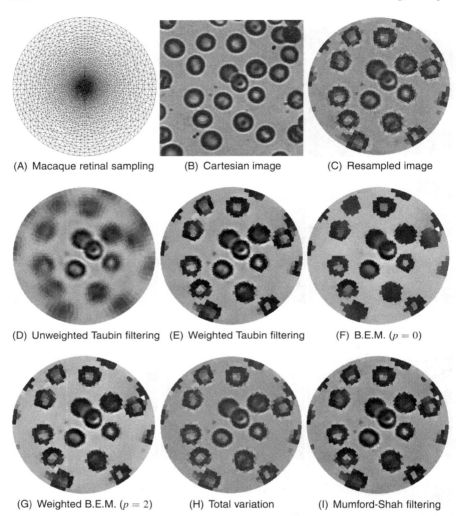

Fig. 5.7 Filtering image data on a biologically sampled image. (**A**) A sampling mesh modeled after the macaque retina. (**B**) Cartesian image. (**C**) Image resampled with the macaque mesh, (**D**)–(**I**) filtering of the data. Note the visual disturbance caused by blurring edges in the poorly sampled peripheral regions in (D)

macaque visual system. Several researchers have taken inspiration from this nonuniform biological sampling of visual space to pursue computer vision approaches or hardware with a similar sampling [158, 277, 321, 328].

We may filter these nonuniform biological samplings of image data in the same framework as before. As with Cartesian sampling, the image sample locations are viewed as nodes, the edge structure is defined by a Delaunay triangulation in the Euclidean plane and the filtered image data **x** is associated with each node (see

[158, 391] for more information). Aside from this new graph, there is *no difference in filtering operation with the standard Cartesian data*. In fact, exactly the same software implementation may be used to perform filtering by simply applying it to the new (non-lattice) graph. For this experiment, the nonuniform sampling structure for the macaque was loaded from the Graph Analysis Toolbox software package [166] that contains an implementation of the filtering techniques discussed in the chapter. Using this same toolbox, a standard Cartesian image of blood cells was imported to the space-variant structure and filtered. Results of the experiment are displayed in Fig. 5.7. Since the same properties of the filtering procedures observed for the Cartesian images apply to the space-variant images as well, only a subset of the filtering methods were employed for this experiment. However, we may make a few observations which are specific to this experiment. The first observation is that the blurring over object boundaries is more visually disturbing in regions of the image which are represented by only a few samples (the periphery in these images). Our second observation is that although the edge weights continue to help avoid blurring over object boundaries, the edge weights may be set in this scenario based on both image data and graph geometry. Therefore, edges connecting nodes which are further apart spatially may be given a lower weight, in addition to a weighting based on intensity difference (see Chap. 4 for more details). In this experiment, edge weights were generated purely from intensity changes in the image.

5.9.2 Three-Dimensional Mesh Filtering

Filtering of geometric data is an important process in computer graphics and the processing of data obtained from various three-dimensional scanners. Is this context, the node data (0-cochain) is a tuple of coordinates assigned to each node. Therefore, the nodal variable **x** in the above algorithms corresponds to an $n \times K$ set of K-dimensional filtered coordinates assigned to each node, with **s** corresponding to the $n \times K$ noisy coordinate values acquired for each node. The output of a filtering procedure is therefore a new set of coordinates for each node. The edge structure of the graph is generally given via a surface extraction preprocessing step. Since the most common method for rendering three-dimensional data requires a list of faces for the surface, the faces are usually extracted via a triangulation process.

We begin this section with a synthetic example of filtering coordinates obtained by generating a circle and adding noise to the coordinates of each point on the circle. Figure 5.8 gives an example of a lowpass filter obtained via the Basic Energy Model with $p = 2$ (i.e., a mean filter) and a lowpass filter given by Taubin's method. By subtracting the lowpass coordinates from the original coordinates, a highpass filter of this ring is obtained.

5.9.2.1 Mesh Fairing

The problem of **mesh fairing** is to produce a smooth three-dimensional mesh from a mesh with noisy coordinate values. In the current framework, the mesh points are

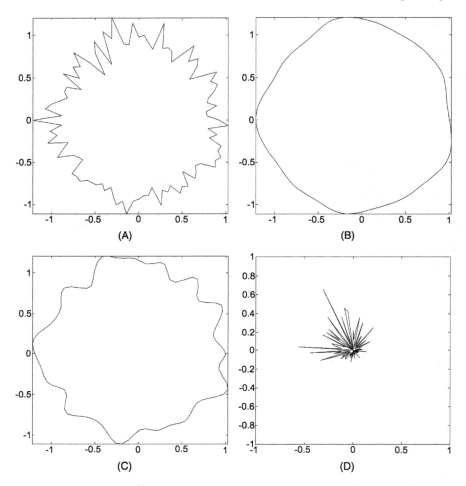

Fig. 5.8 Filtering coordinate data on a ring graph. (**A**) A noisy ring graph produced by adding Gaussian random noise to the radius of nodes arranged in a perfect circle. (**B**) The effect of applying the Basic Energy Model with $p = 2$ (i.e., a mean filter) to the coordinates of the graph in (A). (**C**) The low-pass filter of Taubin [371] applied to the coordinates of the graph in (A). (**D**) A high-pass filter of the coordinates in (C), produced by differencing the low-pass signal of (C) with the original

associated with nodes, the edges are given explicitly by the mesh, and the data tuple $\tilde{s}_i$ associated with each node v_i represents the three-dimensional coordinates of that node. The goal in mesh fairing is to produce filtered data (coordinates) **x**.

Figure 5.9 shows a three-dimensional mesh of a horse. The noise observed in meshes is typically in the direction of the surface normal. For this example, Gaussian noise was added to each of the three coordinates and to each node independently to generate the noisy mesh. Several of the filtering procedures described in this chapter were applied to produce a fairer mesh. In this application, it is possible

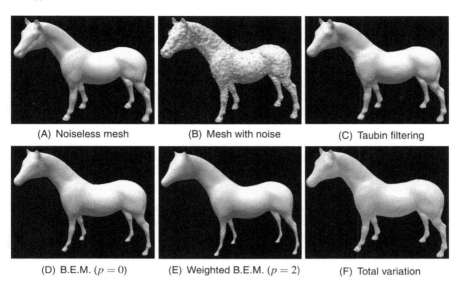

Fig. 5.9 An example of mesh fairing. Gaussian noise was added to each of the coordinates of a three-dimensional mesh and these coordinates were filtered to produce a faired (smoothed) mesh. The nodes and edges are given explicitly by the mesh and the data are the three-dimensional node coordinates. Note how the Basic Energy Models shrink portions of the figure while Taubin's spectral filtering smooths without shrinking by preserving the low frequencies

to see that one of the major benefits of Taubin's spectral approach to filtering is to avoid shrinking of the mesh. Since the Basic Energy Model drives the filtered solution toward a constant value (regardless of the choice of p), the effect observed in mesh fairing is to drive all of the nodes closer together (to the same, constant, location in space). However, by preserving the low frequencies, Taubin's filtering approach avoids the shrinking observed from the other algorithms. The shrinking is particularly noticeable in these figures around the horse's legs, ears and snout. The total variation filtering results in less shrinking than the Basic Energy Model, but smooths the fine details somewhat more than Taubin's filtering algorithm.

5.9.3 Filtering Data on a Surface

In many applications, data is measured at spatial locations along a surface, and to interpret this data properly the analysis must be carried out in a way that respects how the data is distributed along the surface. Examples of such measurements would be data collected from a network of touch sensors on an article of clothing, or samples of the distribution of current along a conductive sheet. Here we consider the application of filtering functional Magnetic Resonance Imaging (fMRI) data measurements of neural activity from positions along the surface of the cerebral cortex of the brain.

We will consider an example taken from an fMRI study that sought to locate brain areas implicated in the processing of vision and, in particular, those areas that are known to be responsible for the processing of *motion* in the visual field. In this example, the neural responses were measured during a visual stimulation consisting of presenting subjects with patterns of moving shapes, or *motion stimuli*, to activate those areas of the visual cortex responsible for processing motion so that these areas could be identified and located within the cortex. Neural "activity" is quantified through statistical analysis of the measured responses, and those measured locations whose statistical significance exceeds a fixed threshold are considered to be the sites of true activations. These activations can be visualized with *activation maps* that depict where on the cortical surface significant activation has been identified. However, random noise in the measurement leads to spurious significant activations by chance, which generates false positives in the activation maps that must be detected and removed for proper interpretation of the data. Because for most experiments it is expected that groups of active locations are nearby in space, spatial filtering of the data helps coalesce locations of true activity while suppressing the significance levels of spurious activity at isolated nodes, forcing them below the significance threshold and thus eliminating them from the final activity map. This spatial prior is often used in fMRI analysis (e.g., [406]).

MRI data is acquired in the form of a stack of images of the brain, thus the measurement consists of a volume of image data. Most techniques for spatial smoothing of fMRI data smooth the data in the original space of the acquired images, i.e., the volumes of image data represented as voxels, and therefore the conventional smoothing can be conveniently enacted by three-dimensional smoothing kernels applied to the volume of image data. Unfortunately much of the spatial structure of the relevant neuronal activity patterns is contained within the surfaces of the brain, such as the cortical gray matter of the cerebral hemispheres where most of the sensory, motor, and higher cognitive functions take place. Smoothing the voxel data in three dimensions is harmful since voxels that are nearby in three dimensions are often sampled from positions on the cortical surface that are far apart when distance is measured *along* the two-dimensional cortical surface—as in the case of two adjacent voxels that sample from opposite, abutting banks of a sulcus. Thus, volumetrically smoothing the voxel data and ignoring the boundary of the cortical surface can mix activity patterns across distant locations of the cortical surface, corrupting the spatial structure of local activity patterns existing along the surface. For this reason, it is advantageous to smooth the data in a way that respects the natural geometry of the cortical surface.

An example of surface smoothing applied to brain activation maps measured with fMRI is presented in Fig. 5.10 (contained in the color plate section at the end of the book). In surface-based fMRI analysis, a mesh representation of the cortical gray matter of the cerebral hemispheres is generated from anatomical MRI data (e.g., [96, 136]), including the two-dimensional exterior and interior boundaries of the gray matter ribbon. For this example data set, the exterior surface is shown in Fig. 5.10(A) and the corresponding interior surface is shown in Fig. 5.10(B). All analysis is restricted to the interior surface of the cortical gray matter, and for ease of visualization the activation maps are typically presented on an "inflated" surface representation (as shown in Fig. 5.10(C)) to reveal the activity buried within the deep sulci of the cortical folds.

5.9 Applications

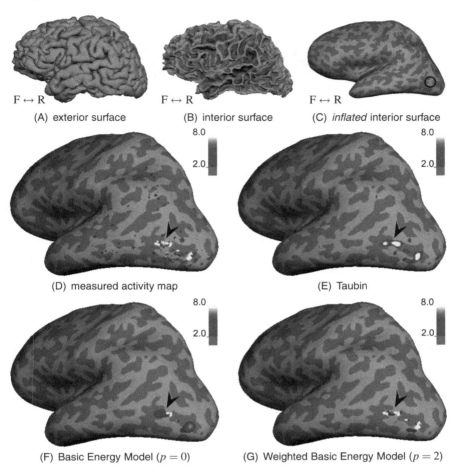

Fig. 5.10 Filtering fMRI data along a cortical surface model. Surface models of the (**A**) exterior surface, (**B**) interior surface, and (**C**) the "inflated" interior surface of the cortical gray matter of the left cerebral hemisphere, with approximate location of area MT indicated by a *circle*. (Surfaces generated with FREESURFER [96, 136].) The *legend* indicates Front–Rear axis of brain. Locations of negative mean curvature (within sulci) are rendered in the *dark gray* and locations of positive mean curvature (within gyri) are rendered in *light gray*. Measured activity map plotted as z-statistics, with color scale provided at *upper right*. The threshold is set to exclude nodes where the activity is not statistically significant, which leads to many isolated points or small clusters of activation appearing in the map—likely false positives due to noise. The results of filtering the data using (**E**) spectral filtering, (**F**) the Basic Energy Model with $p = 0$ and (**G**) $p = 2$, (**H**) Total Variation, and (**I**) the Mumford–Shah algorithm are provided with the same color scale representing the statistical significance. Note that many of the false positives are removed with the filtering. *Arrows* indicate the site of MT activation

The functional activation data from this example is represented on the vertices of the triangular mesh surface representation of the interior surface shown in Fig. 5.10(B), then smoothed with the filtering methods discussed in this chapter, and

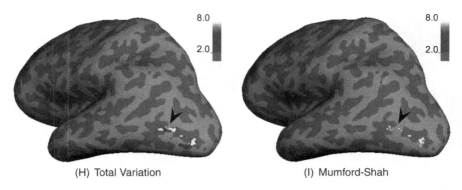

(H) Total Variation (I) Mumford-Shah

Fig. 5.10 (Continued)

the results are visualized on the inflated surface. In the original, unsmoothed data seen in Fig. 5.10(D), we see a cluster of activated nodes in the rear of the brain—which is within the part of the brain that is responsible for vision—accompanied by noisy activations extending up and further into the front of the brain. In order to remove these noisy activations while (ideally) retaining the true activations, surface-based smoothing can effectively highlight the true activity while removing noise.

Each of the filtering methods succeeds in suppressing false activations attributable to noise. The results of spectral filtering of Taubin, the two energy models, and Total Variation shown in Figs. 5.10(E)–(H) highlight two loci of activity in locations near areas where visual motion processing is known to occur known as the "middle temporal" area, or cortical area MT. Beyond removing spurious or noisy activations outside of the visual motion area, the spectral filtering shown in Fig. 5.10(E) also smooths the "true" activity pattern within MT, suggesting that some of the relevant features of the data may be lost along with the false positives. However, this filtering may also aid in eliminating aliasing artifacts in the measurement due to the coarse spatial sampling, and thereby the smoothing process may potentially recover a more faithful representation of the true activation pattern. The results of the basic energy model shown in Fig. 5.10(F) contain a distinct discontinuity that is not salient in the original measurement, which is a sharp feature that violates the expected spatial resolution of the fMRI technique and therefore is likely to be an artifact of the smoothing.

The results of the weighted Basic Energy Model and Total Variation shown in Figs. 5.10(G) and (H) demonstrate both suppression of false positives outside of the presumed true site of activation and retention of most of the structure of the original measured activity map. Therefore, if another form of filtering were desired for the remaining activity cluster, such as additional anti-aliasing filtering, it could be subsequently applied.

The results of the Mumford–Shah algorithm shown in Fig. 5.10(I) drive most of the measured activity below the statistical threshold, and thereby suppresses all but a small island of activity. To gain insight into the relative performance of these filtering methods, the results of Fig. 5.10 are re-plotted in Fig. 5.11 (contained in

5.9 Applications

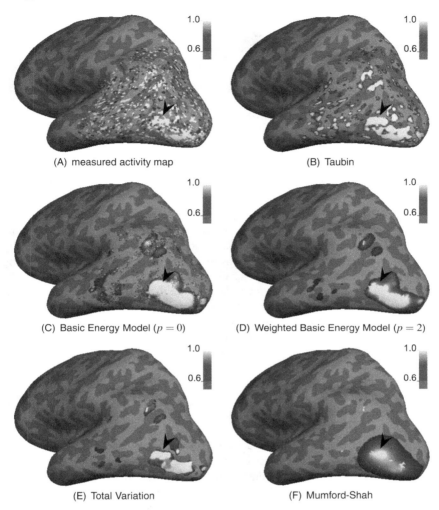

Fig. 5.11 Effect of smoothing methods on sub-threshold fMRI activity. The data of Fig. 5.10 is re-plotted with a color scale that highlights the relative performance and behavior of the filtering methods on activity below the significance threshold. *Reference arrows* are positioned as in Fig. 5.10

the color plate section at the end of the book) but with a lower statistical threshold to examine the sub-threshold patterns of the filtered data. With this color scale, the degree of noise suppression outside of the area of activity is more clear, with the results of the weighted Basic Energy Model shown in Fig. 5.11(D), Total Variation Fig. 5.11(E), and the Mumford–Shah algorithm shown in Fig. 5.11(F) performing best. Additionally, a salient and undesirable feature of the Mumford–Shah algorithm is that it spreads the activity pattern diffusely, losing most of the structure of the original data in this case.

(A) original folded surface   (B) 1,500 iterations   (C) 5,000 iterations

Fig. 5.12 Surface smoothing for cortical inflation. (**A**) Original folded surface. (**B**) Surface after 1,500 iterations of spatial filtering using the Basic Energy Model with $p = 2$. (**C**) Surface after 5,000 iterations of filtering. The high-frequency folds are removed with smoothing, leaving an "inflated" surface in which the regions within the cortical sulci are clearly visible, similar to the explicitly inflated surface presented in Fig. 5.10(C)

In this example, surface data are visualized on the inflated surface representation, which is very common in fMRI studies. Although several tools exist for rapidly computing inflated brain surface, it is instructive to note that the same smoothing operations used to filter the data along the surface can be applied to *filtering the surface mesh vertex coordinates themselves*—as in the previous example on mesh fairing—to smooth out the folding pattern and produce an "inflated" version of the cortical surface representation. Figure 5.12 demonstrates an example of how iterative smoothing using the Basic Energy Model with $p = 2$ can produce an inflated surface representation.

5.9.4 Geospatial Data

A different type of application involving data analysis at discrete locations comes from a parcellation of continuous space into subregions in which measurements are made. Many types of geospatial data fit this description in which a geographical area is parcellated into regions that fit a political, topographic or property description. Examples of this type of data would be soil samples, transportation data, pollution measurements, incidence of infectious disease, or population of a species. Geospatial data is typically managed, analyzed and visualized by software known as a *geographical information system* and this data is typically analyzed with a set of tools known as **spatial statistics** [82, 317].

In this section, we adopt an example of state polling data from the 2008 US Presidential election for the 48 continental states. Each US state is assigned a number equal to the percentage of poll respondents who favored (then candidate) Barack Obama just prior to the 2008 election. Each state is colored brighter if more respondents favored Mr. Obama and darker if fewer respondents favored Mr. Obama. Due to small samples and different polling methodologies, we can assume that there is noise present in this data. A simple model for filtering this polling data would be to assume that a state is more likely to favor a candidate if its neighboring states favor a candidate and less likely to favor a candidate if its neighboring states do not favor the candidate. This model is justified by the concept of **Tobler's Law** or **the First Law of Geography** which asserts that data situated at nearby geographical locations are likely to be correlated [377]. Given this correlation, we may apply a spatial filtering algorithm to the polling data to remove noise. In this example, each state is represented by a node, two states are connected by an (unweighted) edge if they share a border, the measured data **s** is the polling data and the goal is to produce filtered polling measurements **x**. We stress that although the underlying domain (the continental United States) is continuous, the parcellation of this space into political states for which polling measurements are made transforms this problem into the current framework of analyzing data associated with a graph.

Figure 5.13 displays the original polling data and the filtered data. Taubin's filtering method and the weighted Basic Energy Model with $p = 2$ yield similar results. The output of both of these filtering methods is largely unchanged from the original polling data, except that spatial outliers are softened. For example, the weak poll numbers for Mr. Obama in South Carolina and Indiana were improved after filtering because the poll numbers for Mr. Obama in neighboring states were generally stronger. Similarly, the polling numbers in New Hampshire and Maine were balanced after filtering, with the filtered polling numbers showing stronger support for Mr. Obama in New Hampshire and weaker support for him in Maine. The Mumford–Shah filtering produces substantially different results in this application. Specifically, this approach attempts to identify regions within which the polling numbers are expected to be relatively homogeneous. Therefore, the mid-Atlantic states are grouped with most of New England by this algorithm to produce one large voting bloc, the southeastern states are grouped with areas of the midwest to create a second voting bloc, a third voting bloc is produced by the great lakes states and upper midwest and a final voting bloc is produced by the west coast and the southwest. Florida and Maine-New Hampshire are also considered by the Mumford–Shah filtering approach as independent voting blocs. Within these voting blocs, the polling numbers are made more homogeneous (similar to the noisy circle image processing example). Although these voting blocs roughly correspond with meaningful political battle lines in the US in 2008, other parameter settings for this algorithm might produce larger or smaller voting blocs within which the data would be smoothed.

5.9.5 Filtering Flow Data—Brain Connectivity

In this section, we give an example of filtering real flow data along edges. Examples of flow data encountered in practical applications would be traffic networks, communication networks, or migration networks. In fact, if connection strengths between nodes are provided for any directed graph, these strengths could be considered as flow data. When considering flow data, a directed graph may be viewed simply as an undirected graph for which the edge directions represent the directions in which flow is considered to be positive.

Fig. 5.13 Filtering of polling data for the 2008 US Presidential election. Brighter coloring indicates stronger support for (then candidate) Mr. Obama and darker coloring indicates weaker support for Mr. Obama. Each state is represented by a node in the graph where two nodes share an (unweighted) edge if they share a border

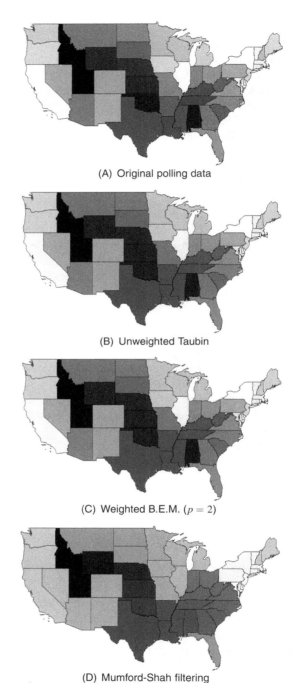

(A) Original polling data

(B) Unweighted Taubin

(C) Weighted B.E.M. ($p = 2$)

(D) Mumford-Shah filtering

5.9 Applications

We address the filtering of the measured strength of brain connectivities between parcellated regions of the cat brain measured by Scannell et al. [329] (and subsequently studied by others [354]). This connectivity network consists of 52 brain regions (nodes) and 818 directed edges. Scannell et al. assigned each edge a connection strength of '1' (weak), '2' (medium) or '3' (strong), which was determined from a compilation of measurements from the cortico-thalamic system of the cat. We may assume that noise was present in these measurements as a result of imprecise measuring devices and as a result of the severe quantization of the data into three categories. The justification for removing noise with the filtering techniques described above also applies to the filtering of flow data. Specifically, in the scalar case, data points within a low-frequency scalar (node) distribution are similar between neighboring nodes. Similarly, data within a low-frequency flow (edge) distribution are similar between neighboring edges in which neighboring edges are either incident on the same node (where similarity means small divergence) or neighboring edges are incident on the same cycle (where similarity means small circulation). Above in Fig. 5.3 we saw that a flow field through a continuous domain straightens out after filtering. Additionally, the noise model justifications for these filtering procedures also applies to the flow case—a zero-mean noise flow distribution will be expected to have zero divergence at all nodes and zero circulation around all cycles.

Several liberties were taken with this data in order to make the filtering operations clearer. First, a random subset of the graph nodes were sampled for presentation purposes to better visualize the results of the filtering. Second, we arbitrarily removed one edge from every pair of nodes connected by two directed edges in the opposite directions. This removal was also made to improve visualization of the results. Figure 5.14 displays the results of our filtering operation on the flow data. The first two figures show the full network and the connections in the subnetwork. The next figures illustrate the measured flow strength (represented by line thickness) and the lowest-frequency eigenvector of the edge Laplacian. The lowest frequency component of the network distributes the flows equally across edges in order to minimize flow divergence at nodes and to minimize flow circulation (curl) around cycles. Note that the eigenvector is *signed*, which is indicated by a change in direction (arrow) for negative flows. Taubin's spectral filtering and the Basic Energy Model ($p = 2$, corresponding to diffusion or generalized mean filtering) were applied to filter the initial flow data. The results of these filtering operations reduce the divergence and circulation of the original measured flow while driving the filtered flow toward the low-frequency eigenvector. The filtering especially dampened the edges contributing to the high-divergence of the leftmost node in the diagram, as well as dampening the edges causing divergence on the central node. In addition to reducing noise, the filtering operation also produced real-valued flows from the initial quantized flows which may allow for a better comparison between edge connectivities.

Since the underlying network is directed, care must be taken when applying these filtering operations that none of the flow signs change, causing direction changes. In this example, this constraint was enforced simply by using a small number of filtering iterations so as to not oversmooth the data.

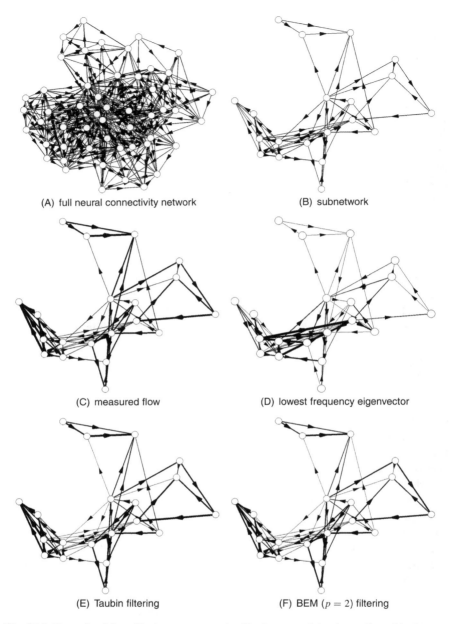

Fig. 5.14 Example of flow filtering on a network of brain connectivity data collected in the cortico-thalamic system of the cat [329]. A subnetwork of the original data was instead processed for visualization purposes. Line thickness represents the measured strength of the connections (the original "flow" data and filtered data). This example illustrates that the lowest frequency component of the network distributes the flows equally across edges in order to minimize flow divergence at nodes and to minimize flow circulation (curl) around cycles. Lowpass Taubin or diffusion ("mean") filtering drive the flows closer toward the flow given by low frequency eigenvector

5.10 Conclusion

In this chapter we reviewed several broad approaches for filtering data defined on arbitrary graphs—even irregular graphs. Although several examples were given, a reader with a particular filtering problem may still be wondering which method to use. The most classic filtering approach is Fourier-based filtering, but such an approach does not permit the preservation of discontinuities, nor does it apply to data defined on irregular (or weighted) graphs. In a more general setting, by far the most common methods are mean filtering or nonlinear anisotropic diffusion since these methods are straightforward to implement, predictable, and run in low-constant linear time. However, mean filtering has a tendency to oversmooth (even with discontinuities permitted), and to drive the data to a single value (the "shrinking" problem in mesh filtering). The filter described by Taubin is, in our opinion, an underutilized method (outside of computer graphics) which solves the second problem with little additional overhead or coding complexity (although it does require the specification of an additional parameter). If more computation is tolerable to provide a better result, then the variational approaches described in this chapter (e.g., median filtering with discontinuities, total variation filtering) are not difficult to implement and produce results that are not oversmoothed, but they do require more computation. Data which fits the assumptions of the Mumford–Shah model—that the data belongs to multiple regions which have smooth internal data—may be filtered well by minimizing the Mumford–Shah energy. If even better results are required, the underlying graph is shift-invariant and computation time is less important, then the variational filters with nonlocal neighborhoods are likely to produce the best results known so far. Finally, the gradient manipulation methods are moderately computationally intense (e.g., requiring a sparse linear system solve), but they provide substantial flexibility for combating data corruption if a model of the expected gradients is known, such as the image processing example of the circle in which it was asserted that the gradients were either large or zero. In the next chapter, we show how these same filtering models may be applied to derive a variety of clustering algorithms.

Chapter 6
Clustering and Segmentation

Abstract Clustering algorithms are used to find communities of nodes that all belong to the same group. This grouping process is also known as *image segmentation* in image processing. The clustering problem is also deeply connected to machine learning because a solution to the clustering problem may be used to propagate labels from observed data to unobserved data. In general network analysis, the identification of a grouping allows for the analysis of the nodes within each group as separate entities. In this chapter, we use the tools of discrete calculus to examine both the *targeted clustering problem* (i.e., finding a specific group) and the *untargeted clustering problem* (i.e., discovering all groups). We additionally show how to apply these clustering models to the clustering of higher-order cells, e.g., to cluster edges.

The clustering problem is to assign a set of labels to a set of cells such that all cells assigned to the same label belong to the same *group*. The most common type of cells to cluster are nodes, and so our discussion will be limited to node clustering, except in Sect. 6.4 where we address the clustering of higher-order cells. Clustering appears in several fields of study, including machine learning and image analysis. In the context of image analysis, the clustering problem is often called *image segmentation*. We will generally use the term *clustering* in this chapter unless we specifically discuss image clustering, in which we will use the term *segmentation*.

Formally, the clustering problem may be formulated by assigning elements b of a label set $\mathcal{L}$ to the set of nodes, $\mathcal{V}$. Specifically, the goal of a clustering algorithm is to produce a segmentation function $\sigma : \mathcal{V} \to \mathcal{L}$. Inspired by image analysis, the term *object* will also be used interchangeably with *cluster* to refer to all nodes that are mapped to the same label through σ.

Clustering algorithms may be broadly categorized along two axes. The first axis ranges from *targeted clustering* to *untargeted clustering*. Targeted clustering algorithms seek to identify a specific set of objects and therefore require some mechanism for training or steering the algorithm to find the desired set of objects. Consequently, targeted clustering algorithms have a specific label set which is drawn from to assign labels to each node. In contrast, an untargeted clustering algorithm seeks to label nodes as belonging to the same group if the group shares some properties, such as features of the data, or locality or connectivity derived from the graph. An untargeted clustering algorithm typically determines the number of labels to be used in order to satisfy its internal assumptions about node homogeneity, although in some cases the number of labels may be fixed. The traditional conception of *the* clustering problem has been the untargeted clustering problem, but interest has been increasing rapidly in targeted clustering algorithms.

All of the algorithms described here model the clustering problem with an objective function (often viewed as an energy function), and the purpose of the clustering algorithm is to find the clustering which optimizes the objective. However, some objective functions are difficult to optimize completely, or even fall into the class of NP-Hard problems. Consequently, the clustering calculated as a solution to one of these difficult objective functions may depend on the initialization to the algorithm. In these cases, a clustering algorithm, which would be untargeted if the objective could be completely optimized, can act as a targeted algorithm by initializing the clustering close to an intended target. We term algorithms of this variety as *semi-targeted* clustering, and treat these algorithms in a separate section.

The second axis for categorizing clustering algorithms ranges from *primal* algorithms (in which the nodes within a cluster are directly labeled) to *dual* algorithms (in which the cells comprising the boundaries of clusters are labeled). Primal algorithms are generally more popular, better developed, and easier to generalize. Dual algorithms work only under limited circumstances and depend on an embedding, but they constitute an important class of algorithms in image analysis that are of a fundamentally different character than the primal algorithms.

Several clustering algorithms in this chapter assume that each node is associated with some data (i.e., an attributed graph) and it is this data which is to be clustered. However, the predominant methodology is to transform this data into graph structure via the weights (using the functions in Chap. 4) and then apply the clustering algorithms to the weighted graph. Consequently, the clustering algorithms utilize both the network connectivity structure and the weights to determine a clustering and may therefore be applied to produce a clustering of any general network (even when the nodes are not associated with data).

This chapter is organized to consist of three main sections that describe the targeted, untargeted, and semi-targeted classes of clustering algorithms, with subsections detailing the primal and dual versions of these algorithms. A smaller fourth section addresses the extension of these algorithms to the clustering of higher-order cells. We conclude with a section providing some example applications of the presented algorithms.

6.1 Targeted Clustering

Targeted segmentation algorithms require information to be input about the desired output object or set of objects to cluster. This prior information can take different forms, such as a partial labeling of the nodes or a probability, assigned to a set of nodes, of belonging to each label. In the field of image processing, a targeted segmentation algorithm might have the goal of segmenting a particular tumor in a medical image, or an incoming missile in a military application. Such an algorithm could also be used interactively to extract an object from a photograph for editing. In a World Wide Web application, the goal of a targeted clustering algorithm might be to extract a list of websites in the same cluster as a target website. An example in social networking would be to extract all members of the group in which a particular individual is a member. In the context of machine learning, targeted clustering algorithms are related to both *supervised learning* and *semi-supervised learning* algorithms, with the difference that these algorithms must generalize to all unseen data (nodes). However, in the context of machine learning, the targeted clustering algorithms described here may be viewed as examples of *transductive* learning algorithms that simply focus on labeling a known set of data points. Section 6.5.3 contains more information about this view of clustering as a machine learning algorithm.

We begin by addressing primal algorithms for targeted segmentation since primal algorithms comprise the majority of existing techniques for both targeted and untargeted methods. Additionally, primal algorithms are generally much more straightforward to describe and implement than dual algorithms.

6.1.1 Primal Targeted Clustering

The basic components of a targeted segmentation algorithm are a known label set comprised of a finite number of labels, an energy for which the extrema describe "good" clusters, and additional information about how the labels relate to the node set. Examples of additional information exploited in a primal algorithm include:

1. A method for assigning membership probabilities for each label to a subset of nodes.
2. Known labels for a subset of the nodes.
3. A set of edges which the boundary is known to cross.

We will address each of these types of information sequentially.

In all of the subsequent discussion on targeted primal clustering algorithms, we can formulate our goal as solving for a probability $x_{i,b}$ that node v_i belongs to label b. (Without loss of generality, we assume that each label is represented by an integer from $0 \le b < |\mathcal{L}|$.) Unless otherwise noted, the segmentation function $\sigma(v_i) = b$ is obtained via choosing the most likely label for each node, i.e.,

$$\sigma(v_i) = \underset{b}{\operatorname{argmax}}\{x_{i,b}\}. \tag{6.1}$$

Consequently, our focus will be on finding the membership probabilities $x_{i,b}$ for each node v_i and label b, since this set of membership probabilities defines the clustering via (6.1).

6.1.1.1 Probabilities Assigned to a Subset

Consider a set of nodes $\mathcal{V}_S \subseteq \mathcal{V}$, such that for each node $v_i \in \mathcal{V}_S$, we have a prior probability that v_i is assigned to label b, given by $\bar{s}_{i,b}$. We may now apply *any filtering technique* to produce a clustering (see Chap. 5 for a discussion on filtering on cell complexes). The approach for applying a filtering technique to compute a targeted clustering is to begin by extending the vector of priors $\bar{\mathbf{s}}$ beyond the subset $\mathcal{V}_S$ to priors $\mathbf{s}$ defined over all nodes in the complex, $\mathcal{V}$, via setting entries corresponding to the nodes not in the subset $\mathcal{V}_S$ to zero, i.e.,

$$s_{i,b} = \begin{cases} \bar{s}_{i,b} & \text{if } v_i \in \mathcal{V}_S, \\ 0 & \text{otherwise.} \end{cases} \tag{6.2}$$

Any filtering technique may then be applied to produce segmentation probabilities by treating the prior probabilities $\mathbf{s}$ as noisy data in a filtering technique. We saw several approaches for filtering noisy data in Chap. 5 which we now apply to find clustering probabilities $\mathbf{x}$. The first method that we apply from Chap. 5 is Taubin's method, in which we set the initial conditions $x_{i,b}^{[0]} = s_{i,b}$ and produce the final values of $x_{i,b}$ iteratively via Taubin's filtering. Specifically, to apply Taubin's filtering method (see Chap. 5) to find the $\mathbf{x}$ for each label, we employ the iteration rule

$$\mathbf{x}_b^{[2k+1]} = \mathbf{x}_b^{[2k]} - \lambda \mathbf{L}\mathbf{x}_b^{[2k]} = \mathbf{x}_b^{[2k]} - \lambda \mathbf{A}^\mathsf{T} \mathbf{A} \mathbf{x}_b^{[2k]}, \tag{6.3}$$

$$\mathbf{x}_b^{[2k+2]} = \mathbf{x}_b^{[2k+1]} + \mu \mathbf{L}\mathbf{x}_b^{[2k+1]}. \tag{6.4}$$

Similarly, we could minimize the Basic Energy Model from Chap. 5,

$$\mathcal{E}_{\text{BEM}}[\mathbf{x}_b] = \mathbf{1}^\mathsf{T} (\mathbf{G}^{-1})^p |\mathbf{A}\mathbf{x}_b|^p = \sum_{e_{ij}} w_{ij}^p |x_{i,b} - x_{j,b}|^p, \tag{6.5}$$

by applying an iterative mode ($p = 0$), median ($p = 1$), mean ($p = 2$) or minimax ($p = \infty$) filter to the initial probabilities. In other words, for each label b, the label probabilities $x_{i,b}$ (for node v_i) are filtered to produce final label probabilities and then each node is assigned the label for which is has the greatest probability (i.e., via (6.1)). For each label, the solution is initialized to $\mathbf{x}_b^{[0]} = \mathbf{s}_b$. Chapter 5 provides justification for why the mode, median, mean and minimax filters minimize the energy for the various values of p.

Figure 6.1 shows an example of how these filters may be applied to targeted image segmentation when the probabilities are assigned based on an intensity model of the target object. In this example, the image intensities inside the circle and outside the circle were both drawn from a uniform distribution with the same variance, but with a different mean inside and outside. In this case, the variance of the distributions was 2.5 times greater than the difference in mean between the distributions. The foreground priors were generated by assuming a Gaussian distribution for the intensities inside the circle with a mean equal to the minimum intensity in the image.

6.1 Targeted Clustering

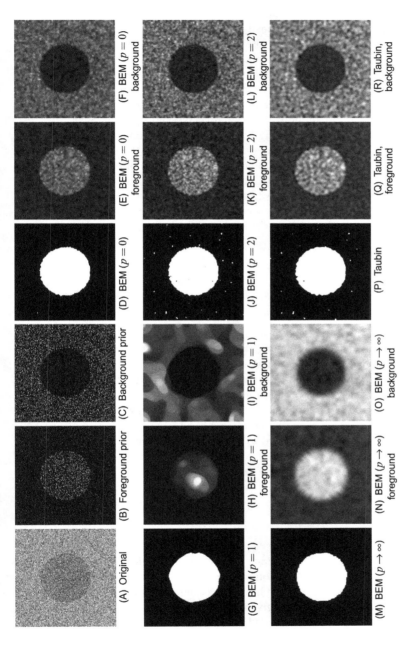

Fig. 6.1 Targeted image segmentation using the Basic Energy Model and Taubin spectral filtering method. Each algorithm was initialized with known foreground and background intensity priors (obtained by modeling the foreground and background as a Gaussian distribution with a mean equal to the minimum/maximum intensity values respectively). These priors were filtered with each algorithm to produce foreground and background probabilities (a few iterations of each minimization was used for the BEM models). Each node was then assigned to the label (foreground or background) for which it has a higher probability

With this Gaussian prior model, each pixel in the foreground prior image was assigned the probability that it was drawn from the Gaussian model. The background priors were produced in the same manner, except that the mean of the Gaussian was chosen to be equal to the highest intensity pixel in the image. By simply smoothing these foreground and background priors (using the filtering methods described in Chap. 5) we are able to obtain a smoothed "probability" that each pixel belongs to the foreground or background, which may then be used to generate the final label for each pixel by comparing the relative foreground and background "probabilities". In this manner, *any* filtering algorithm may be used as a clustering algorithm *if* it is possible to assign a prior likelihood that each node belongs to a particular label.

In Chap. 5 we noted that the danger with variational problems of the form in the Basic Energy Model is that there is a trivial optimum—the solution where $\mathbf{x}_b$ is constant. In practice we could simply take a few iterations of our mode, median, mean or minimax filtering of $\mathbf{x}_b$ (as was done to generate the results in Fig. 6.1). An alternative to this iterative approach is to consider the Extended Basic Energy Model in which the prior information is represented by a second term that the solution must balance. In the context of clustering, the Extended Basic Energy Model is given by

$$\mathcal{E}_{\text{EBEM}}[\mathbf{x}_b] = \mathbf{1}^\mathsf{T}(\mathbf{G}^{-1})^p |\mathbf{A}\mathbf{x}_b|^p + \lambda \sum_a \mathbf{s}_a^\mathsf{T} |\mathbf{x}_b - \delta(a,b)|^p$$

$$= \sum_{e_{ij}} w_{ij}^p |x_{i,b} - x_{j,b}|^p + \lambda \sum_{v_i} \sum_a s_{i,a} |x_{i,b} - \delta(a,b)|^p, \quad (6.6)$$

where $\delta(a,b)$ is a Kronecker delta function. It was shown in Chap. 5 that the strength of the regularization parameter λ is inversely related to the number of iterations used for smoothing in the Basic Energy Model.

We may continue with the application of filtering methods to the clustering problem by considering the Total Variation model from Chap. 5. One benefit in image processing of the Total Variation model over the Basic Energy Model is that Total Variation has less tendency to exhibit gridding artifacts. In the context of clustering, the Total Variation Model is described by the energy functional

$$\mathcal{E}_{\text{TV}}[\mathbf{x}_b] = \mathbf{1}^\mathsf{T}\bigl(|\mathbf{A}^\mathsf{T}||\mathbf{A}\mathbf{x}_b|^2\bigr)^{\frac{p}{2}} + \lambda \sum_a \mathbf{s}_a^\mathsf{T} |\mathbf{x}_b - \delta(a,b)|^p$$

$$= \sum_{v_{i,b}} \left(\sum_{e_{ij}} |x_{i,b} - x_{j,b}|^2\right)^{\frac{p}{2}} + \lambda \sum_{v_i} \sum_a s_{i,a} |x_{i,b} - \delta(a,b)|^p. \quad (6.7)$$

As before, this model is used to generate a clustering by finding a solution for each label b and then comparing these solutions to produce a final labeling by assigning each node, v_i, to the label for which it has the maximum solution. More information on the optimization of this model is given in Chap. 5.

All of these models may be included to incorporate implicit boundaries via an edge weighting. These models are written with edge weights in Chap. 5 and we do not repeat this material here. The same weighting functions that were used to weight edges for filtering applications may also be applied for clustering.

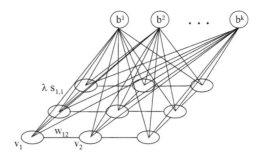

Fig. 6.2 Mathematically, the use of label priors is equivalent to using k labeled, "phantom" nodes that correspond to each label and are connected to each node. Despite the abuse of notation, note the labels b are distinguished with superscript index indicating the different labels

6.1.1.2 Known Labels for a Subset of Nodes

Many targeted clustering applications permit some method of assigning to each node a prior probability quantifying the likelihood that the node belongs to each label. In these situations, the targeted segmentation problem may be solved via any of the filtering methods reviewed in the previous section. However, another common form of targeting information is to know the labeling for a small subset of nodes. The nodes with a known label within this subset are often called **seeds** in the image processing literature. Seeded nodes can also take advantage of the filtering approaches to clustering discussed above by simply assigning the x_i values for each seed node and performing the optimization with respect to this assignment. One major advantage of targeting with seeded nodes is that a global, nontrivial optimum may be found for each of the variational models (i.e., the Extended Basic Energy Model or the Total Variation Model) even if $\lambda = 0$. That is, the initial assignment of seed notes comprises a constraint that the solution is explicitly forced to satisfy. Therefore, if prior probabilities are not available or not reliable, they can be ignored by setting $\lambda = 0$. In this way, each model can be used completely parameter-free when seeds are available.

Formally, define the set of seeded or "marked" nodes $\mathcal{V}_M \subset \mathcal{V}$ for which $\sigma(\mathcal{V}_M)$ is known. These labels are assumed to have been obtained via a different process such as user interaction or an automatic seeding. Using this information, we can fix $x_{i,b}$, $\forall v_i \in \mathcal{V}_M$ via

$$x_{i,b} = \begin{cases} 1 & \text{if } \sigma(v_i) = b, \\ 0 & \text{if } \sigma(v_i) \neq b. \end{cases} \quad (6.8)$$

Fixing these values as Dirichlet boundary conditions on the set of seeded nodes allows for optimization of the above models to produce a nontrivial solution for $\mathbf{x}_b$ if, for each label b, $\sigma(v_i) = b$ for some seed node $v_i \in \mathcal{V}_M$ (i.e., each label is associated with at least one seed). These seeds could also be used with Taubin's method applied to segmentation by initially fixing the values in $\mathcal{V}_M$ and not updating the corresponding values of $\mathbf{x}_b$ during the iterations. See Appendix B for more information on optimization in the presence of Dirichlet boundary conditions.

Although we distinguish between the prior probability method and seeded method for producing a targeted clustering, it is possible to view the prior probability method as equivalent to the seeded method. Specifically, the prior likelihood term in the Extended Basic Energy Model or the Total Variation Model may also be obtained via seeding. The "seed" in this case is a "phantom seed" representing each label, b, which is attached to node v_i with weight $\lambda s_{i,b}$. This interpretation of the prior term is represented in Fig. 6.2 and has previously appeared in the literature [53, 159]. Since the method of specifying a targeted clustering via prior terms is equivalent to seeding (with the phantom seeds construction), we may henceforth treat only the case of the seeded model, since it is understood that all of the results apply also to the incorporation of prior terms. Consequently, in the context of discussing seeded targeted clustering algorithms, we will employ the term "Basic Energy Model" to refer either to the "Basic Energy Model" or "Extended Basic Energy Model" since the additional term may be viewed simply as another form of seeding.

Incarnations of the seeded Basic Energy Model with various values of p have been heavily utilized in the literature. We review several targeted clustering algorithms that can be interpreted as special cases of the Basic Energy Model, along with one algorithm that can be interpreted as a seeded Total Variation Model.

Max-flow/Min-cut

It was shown by Sinop and Grady [350] that when $p = 1$ the solution $\mathbf{x}_b$ given by the Basic Energy Model (6.6) may be found by computing a max-flow/min-cut between the seeds (both real and "phantom") labeled '1' and the seeds labeled '0'. Consequently, the use of max-flow/min-cut to perform targeted node clustering may be viewed as an instance of employing the Basic Energy Model for targeted clustering with norm $p = 1$. Since max-flow/min-cut may be used to optimize the Basic Energy Model when $p = 1$, this energy model may be interpreted as minimizing the boundary length between the labeled regions. In this case, the seeds are identified with the source/sink terminals in the traditional max-flow/min-cut problem.

Max-flow/min-cut techniques have a long history in clustering problems [246, 407, 415]. Appendix B and Refs. [55, 56, 173] contain more details on this optimization. In the context of image segmentation, this seeded max-flow/min-cut model has been known as "graph cuts" [53].

Random Walker

Taking $p = 2$ in the Basic Energy Model (6.6) allows the solution at node v_i, $x_{i,b}$, to be interpreted as the probability that a random walker leaving node v_i arrives at a seed (real or "phantom") labeled b before arriving at a seed not labeled b [161] (see Chap. 3). Consequently, the clustering algorithm that employs $p = 2$ is known as the "random walker" algorithm in the image segmentation literature. Alternately, the clustering obtained by this model is equivalent to the clustering obtained by computing the effective resistance between each node and the seeds of each label (viewed as a single node) and assigning the node to the label having the smallest effective resistance [161] (for more information about effective resistance, see Chap. 3). The

6.1 Targeted Clustering

Table 6.1 Targeted clustering using the Basic Energy Model with seeds. Different values of the norm parameter p give different interpretations to the model when applied to clustering

Choice of p	1	2	∞
name	Max-flow (Graph cut)	Random walk	Geodesic
objective function	$\sum_{e_{ij} \in \mathcal{E}} w_{ij}\|x_i - x_j\|$	$\sum_{e_{ij} \in \mathcal{E}} w_{ij}^2\|x_i - x_j\|^2$	$\max_{e_{ij} \in \mathcal{E}} w_{ij}\|x_i - x_j\|$
objective function interpretation	boundary cut	effective conductance	minimum Lipschitz extension
optimization method	maximum flow	solution of a sparse linear system	shortest path
uniqueness	not unique	unique	not unique

earliest adoption of this model for clustering may have been Kodres [237] who used it to determine how to design a circuit layout. This clustering algorithm has also been applied to machine learning [424] and extended to directed graphs [349].

Geodesic Segmentation

When $p = \infty$, it was shown in [350] that the problem can be recognized as a discrete formulation of the minimal Lipschitz extension [13]. Additionally, it was shown in [348] that a minimum of the Basic Energy Model (6.6) may be given by

$$x_{i,b} = d_{i,b}/(d_{i,b} + d_{i,\bar{b}}), \tag{6.9}$$

where $d_{i,b}$ is used to indicate the weighted length of the shortest path from v_i to any seed labeled b and $d_{i,\bar{b}}$ is used to indicate the weighted length of the shortest path from v_i to any seed not labeled b. This clustering approach is often called **geodesic segmentation** in the image segmentation literature. Using shortest paths in this way for image segmentation have been popularized by several groups [8, 18, 94, 128] (albeit without explicit reference to the energy minimization interpretation).

P-Brush

These three choices of the norm parameter p in the Basic Energy Model were compared by Sinop and Grady [350] to find that a smaller value of p produces a clustering that has less dependence on the location of the seeds, but the clustering obtained by using a larger value of p has less dependence on the number of seeds. Table 6.1 compares the clustering algorithms generated by various values of p. The use of the Basic Energy Model (6.6) for clustering with fractional values of p was recently examined in [106], which found that the clustering algorithms obtained by minimizing fractional values of p effectively interpolate between the clustering algorithms corresponding to integer values of p. Additionally, the solutions given by the Basic Energy Model are pointwise continuous with respect to changes in p. This clustering algorithm with fractional values of the norm parameter p was called **P-Brush**.

Power Watershed

The classical watershed algorithm from mathematical morphology [320, 337] may also be seen as an instance of the Basic Energy Model with only a slight modification. Couprie et al. [90] modified the Basic Energy Model from (6.6) to produce

$$\mathcal{E}_{\text{PW}}[\mathbf{x}_b] = \mathbf{1}^\mathsf{T}(\mathbf{G}^{-1})^q |\mathbf{A}\mathbf{x}_b|^p = \sum_{e_{ij}} w_{ij}^q |x_{i,b} - x_{j,b}|^p. \tag{6.10}$$

Once again, we assume that some nodes are seeded (alternately, a prior term is included as in the Extended Basic Energy Model). The only difference between (6.10) and the Basic Energy Model in (6.6) is the exponent, q, on the edge weights. It was shown by Allène et al. [4] that the $p = 1$ model above (max-flow/min-cut) is also optimized by a watershed computation for a value of q above some constant. Therefore, as $q \to \infty$, the model in (6.10) becomes the watershed clustering algorithm when $p = 1$. Viewed differently, Allène et al. [4] showed that as the power of the weights increases to infinity, then the max-flow/min-cut algorithm produces a clustering corresponding to the maximum spanning forest (MSF) used in a watershed computation [4]. Interpreted from the standpoint of the Welsch weighting function in Chap. 4, it is clear that we may associate $q = \frac{1}{\alpha}$ to understand that the watershed equivalence arises when the weighting function is employed using a particular range of parameter values. An important insight from this connection is that *when the value of α is sufficiently small, we can replace the expensive max-flow computation with an efficient maximum spanning forest computation.*

Algorithm 6.1 Power Watershed algorithm, optimizing $q \to \infty$, $p \geq 1$

Data: A weighted graph $\mathcal{G}(\mathcal{V}, \mathcal{E})$ and a subset of foreground seeds $\mathcal{V}_{\text{FG}}$ and background seeds $\mathcal{V}_{\text{BG}}$
Result: A solution $\mathbf{x}$
Set $\mathbf{x}_{\text{FG}} = 1$, $\mathbf{x}_{\text{BG}} = 0$ and all other $\mathbf{x}$ values as unknown, mark all edges as unprocessed.
Sort the edges of $\mathcal{E}$ by decreasing order of weight.
while *any node has an unknown potential* **do**
  Find an edge (or a plateau) E_{MAX} in $\mathcal{E}$ which is both of maximal weight and
  safe; denote by $\mathcal{S}$ the set of nodes connected by E_{MAX}.
  if $\mathcal{S}$ *contains any nodes with known potential* **then**
    Find $\mathbf{x}_\mathcal{S}$ minimizing (6.10) (using the input value of p) on the subset $\mathcal{S}$ with
    the weights in E_{MAX} set to $w_{ij} = 1$, all other weights set to $w_{ij} = 0$ and the
    known values of $\mathbf{x}$ within $\mathcal{S}$ fixed to their known values. Consider all $\mathbf{x}_\mathcal{S}$
    values produced by this operation as known.
  else
    Merge all of the nodes in $\mathcal{S}$ into a single node, such that when the value of
    $\mathbf{x}$ for this merged node becomes known, all merged nodes are assigned the
    same value of $\mathbf{x}$ and considered known.

6.1 Targeted Clustering

Table 6.2 The targeted clustering algorithms obtained by minimizing the Power Watershed model in (6.10) for various choices of p and q

p	q			
	0	1	2	∞
1	collapse to seeds	Max-flow	Max-flow	Watershed
2	ℓ_2-norm Voronoï	Random walker	Random walker	Power watershed, $p = 2$
∞	ℓ_1-norm Voronoï	Geodesic	Geodesic	Power watershed, $p = \infty$

Couprie et al. [90] went on to explore the clustering model in (6.10) when $q \to \infty$ for any value of p. Since this family of watersheds was characterized by the exponent p, they termed this clustering algorithm the **power watershed**. Algorithm 6.1 gives an algorithm for finding the solution to $\mathbf{x}_b$ that optimizes the Power Watershed model in (6.10). In Algorithm 6.1, if $\mathcal{E}_{\text{MSF}}$ is a set of edges forming a subset of an MSF, then an edge e_i is considered *safe* if $\mathcal{E}_{\text{MSF}} \cup e_i$ is also a subset of an MSF. Note that this algorithm applies only to two labels. In a multilabel targeted clustering problem, the potential function $\mathbf{x}_b$ for each target label b would be set to the value '1' for all nodes assigned to the "foreground" ('1') or to the value '0' for all nodes set to the "background", and the final labeling for node v_i assigned by choosing the label with the largest value (as expressed in (6.1)). Table 6.2 gives a description of the segmentation algorithms obtained by minimizing the Power Watershed energy for various pairs of values for p and q.

Continuous Max-flow

The seeded Total Variation Model has also been applied in the context of image processing. When $p = 1$, this model is equivalent to the **continuous max-flow** formulation of [358] that was subsequently applied to image segmentation [11]. Additionally, due to the interpretation of (6.7) as a minimization of total variation, the minimization of (6.7) has also been applied to image segmentation under the name "total variation segmentation" or "TVSeg" [386]. Fast algorithms for this minimization are given in [11] and [68, 98, 385]. When $p = 2$, the Basic Energy Model and the Total Variation Model are equivalent (i.e., the "random walker" algorithm). To the knowledge of the authors, the cases of $p = \infty$ or the separation of the exponent onto the weights (yielding a power watershed-like algorithm) have not been explored in conjunction with the Total Variation Model for targeted clustering.

6.1.1.3 Negative Weights

A less direct method for specifying a clustering target is to assign *negative weights* to some of the edges which are known to lie between clusters. Negative weights are generally sufficient without seeds or prior probabilities to produce a nontrivial solution of any of the above models, since the value of the objective function can dip below zero. Negative weights can be used to encode *repulsion* between two nodes, causing the difference between the values of their membership probabilities $\mathbf{x}_b$ values to grow [413, 414]. However, negative weights can cause difficulties in optimization since the objective function may become unbounded, yielding no useful solution. To avoid these unbounded scenarios, restrictions on $\mathbf{x}_b$ must be imposed, such as requiring that $0 \leq \mathbf{x}_b \leq 1$ [257]. Optimization of the Basic Energy Model, the Power Watershed Model or the Total Variation Model are substantially more difficult problems to solve with negative weights, with limited benefit demonstrated so far. Consequently, there has been less attention devoted to employing repulsion (via negative weights) for specifying a clustering target.

We have now covered the most prominent variational models applied to primal image segmentation. All of these targeted clustering algorithms were derived from filtering algorithms applied to prior knowledge (likelihood priors and/or seeds) of the segmentation labels. In the next section, we discuss targeted segmentation via the *dual* complex. Adopting a dual viewpoint departs from the previous discussion on filtering.

6.1.2 Dual Targeted Clustering

Instead of clustering by labeling nodes on the primal graph, clustering on the dual graph seeks to find the set of edges defining the *boundary* of the labeled regions. Finding this set of edges has two disadvantages over the primal algorithms we have studied.

1. Cycles in the dual complex map to cutsets in the primal complex for only a limited set of complexes (e.g., planar graphs in two dimensions).
2. The dual complex *depends on the dimensionality of the embedding*, and therefore these algorithms may require modification if the embedding changes. For example, embedding in $\mathbb{R}^2$ can be substantially different than embedding in $\mathbb{R}^3$, since edges are dual to edges in two dimensions and edges are dual to faces in three dimensions.

Nonetheless, there are advantages of using a dual algorithm. Some dual algorithms are faster, and sometimes the inputs to the targeted segmentation algorithm are easier to specify in a dual lattice (i.e., boundary locations). Additionally, some objective functions are easier to express as functions of the dual elements (e.g., edges, faces) rather than the primal elements (e.g., nodes). A computational advantage of dual algorithms is that the number of boundary edges is typically quite small compared to the number of internal labeled nodes. Consequently, the boundary may be represented more efficiently, and perturbations of the boundary from an initialization may be accomplished efficiently. Due to the natural embedding of images (as two-dimensional or three-dimensional lattices), most of the work on dual clustering algorithms has appeared in this literature (in which boundary-focused algorithms are sometimes called a "boundary parameterization" or a "Lagrangian representation" [339]).

6.1 Targeted Clustering

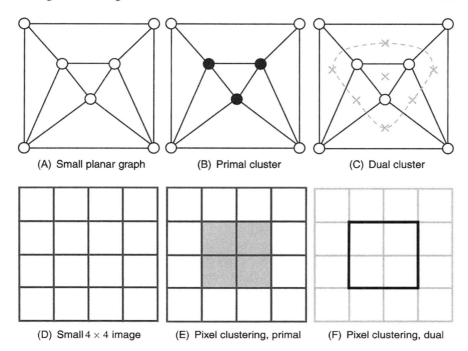

Fig. 6.3 A comparison of primal clustering with dual clustering. (**A–C**) A small planar graph in which the primal clustering is represented as a binary labeling of each node (indicating the cluster membership of each node) and the dual representation of the clustering is as a closed boundary of edges which separate the inside of the cluster from the outside. (**D–F**) A set of pixels from a small 4 × 4 image. A primal clustering of these pixels is represented by assigning a binary labeling to each pixel to indicate cluster membership. In contrast, the dual representation of the clustering is obtained by identifying the edges *between* the pixels inside the cluster and those pixels outside the cluster

Figure 6.3 contrasts primal clustering with dual clustering on a small planar graph. A primal clustering algorithm assigns a *label* to each node in the primal graph, while a dual clustering algorithm identifies boundary edges in the dual graph. The first row of the figure gives an example of a small planar graph in which the primal clustering is represented as a binary labeling of each node (indicating the cluster membership of each node) while the dual representation of the clustering is as a closed boundary of edges which separate the inside of the cluster from the outside. The second row of the example shows a set of pixels from a small 4 × 4 image. A primal clustering of these pixels is represented by assigning a binary labeling to each pixel to indicate cluster membership. In contrast, the dual representation of the clustering is obtained by identifying the edges *between* the pixels inside the cluster and those pixels outside the cluster. In the image processing literature, these dual edges between the pixels have sometimes been called "cracks" or "bels" (for "boundary elements") [129].

A common algorithm for targeted clustering on the dual graph in two dimensions is known as intelligent scissors or live wire [129, 287]. This algorithm interactively builds an open contour on the dual graph up to the last step, at which time the contour is closed to enclose a region of nodes. In the intelligent scissors algorithm, a series of nodes in the dual graph are sequentially input into the algorithm in pairs (e.g., interactively) and a shortest path is computed between the pairs of nodes. Formally, assume that we have computed appropriate edge weights between nodes on the primal graph (in the same manner as in the primal algorithms, see Chap. 4). Define an indicator vector $\mathbf{y}^s$ of edges representing a line segment (chain of dual edges) s of the cluster boundary consisting of the set of dual edges such that

$$y_i^s = \begin{cases} 1 & \text{if dual edge } e_i \text{ is a member of the boundary,} \\ 0 & \text{otherwise.} \end{cases} \quad (6.11)$$

Therefore, the edges indicated by $\mathbf{y}$ serve to "surround" the target cluster of nodes in the primal graph. Given two dual nodes v_i and v_j that are known to lie on the desired boundary (produced either interactively by a user or automatically), the intelligent scissors/live wire algorithm finds a solution to

$$\min_{\mathbf{y}^s} \mathbf{w}^T \mathbf{y}^s, \quad (6.12)$$

$$\text{s.t.} \quad \mathbf{A}^T \mathbf{y}^s = \mathbf{p},$$

where $\mathbf{w}$ is the vector of dual distance edge weights (equal to the primal affinity edge weights in the two-dimensional case, since in two dimensions edges are dual to edges—see Chap. 2 for more details) and the boundary vector $\mathbf{p}$ is defined as

$$p_k = \begin{cases} +1 & \text{if } k = i, \\ -1 & \text{if } k = j, \\ 0 & \text{otherwise.} \end{cases} \quad (6.13)$$

The optimization in (6.12) may be performed quite efficiently using Dijkstra's shortest path algorithm. The constraints in (6.12) demonstrate the use of the incidence matrix $\mathbf{A}^T$ as the boundary operator, since we may interpret the constraint that $\mathbf{A}^T \mathbf{y}^s = \mathbf{p}$ as the requirement that the set of edges represented by $\mathbf{y}^s$ has endpoints given by v_i and v_j. After $\mathbf{y}^s$ has been computed, a new dual node is input to the intelligent scissors/live wire algorithm to define a new $\mathbf{p}$ vector. A series of $\mathbf{y}^s$ line segments are computed using a sequential set of points which are then combined to form the final output $\mathbf{y} = \sum_s \mathbf{y}^s$. An important implementation detail in the intelligent scissors/live wire algorithm is that because $\mathbf{y}$ is strictly binary valued, the orientation of edge traversal must be encoded in the incidence matrix. Consequently, the incidence matrix must be modified to represent each edge twice with opposite orientation (see Chap. 4 for more details on this over-representation). In practice, the use of Dijkstra's algorithm obviates these details—Dijkstra's algorithm implicitly solves (6.12). Therefore, all that is necessary for a practical implementation is to use Dijkstra's algorithm to compute the shortest path between each successive series of node pairs on the boundary until the boundary is closed. A second implementation detail is that the edges connecting the dual nodes to the "outside face" (e.g., the border of the image) must be assigned some weight to allow for the contour to close around a set of nodes on the border. One approach to weighting the edges on the outside face is to assign these edges a weight equal to the average weight inside the complex.

A very recent clustering algorithm that employs a dual formulation is the algorithm of Schoenemann et al. [334] which uses a dual formulation as a way of encoding the *curvature* of the cluster boundary. The dual formulation is employed because the discretization of curvature employed by Schoenemann et al. was based on the work of Bruckstein et al. [61], who showed how angles between successive line segments of a polygon could be used to approximate curvature of the polygon. By associating the dual edges with polygonal line segments, Schoenemann et al. [334] produced an optimization method that optimized the cluster boundary to have small curvature.

6.1.2.1 Dual Algorithms in Three Dimensions

Since the dual complex changes with embedding and dimensionality, any dual algorithm must be altered to accommodate these changes. In order to apply the shortest-path targeted clustering algorithm to a 3-complex, we now transition from minimal paths to minimal surfaces. Unless otherwise stated, we will assume for simplicity that our 3-complex is a three-dimensional, 6-connected lattice. Fortunately, the dimensionality of the minimal path problem may be increased simply by using the dimension-appropriate incidence matrix (acting as the boundary operator) and boundary vector **p**. This dimension-increased shortest path problem is therefore stated as: *Given the boundary of a two-dimensional surface (i.e., a closed contour or series of closed contours), find the minimal two-dimensional surface with the prescribed boundary.* We note that this problem may be considered as a discrete instance of Plateau's problem [363].

In this three-dimensional problem, the boundary operator is the edge–face incidence matrix defined in Chap. 2. Instead of the lower-dimension boundary vector, **p**, we can now employ the vector **r** as a signed indicator vector of a closed contour with an associated ordering of vertices obtained via a traversal along the edges comprising the contour. Given a contour represented by an ordering of vertices $(v_a, v_b, v_c, \ldots, v_a)$ such that each neighboring pair of vertices is contained in the edge set, the contour may be represented with the vector

$$r_i = \begin{cases} +1 & \text{if the vertices comprising edge } e_i \text{ are contained in the contour} \\ & \text{with coherent orientation,} \\ -1 & \text{if the vertices comprising edge } e_i \text{ are contained in the contour} \\ & \text{without coherent orientation,} \\ 0 & \text{otherwise.} \end{cases} \quad (6.14)$$

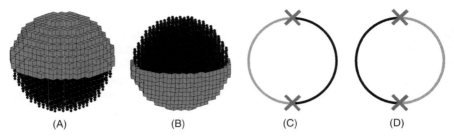

Fig. 6.4 The discrete minimal surface given a boundary may not be unique. For example, if the surface boundary is given as a closed contour at the equator of a sphere, then either the upper (**A**) or lower (**B**) hemisphere is a valid minimum solution. This same lack of uniqueness may also appear in the shortest path problem. Analogously, if two endpoints were placed at antipodal points of a circle, the shortest path may be returned as either the left (**C**) or right (**D**) path around the circumference of the circle. However, in applications with real data, both the shortest path and minimal surface are typically unique for a given input

Therefore, the discrete minimal surface problem is

$$\min_{\mathbf{z}} \mathcal{Q}[\mathbf{z}] = \sum_i w_i z_i,$$

subject to $\quad \mathbf{Bz} = \mathbf{r},$

(6.15)

where **z** is an indicator vector indicating whether or not a face (in the dual complex) is in the minimal surface and w_i is used to indicate the weights of a face. The surface represented by the solution to (6.15), **z**, is a (discrete) minimal surface. Since the faces in the dual lattice correspond to edges in the primal lattice (where the image data is represented), any of the weighting functions in Chap. 4 may be used to produce the set of face weights.

The discrete minimal surface problem was extensively treated by Sullivan [364], who showed that a fast algorithm exists for its optimization. However, the equality constraints in (6.15) are pre-unimodular matrix (see Appendix B and [162]). Therefore, a generic linear programming solver could be used to produce an optimal integer solution of (6.15) even though linear programming uses a real-valued relaxation of the variable **z**. For more details on this optimization, see Refs. [162, 364].

The solution to the discrete minimal surface problem may not be unique. However, the shortest path problem may also have a solution which is not unique. Figure 6.4 illustrates this issue. For example, a closed contour located precisely at the equator of the sphere in Fig. 6.5(A) could result in a solution indicating either the upper or the lower hemisphere. This is analogous to the one-dimensional case where multiple solutions exist that give shortest paths between two antipodal points on a circle.

Figure 6.5 gives three examples drawn from three-dimensional image segmentation to illustrate the properties of this discrete minimal surface algorithm. Firstly, we use the algorithm to find the surface of a black sphere in a white background,

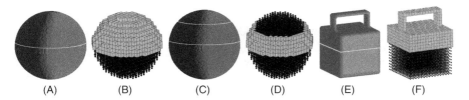

Fig. 6.5 Minimal surface segmentation of synthetic three-dimensional images. Renderings of the original object (with the input contours) are shown, along with the solutions. The input volumes consisted of black voxels indicating voxels belonging to the object and white voxels indicating background. Black voxels are represented in the figure by *small black spheres*. The *white stripe* in each of the rendered views shows the input contour(s). In the solution visualizations, black dots are plotted at the center of the black (object) voxels and faces are shown to indicate the computed surface. (**A**, **B**) A sphere with an input contour along a parallel. Note that, unlike shortest paths (which require two endpoints), a single boundary contour input is sufficient to define a solution. (**C**, **D**) A sphere with two input contours at parallels of different heights. (**E**, **F**) A lunchbox shape with a handle on the top and a contour input around the middle of the object. The algorithm will correctly find minimal surfaces with topological changes

given an initial contour around one parallel. Secondly, we find the surface of the same sphere using a boundary consisting of contours around two parallels (i.e., a contour given on two slices). Finally, we segment a "lunch box" shape given a contour around the middle of the object. This experiment shows that the algorithm correctly handles changes in object genus without any special handling. In contrast to the shortest path problem in which two points are necessary to define a path, Fig. 6.5(A) shows that a single closed contour is sufficient to define the boundary of a surface. The applications in Sect. 6.5.1 illustrate the use of this segmentation algorithm on real three-dimensional image data.

6.2 Untargeted Clustering

Untargeted clustering is the traditional clustering problem in which the goal is to divide a graph into a hierarchy of clusters where the number of clusters is unknown. This problem is not well defined since there is no agreed-upon criteria for determining what constitutes a "good" cluster. Even for those algorithms with a well-defined, seemingly simple criteria for a good clustering, the problem is generally NP-Hard (e.g., k-means is NP-Hard [5, 269]). Consequently, nearly every untargeted clustering algorithm defines its own meaning of what constitutes a good clustering, and then supplies a heuristic that produces a suboptimal solution to the stated objective.

Untargeted clustering has a huge range of uses, including data discovery, compression [225, 332], parallelization [392], image processing [216, 303, 382], the efficient solution of PDEs (via domain decomposition) [351, 381], sparse matrix ordering (nested dissection) [229, 281] and identification of neural substructures [41, 190]. The collection of techniques for untargeted clustering is far too vast to provide a comprehensive review here (for a recent review on graph clustering algorithms see

Ref. [330]). Instead, we present only a few techniques that fit well into the theme of this book. In this section, we mainly address *primal* untargeted segmentation algorithms (which comprise almost all untargeted clustering algorithms), although dual untargeted algorithms are treated briefly at the end of this section.

We begin by noting that any of the *targeted* clustering algorithms can be used to produce an untargeted clustering algorithm. Specifically, a targeted clustering algorithm may be converted to an untargeted clustering algorithm by the following steps.

1. Select the number of clusters K.
2. Randomly select K nodes and assign these nodes each a different label.
3. Apply the targeted segmentation algorithm with these seeds.
4. Move the seed location to the "center" of each labeled cluster.
5. Continue applying the targeted segmentation and moving the seed locations until all of the seed locations no longer move.

Essentially, this procedure is an adaptation of Lloyd's algorithm for k-means clustering [263, 264] to any of the targeted clustering methods. Note that convergence of this procedure may not be guaranteed, and therefore in practice it may be beneficial to limit the number of iterations.

One common approach for designing untargeted clustering algorithms is to define an algorithm that partitions the graph into two clusters and then recursively applies the partitioning on each cluster until some measure of partition quality is met and the recursion exits. Although there exist good reasons for theoretical and practical concerns with such a recursive approach (see Ref. [347]), the recursion approaches have the advantage of not requiring prior knowledge of the total number of clusters and they do work reasonably well in practice. We will now review a standard approach for dividing a graph into two clusters with the understanding that this partitioning may be applied recursively to obtain multiple clusters.

6.2.1 Primal Untargeted Clustering

In all of the subsequent discussion on primal bipartitioning algorithms, we can formulate our goal as solving for a 0-cochain **x**, with coefficients $x_i \in \mathbb{R}$, that determine whether node v_i belongs to label '0' or '1'. The segmentation function is defined here as $\sigma(v_i) \mapsto \{0, 1\}$ obtained by thresholding the cochain **x**, i.e.,

$$\sigma(v_i) = \begin{cases} 1, & \text{if } x_i \geq \theta, \\ 0, & \text{otherwise,} \end{cases} \quad (6.16)$$

where the threshold θ is set manually, automatically or in some application-dependent manner (see Ref. [352] for some standard possibilities for choosing a threshold θ). Our initial focus will be on the production of the inclusion cochain **x** for each node before continuing to a discussion of how to set the threshold θ.

6.2 Untargeted Clustering

If we reconsider the Basic Energy Model (6.6) in the context of untargeted bipartitioning, we see that the absence of the targeting information causes the energy to have a trivial minimum at $\mathbf{x} = k$ for some constant k. One approach for addressing this trivial minimum is to add an extra term that attempts to expand the cluster. This additional force is sometimes known in the image processing literature as a *balloon force* [83, 84]. Specifically, consider the Untargeted Basic Energy Model

$$\begin{aligned}\mathcal{E}_{\text{UBEM}}[\mathbf{x}] &= \mathbf{G}_1^{-1}|\mathbf{Ax}|^p - \lambda \mathbf{G}_0^{-1}|\mathbf{x}|^q \\ &= \sum_{e_{ij}} w_{ij}|x_i - x_j|^p - \lambda \sum_{v_i} w_i |x_i|^q,\end{aligned} \quad (6.17)$$

for some $\lambda > 0$. This energy can be made arbitrarily low by setting $\mathbf{x} = k$ for some constant value k. A variety of strategies for avoiding this problem may be adopted. For each combination of p and q, we discuss how this problem has been overcome in the past. The gradient of $\mathcal{E}_{\text{UBEM}}[\mathbf{x}]$ equals zero when $\mathbf{x}$ satisfies

$$p\mathbf{A}^T \mathbf{G}_1^{-1} |\mathbf{Ax}|^{p-1} = \lambda q \mathbf{G}_0^{-1} |\mathbf{x}|^{q-1}. \quad (6.18)$$

Although several combinations of p and q values have not been investigated in the literature, there are some cases which lead to algorithms that have had an important impact on the automatic clustering community. Optimization of the Untargeted Basic Energy Model produces a solution which may be thresholded to provide a bipartition into two clusters. The standard approach of using the Untargeted Basic Energy Model to produce more than two clusters is to recursively bipartition these clusters until some measure of the partition quality (usually the value of $\mathcal{E}_{\text{UBEM}}[\mathbf{x}]$) fails to be satisfied.

We begin by examining the Untargeted Basic Energy Model for $p = 2$ and $q = 2$. In this case, then the minimum taken in (6.18) is given by

$$\mathbf{A}^T \mathbf{G}_1^{-1} \mathbf{Ax} = \mathbf{Lx} = \lambda \mathbf{G}_0^{-1} \mathbf{x}. \quad (6.19)$$

When $\mathbf{G}_0^{-1} = \mathbf{I}$ then the solution $\mathbf{x}$ is an eigenvector of $\mathbf{L}$. As an eigenvector problem, we can view (6.19) as a minimization of the Rayleigh quotient

$$\lambda = \frac{\mathbf{x}^T \mathbf{L} \mathbf{x}}{\mathbf{x}^T \mathbf{G}_0^{-1} \mathbf{x}}. \quad (6.20)$$

The first eigenvector of $\mathbf{L}$ is $\mathbf{x} = k$ for some constant k. However, since $\mathbf{L}$ is symmetric, the eigenvectors will be orthogonal. Therefore, by taking $\mathbf{x}$ to be the eigenvector corresponding to the second smallest eigenvalue we have an optimum to the Untargeted Basic Energy Model (6.17) when optimized in the space orthogonal to $\mathbf{x} = k$. Consequently the problem of an unbounded solution for the Untargeted Basic Energy Model in (6.17) is resolved by adopting this eigenvector because the optimization has been effectively performed in the space orthogonal to the problematic (constant) solution. The second smallest eigenvalue is often called the **Fiedler value** (although Fiedler originally called it the "algebraic connectivity" [133]) and the corresponding eigenvector is known as the **Fiedler vector**. There are several reasons to support the Fiedler vector as a method for graph clustering: (i) A smaller value of λ represents less perturbation from the targeted model, allowing the smoothness term to have the greatest effect; (ii) It was shown by Fiedler [134, 135] that labeling nodes above a threshold as foreground and below the threshold as background guarantees that both foreground and background components are connected [134, 135]; and (iii) The Fiedler value can be used to bound the isoperimetric constant of a graph [81]. Automatic clustering by thresholding the Fiedler vector is called **spectral clustering**, which has been rediscovered several times in the clustering literature [114, 184, 310]. We can view spectral clustering from the continuous calculus perspective as an instance of the *Helmholtz equation*, e.g.,

$$\nabla^2 x = \lambda x. \tag{6.21}$$

Recall that the Helmholtz equation describes the harmonic frequencies of an ideal membrane. Consequently, we can interpret the solution to (6.19) as the lowest frequency nontrivial harmonic which is then thresholded to produce a partitioning. Figure 6.6 shows two examples of graphs which are mapped to heights that correspond to the values of the lowest frequency eigenvector (i.e., the Fiedler vector). The first example shows an elongated lattice for which the harmonic reflects over the principle axis of symmetry. The second example shows a more separated graph for which each cluster isolates well as one component of the harmonic. An additional geometric interpretation of the spectral clustering approach is as a mapping of the nodes to the real line in order to both locate each node (on the real line) at the average location of its neighbors as well as maintain a unit average distance between all pairs of nodes [14].

We may again consider the spectral clustering algorithm (6.19) in the case that $\mathbf{G}_0^{-1} = \mathbf{D} = \text{diag}(\mathbf{d})$ where $\mathbf{d}$ is the vector of node degrees. In this case, the solution is the *generalized eigenvector* of $\mathbf{L}$ and $\mathbf{D}$. This variant of the spectral clustering algorithm (6.19) first appeared in the image processing literature as **Normalized Cuts** [345] in which the authors claimed that using $\mathbf{G}_0^{-1} = \mathbf{D}$ significantly improved the quality of spectral clustering applied to image segmentation, presumably because of the large range of weight values in a weighted graph derived from an image. The use of $\mathbf{G}_0^{-1} = \mathbf{D}$ for clustering has been additionally supported both theoretically and empirically by Coifman et al. [85].

We now examine the Untargeted Basic Energy Model (6.17) when $p = 2$ and $q = 1$. In this case, the minimum of the Untargeted Basic Energy Model (given by (6.18)) is taken when $\mathbf{x}$ satisfies

$$\mathbf{A}^T \mathbf{G}_1^{-1} \mathbf{A} \mathbf{x} = \mathbf{L} \mathbf{x} = \lambda \mathbf{g}_0^{-1}, \tag{6.22}$$

where $\mathbf{g}_0^{-1}$ represents the vector consisting of the diagonal elements of $\mathbf{G}_0^{-1}$, i.e., $\mathbf{G}_0^{-1} = \text{diag}(\mathbf{g}_0^{-1})$. Unfortunately, because $\mathbf{L}$ is singular the solution of this problem is undefined. This singularity corresponds to the unbounded solution $\mathbf{x} = k$ discussed above for the Untargeted Basic Energy Model. The typical solution to this

6.2 Untargeted Clustering

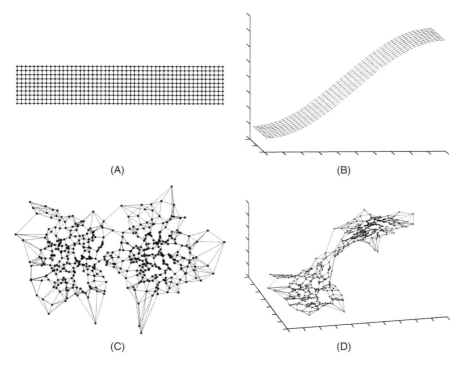

Fig. 6.6 The first nontrivial harmonic of two example graphs. From a physics standpoint, we may consider the graphs as a network of springs for which the Fiedler vector provides the lowest frequency harmonic. Additionally, the lattice graph may be viewed as a finite differences approximation to a membrane. (**A, C**) Example graphs, (**B, D**) Corresponding embeddings in which the first nontrivial harmonic is mapped to the height of each node

singularity employed by spectral partitioning is to perform the optimization of **x** in the space orthogonal to $\mathbf{x} = \mathbf{k}$. However, an alternative approach to this problem was suggested in [168, 169] that a single reference node, v_r, be chosen such that $x_r = 0$ is fixed. By fixing a reference node to one partition, the problem in (6.22) takes a unique solution which may then be thresholded to produce the clustering into two partitions. Note that the clustering obtained in this manner does not depend on the value of λ and we therefore ignore this value. As with spectral clustering above, both $\mathbf{G}_0^{-1} = \mathbf{I}$ and $\mathbf{G}_0^{-1} = \mathbf{D}$ have been employed with the choice of $\mathbf{G}_0^{-1} = \mathbf{D}$ generally producing better clusters [169]. Additionally, it was proved by Grady and Schwartz [169] that the partitioning performed in this way guaranteed that the partition connected to the reference node is connected. Since the solution to (6.22) may be interpreted as the steady-state DC potentials for a resistive network with $\mathbf{c}_0$ representing currents injected into each node, the reference node was called the **ground node** [169]. Figure 6.7 shows the equivalent circuit for which the solution to (6.22) gives the steady-state electrical potentials. We can view (6.22) from the continuous

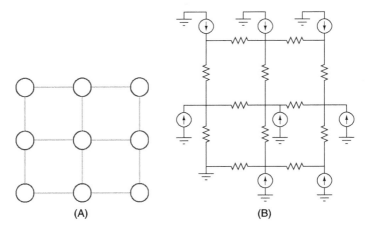

Fig. 6.7 An example of a simple graph (**A**), and its equivalent circuit (**B**). Solving (6.22) (using the node in the lower left as ground) for the graph depicted in (A) is equivalent to connecting this node to ground in the circuit depicted in (B) and then measuring the potential values at each node

calculus perspective as an instance of the *Poisson equation*,

$$\nabla^2 x = \lambda c_0, \qquad (6.23)$$

subject to the internal Dirichlet boundary condition that $x_r = 0$. This algorithm was called the **isoperimetric algorithm** by Grady and Schwartz [169] due to the connection with a relaxed version of the isoperimetric ratio. See Chap. 8 for more information about the isoperimetric ratio of a graph and Appendix B for more information about the optimization of a ratio. The primary advantages of this isoperimetric algorithm over spectral clustering are that solving a symmetric positive definite linear system of equations is more efficient than solving a (generalized) eigenvector problem, the solution to the eigenvector problem may not be unique, and the spectral algorithm overly tries to produce balanced clusters [179]. See Ref. [169] for more details on this comparison.

The Untargeted Basic Energy Model has also been studied for the case of $p = 1$ and $q = 1$ with the further restriction that the solution **x** is binary. This approach has the advantage that no threshold of **x** needs to be chosen and that it is possible to choose an optimal eigenvalue λ automatically rather than manually. See Kolmogorov et al. [242] for examples of this model in the context of image segmentation and more information about its optimization.

6.2.2 Dual Untargeted Clustering

Dual untargeted clustering algorithms have received little, if any, treatment in the literature. A primal algorithm has the advantage that it is independent of the embedding. This independence was forfeited in the targeted clustering problem in return for a unique method of targeting a cluster (i.e., by providing a partial boundary). Therefore, in the untargeted clustering problem, it is worth briefly examining whether anything is gained by a dual formulation. One possible advantage of a dual formulation is that it is easy to use the boundary operator to impose the constraint that a solution is a closed boundary. For example, if our dual graph were planar, then we could express the space of closed two-dimensional contours as any contour satisfying the constraint that

$$\mathbf{A}^T \mathbf{y} = \mathbf{0}, \qquad (6.24)$$

where $\mathbf{y}$ is a 1-cochain describing the set of edges in the boundary. An attractive aspect of the condition in (6.24) is that the $\mathbf{A}^T$ matrix is totally unimodular, which means that an integer solution may be obtained under this condition even if the optimization were over a linear functional of real-valued $\mathbf{y}$ variables. A second attractive aspect of the condition (6.24) is that the nullspace of $\mathbf{A}^T$ is known, allowing for an easy change of variables to create an unconstrained optimization problem. However, despite the ease of the dual constraint to describe a closed contour, this form seems to have been explored only very recently in the literature [293].

6.3 Semi-targeted Clustering

In the introduction to this chapter we commented that some clustering algorithms establish an objective function for the clustering which cannot be feasibly optimized. In these circumstances the clustering optimization may be initialized, leading to a solution similar to the initialization. Consequently, even though the objective function may describe an untargeted clustering problem, the dependence of the optimization on the initialization may lead to a clustering that is somewhat targeted. In this section, we consider a primal semi-targeted clustering algorithm, the k-means algorithm and its generalization to the Mumford–Shah model.

Traditionally, the Mumford–Shah model has appeared only in the field of image processing. Therefore, the connections between k-means and the Mumford–Shah have received little attention in the literature. However, the recent work on formulating (and optimizing) the Mumford–Shah model on an arbitrary graph [97, 121, 163, 418] has made it more clear how to interpret the Mumford–Shah model as a generalization of the k-means model. In the next section, we adopt this approach explicitly by building from the k-means model to the full, piecewise smooth, Mumford–Shah model.

6.3.1 The k-Means Model

The k-means model is probably the most common method for the untargeted clustering of data. Given a predetermined number of labels k, the k-means algorithm seeks to find a clustering that is a partitioning of the node set into k disjoint subsets of nodes, defined by the sets $\mathcal{R}_1, \mathcal{R}_2, \ldots, \mathcal{R}_k$, such that the intersection of any two sets is empty and the union of all sets is $\mathcal{V}$. We can denote this partitioning with the partitioning function Z, which maps the nodes into the collection of partitions $\mathcal{P} = \{\mathcal{R}_1, \mathcal{R}_2, \ldots, \mathcal{R}_k\}$, i.e., $Z : \mathcal{V} \to \mathcal{P}$. The optimum clustering in the k-means algorithm is defined as the optimization of

$$\mathcal{E}_{\text{KM}}[Z] = \sum_i \sum_{v_j \in \mathcal{R}_i} \|\tilde{c}_{\mathcal{R}_i} - \tilde{s}_j\|^2, \qquad (6.25)$$

where $\tilde{s}_j$ is the **data tuple** at node v_j and $\tilde{c}_{\mathcal{R}_i}$ is the mean data tuple inside $\mathcal{R}_i$ representing the *centroid* of the data within region $\mathcal{R}_i$.[1] Therefore, the k-means algorithm attempts to find the k-way partitioning of the data such that the data tuples in each partition have minimal within-cluster sum-of-squares difference from the mean tuple. Note that here we use tuple-valued data to handle cases in which multivariate data is available (e.g., RGB or colorscale image data), however in the simple univariate case these tuples can be replaced with scalars.

The classical method for optimizing the k-means model is Lloyd's algorithm [263, 264], which consists of the following steps.

1. Randomly partition the data nodes into k regions.
2. Repeat until convergence:
   a. Update each centroid tuple, $\tilde{c}_{\mathcal{R}_j}$, to the mean of the data tuples of all nodes in $\mathcal{R}_j$.
   b. Assign each point, v_i, to the label for which its data tuple $\tilde{s}_i$ is closest to the label centroid, i.e., $\sigma(v_i) = \operatorname{argmin}_j \{\|\tilde{c}_{\mathcal{R}_j} - \tilde{s}_i\|_2\}$.

Since the output of Lloyd's algorithm depends on the initialization, the algorithm is typically run to convergence multiple times with different initializations in order to find a good partitioning.

The k-means algorithm is fast and simple to implement, which accounts for its tremendous popularity. However, some data clustering problems have a *spatial* component as well as a data component. For example, the k-means algorithm above would cluster data tuples of an image (e.g., RGB image data) without accounting for the spatial arrangement of the pixels. However, based on the low-frequency models established for data in Chap. 5 and cluster labels developed above, neighboring nodes should be more likely to take the same label. Under this model, an outlier gray pixel inside a uniform region of black pixels is likely to be an artifact and can reasonably be assigned to the same label as the surrounding black pixels, while an outlier gray pixel inside a uniform region of white pixels is likely to be a artifact and can reasonably be assigned to the same label as the surrounding white pixels. However, the definition of the k-means algorithm above has no mechanism to

---

[1] In the rest of this chapter we treated the data as univariate in order to simplify the exposition, with the understanding that all of the machinery could also be applied to multivariate data. However, since k-means is almost exclusively applied to multivariate data we have adopted a multivariate view of data in this section. Therefore, it is assumed that each node (data point) is associated with a tuple of data, rather than a scalar.

6.3 Semi-targeted Clustering

account for the neighborhood structure of the data values.[2] Consequently, we can modify the k-means algorithm by adding a spatial regularization term which penalizes cases where neighboring nodes (defined by an edge set) are assigned different labels. This penalty may be viewed as encoding the *boundary length* of each region $\mathcal{R}_j$. Additionally, the penalty could be weighted using any of the weighting methods where the weights are derived from the data as described in Chap. 4, but this direction has not been pursued in the literature and we therefore limit our exposition to the unweighted (i.e., the unity-weighted) case with the understanding that boundary length could be trivially modified to incorporate data weights. The energy functional describing this modified k-means algorithm is given by

$$\mathcal{E}_{\text{MPCV}}[Z] = \sum_i \sum_{v_j \in \mathcal{R}_i} \|\tilde{c}_{\mathcal{R}_i} - \tilde{s}_j\|^2 + \nu \sum_{e_{ij} \in \mathcal{E}} \delta(\sigma(v_i) - \sigma(v_j)), \qquad (6.26)$$

where $\delta(\cdot)$ represents the Kronecker delta function. The value of the trade-off parameter ν is a free parameter that can be used to weight the relative importance of spatial coherency in the k-means algorithm. Unfortunately, the spatial regularization term in (6.26) makes this energy quite a bit more difficult to optimize than the standard k-means energy of (6.25) when $k > 2$. Recent work has addressed approximation methods for optimizing the boundary for multiple labels [17, 122]. Therefore, we simplify the energy functional of the modified k-means algorithm (6.26) by restricting the number of classes to two (i.e., $k = 2$) and optimizing instead

$$\mathcal{E}_{\text{CV}}[\mathcal{R}] = \mathbf{r}^\mathsf{T} \|\tilde{c}_{\mathcal{R}} - \tilde{\mathbf{s}}\|_2^2 + (\mathbf{1} - \mathbf{r})^\mathsf{T} \|\tilde{c}_{\bar{\mathcal{R}}} - \tilde{\mathbf{s}}\|_2^2 + \nu(\mathbf{1}^\mathsf{T}|\mathbf{A}\mathbf{r}|), \qquad (6.27)$$

where $\tilde{\mathbf{s}}$ represents the $|\mathcal{V}| \times 1$ vector of tuples $\tilde{s}_j$, or a tuple-valued vector over the nodes. Since there are only two labels, we can represent them by the set $\mathcal{R}$ and $\bar{\mathcal{R}}$, in which case $\mathbf{r}$ is a node vector (0-chain) acting as an indicator function for the set $\mathcal{R}$. Similarly, we use $\tilde{c}_{\mathcal{R}}$ to represent the data centroid tuple in set $\mathcal{R}$ and $\tilde{c}_{\bar{\mathcal{R}}}$ to represent the data centroid tuple in the complement set, $\bar{\mathcal{R}}$. Thus the expression $\|\tilde{c}_{\mathcal{R}} - \tilde{\mathbf{s}}\|_2^2$ is used to represent a $|\mathcal{V}| \times 1$ vector where each entry j equals $\|\tilde{c}_{\mathcal{R}} - \tilde{s}_j\|_2^2$. In the image processing literature this special case energy model is known by different names as the *piecewise constant Mumford–Shah* functional [289] or the **Chan–Vese model** [70], although it is important to note that both of these models were formulated in a continuous setting, and the first definitions of the models on a general graph appeared much more recently [97, 121, 418]. The expression of this model for multiple classes (i.e., $k > 2$), described in (6.26), has been known as the *multi-phase Chan–Vese model* and is given by the energy $\mathcal{E}_{\text{MPCV}}$ defined above.

The two-class model in (6.27) may be optimized in a manner similar to Lloyd's algorithm for optimizing k-means. The only difference with Lloyd's algorithm is

---

[2]Some authors have tried to incorporate spatial location into k-means by using the pixel coordinates as part of the feature vector in the application of k-means. This device can mitigate the problem described here in certain circumstances, but does not generalize to applications in which the network has no embedding or when the embedding is complicated, as in the gene expression example in Sect. 6.5.4 or the geospatial example in Sect. 5.9.4.

that the labeling step is more complicated, because the boundary term must now be accounted for. Specifically, by viewing **r** as a labeling vector, then (6.27) can be considered as a min-cut problem since

$$\text{Boundary}(\mathbf{r}) = \nu \mathbf{1}^T |\mathbf{A}\mathbf{r}| = \sum_{e_{ij} \in \mathcal{E}} \nu |r_i - r_j|, \quad (6.28)$$

and

$$\mathbf{r}^T \|\tilde{c}_\mathcal{R} - \tilde{\mathbf{s}}\|_2^2 = \sum_{v_i \in \mathcal{V}} r_i \|\tilde{c}_\mathcal{R} - \tilde{s}_i\|_2^2, \quad (6.29)$$

which may also be viewed as a min-cut (a unary term) via the construction

$$\text{Cut}(\tilde{c}_\mathcal{R}) = \sum_{v_i \in \mathcal{V}} \|\tilde{c}_\mathcal{R} - \tilde{s}_i\|_2^2 |r_i - 0|, \quad (6.30)$$

where $\mathcal{E}_0$ is a set of auxiliary edges connecting each node to a phantom terminal (as in Fig. 6.2). Similarly,

$$\text{Cut}(\tilde{c}_{\bar{\mathcal{R}}}) = \sum_{e_{ij} \in \mathcal{E}_1} \|\tilde{c}_{\bar{\mathcal{R}}} - \tilde{s}_i\|_2^2 |1 - r_i|. \quad (6.31)$$

Therefore, the optimization of the two-class model in (6.27) is enacted by the following steps.

1. Randomly initialize the set $\mathcal{R}$ and compute the centroid $\tilde{c}_\mathcal{R}$ of $\mathcal{R}$, and the centroid $\tilde{c}_{\bar{\mathcal{R}}}$ of set $\bar{\mathcal{R}}$.
2. Repeat until convergence:
   a. Find the labeling vector **r** that minimizes

   $$\mathcal{E}[\mathcal{R}] = \text{Boundary}(\mathbf{r}) + \text{Cut}(\tilde{c}_\mathcal{R}) + \text{Cut}(\tilde{c}_{\bar{\mathcal{R}}})$$

   from the expressions above using a max-flow/min-cut algorithm.
   b. Assign each centroid, $\tilde{c}_\mathcal{R}$ and $\tilde{c}_{\bar{\mathcal{R}}}$, to the mean of the data vectors of all nodes in $\mathcal{R}$ and $\bar{\mathcal{R}}$, respectively.

The k-means model was generalized to incorporate spatial information by introducing a spatial regularization term. However, in cases where the spatial arrangement of the nodes is relevant, this generalized k-means approach seems limited by the fact that the data is modeled as a single centroid for every node in the cluster (i.e., the centroid value is uniform *everywhere in space* within a cluster). Consequently, this generalized k-means model may be further generalized by allowing each node to have its own idealized "centroid" (i.e., to allow the term corresponding to the centroid to vary spatially). To avoid confusion, we term this spatially varying version of the region centroid as a *pseudocentroid*, in which every node in the graph has a both pseudocentroid for $\mathcal{R}$ and a pseudocentroid for $\bar{\mathcal{R}}$. However, to be meaningful, the pseudocentroid cannot be allowed to vary arbitrarily and therefore we can impose

6.3 Semi-targeted Clustering

a smoothness penalty on the pseudocentroids for each region. Specifically, for the two-class problem, let

$$\mathcal{E}_{\text{smooth}}[\mathcal{R}] = \sum_{e_{ij} \in \mathcal{E}} r_i \|\tilde{c}_{i\mathcal{R}} - \tilde{c}_{j\mathcal{R}}\|_2^2 + \sum_{e_{ij} \in \mathcal{E}} (1 - r_i) \|\tilde{c}_{i\bar{\mathcal{R}}} - \tilde{c}_{j\bar{\mathcal{R}}}\|_2^2, \qquad (6.32)$$

where now $\tilde{c}_{i\mathcal{R}}$ represents the pseudocentroid of node v_i in set $\mathcal{R}$. Therefore, the pseudocentroid at each node should change smoothly between nodes within a region. This model may be called the *piecewise-smooth Mumford–Shah model*, even though the formulation of Mumford and Shah [289] was in a continuous space. This form of the model on a graph appeared in Ref. [163]. By allowing the pseudocentroids to vary with each node, the output of our minimization is both a clustering *and* and idealized form of our data. This idealized form of the data was used in Chap. 5 for filtering noisy data, while the focus of this chapter is on the clustering. This smoothness term could be penalized differently for smoothness changes across different edges by employing any of the affinity edge weighting functions derived from data which were detailed in Chap. 4. However, since this smoothness penalty has not been adjusted for edge weighting in the literature, we will continue the exposition assuming an unweighted (i.e., a unity weighted) edge set.

When we consider each region to be represented by a single centroid, then it is easy to compare the data at every node to a cluster centroid to see how well the node might fit in the cluster. However, when we allow each cluster to be represented by several pseudocentroids (in fact, each node is represented by its own pseudocentroid), then it is less clear how we might compare the data at nodes which are not inside a particular cluster with the pseudocentroids of that cluster, since each cluster is no longer represented by a single centroid. Therefore, the piecewise smooth Mumford–Shah model requires an estimate of the pseudocentroid for each cluster at each node. This estimation was performed by Grady and Alvino [163] by smoothly extending the pseudocentroids outside of each cluster to all nodes by solving a Laplace equation. Figure 6.8 shows an example of these pseudocentroids in the context of image segmentation.

Once a cluster pseudocentroid is generated for each cluster and node, then the optimization of the piecewise smooth Mumford–Shah model is the same as the piecewise constant model above. At each iteration of the model, it is necessary to estimate the cluster pseudocentroid for each pixel and then solve a max-flow/min-cut problem to find the optimal clustering (for the two-class case). See Ref. [163] for more details. This clustering model could also be extended to multiple classes by recursive bisection or by the approximation approaches taken by El-Zehiry et al. [122] and by Bae and Tai [17]. Figure 6.9 shows the clustering and idealized data values obtained for a synthetic image.

Among the class of semi-targeted algorithms, we have so far considered only the k-means algorithm and the more generalized Mumford–Shah algorithm, which both operate on the primal graph. Dual semi-targeted algorithms are not considered here because the authors are unaware of any algorithms of this type. However, in the image processing literature there is a vast amount of work on *active contour* methods

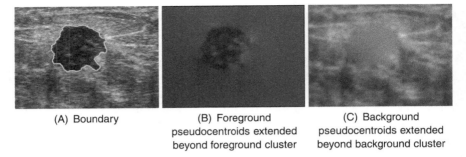

Fig. 6.8 Optimization of the piecewise smooth Mumford–Shah using max-flow requires specification of values for the cluster pseudocentroid estimation function $\tilde{c}_{i\mathcal{R}}$ in the region *outside* the cluster region $\mathcal{R}$. Using a Laplace equation regularizer allows us to estimate the values for the foreground $\tilde{c}_{i\mathcal{R}}$ tuples and the background $\tilde{c}_{i\bar{\mathcal{R}}}$ tuples for the entire domain (i.e., the graph representing the image), as shown above. See Ref. [163] for more details

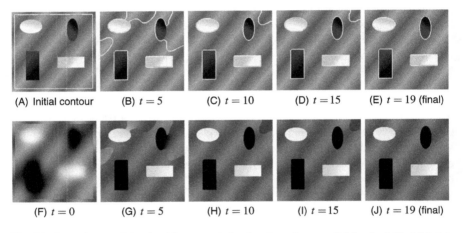

Fig. 6.9 Several steps of the algorithm to optimize the piecewise smooth Mumford–Shah Model. (**A–E**) Partitioning evolution from initialization to stabilization. *Blue contours* indicate boundary location. (**F–J**) Corresponding pseudocentroid reconstruction of the piecewise smooth estimate of the image data, given a contour. For each image, the iteration step t is provided below

(e.g., [222, 409]) which are very similar in spirit. These methods parameterize a cluster boundary as a polygon with a series of control points which are evolved to (locally) optimize an objective function. However, these methods do not fit into the scope of the present work since they require an embedding, the control points are allowed to vary anywhere in the embedded space and the methods are not explicitly formulated on a dual complex.

6.4 Clustering Higher-Order Cells

In the previous sections we described variational models for targeted and untargeted clustering of the nodes in a graph. We end this chapter by briefly recasting these models to apply to the clustering of higher-order cells. We will focus on recasting these models in terms of edge clustering with the understanding that the same methods could be further applied to the clustering of p-cells for any p (as in Chap. 5).

6.4.1 Clustering Edges

We will consider targeted, untargeted and initialized primal clustering formulations. Formally, the clustering problem on edges may be formulated by assigning labels b of a label set $\mathcal{L}$, with $b \in \mathcal{L}$, to the set of edges, $\mathcal{E}$. As in the case of nodes, the goal of a clustering algorithm is to produce a segmentation function $\sigma : \mathcal{E} \to \mathcal{L}$. In the following discussion, we assume that a cycle set has been identified and this cycle set constitutes a basis for the cycle space, i.e., that the edge Laplacian is full rank. Additionally, we assume that $\mathbf{G}_1 = \mathbf{I}$.

6.4.1.1 Targeted Edge Clustering

The basic components of a targeted edge clustering algorithm are the same as those components of a targeted node clustering algorithm. Specifically, additional information is required that associates some or all edges to particular labels. Examples of additional information in a targeted edge clustering algorithm could include:

1. A method for assigning probabilities to a subset of the edges that these edges belong to each label.
2. Known labels for a subset of the edges.

In all of the subsequent discussion on targeted edge clustering algorithms, we can formulate our goal as solving for a membership probability $y_{i,b}$ that edge e_i belongs to label b. Unless otherwise noted, the segmentation function $\sigma(e_i) = b$ is obtained via $\sigma(e_i) = \mathrm{argmax}_b\{y_{i,b}\}$, similar to the case with segmenting nodes. Consequently, our focus will be on the calculation of $y_{i,b}$ for each edge and label.

We begin by rewriting the weighted version of the Basic Energy Model in (6.5) to apply to clustering edges. As in the filtering case treated in Chap. 5, we replace the gradient of the node data by both the curl and the divergence of the flow data along each edge, since this replacement preserves the character of the operator as penalizing high-frequency functions (with respect to the edge Laplacian). Specifically, we may rewrite (6.5) for edges as

$$\mathcal{E}_{\mathrm{BEM}}[\mathbf{y}_b] = \mathbf{1}^\mathsf{T}\left|\mathbf{G}_2^{-1}\mathbf{B}\mathbf{y}_b\right|^p + \mathbf{1}^\mathsf{T}\left|\mathbf{G}_0\mathbf{A}^\mathsf{T}\mathbf{y}_b\right|^p. \tag{6.33}$$

Once again, we can avoid a trivial minimum of this energy by incorporating a pre-specified set of labels for some edges (similar to the use of *seeds* in the node clustering case). Formally, define the set of seeded or "marked" edges $\mathcal{E}_M \subset \mathcal{E}$ for which $\sigma(\mathcal{E}_M)$ is known. These labels are assumed to have been obtained via a different process such as user interaction. Using this information, we can fix $y_{i,b}$ for all edges $e_i \in \mathcal{E}_M$ via the assignment

$$y_{i,b} = \begin{cases} 1 & \text{if } \sigma(e_i) = b, \\ 0 & \text{if } \sigma(e_i) \neq b. \end{cases} \quad (6.34)$$

These fixed values allow us to solve for a nontrivial minimum of the edge-focused Basic Energy Model in (6.33), which can be computed via the optimization methods in Appendix B.

As observed for node clustering, the trivial minimum of the Basic Energy Model could also be avoided by adding a term encoding the prior of each edge to belong to each label. We may continue to apply the targeted clustering models of Sect. 6.1 to flow data. When considering the Basic Energy Model expressed in (6.33) we treated the prior probabilities as initialized with an initial condition then iteratively updated to minimize the target energy. However, unless pre-labeled seeds are known, an optimal solution of (6.33) is trivially zero. Therefore, we may take a different approach to incorporating the prior probabilities by separating the smoothness constraints and priors into two different terms whose influence on the solution is controlled with a parameter. This edge-focused Extended Basic Energy Model is expressed by

$$\mathcal{E}_{\text{EBEM}}[\mathbf{y}_b] = \mathbf{1}^T \left| \mathbf{G}_2^{-1} \mathbf{B} \mathbf{y}_b \right|^p + \mathbf{1}^T \left| \mathbf{G}_0 \mathbf{A}^T \mathbf{y}_b \right|^p + \lambda \sum_a \mathbf{s}_a^T \left| \mathbf{y}_b - \delta(a-b) \right|^p, \quad (6.35)$$

where $\mathbf{s}_a$ represents the prior likelihoods that each edge belongs to label a, $\delta(a-b) = 1$ if $a = b$ and $\delta(a-b) = 0$ otherwise. At the cost of introducing a free parameter λ we now have a model that produces a nontrivial minimum. Additionally, we may also introduce a set of seeds in the same manner as before, which allows us to find a nontrivial solution even when $\lambda = 0$ (i.e., prior probabilities are unknown).

6.4.1.2 Untargeted Edge Clustering

Building on the untargeted Basic Energy Model of (6.17), we now consider the edge-focused untargeted Basic Energy Model in the context of untargeted bipartitioning of an edge set. Specifically, consider the energy

$$\mathcal{E}_{\text{UBEM}}[\mathbf{y}] = \mathbf{1}^T \mathbf{G}_2^{-1} |\mathbf{B} \mathbf{y}|^p + \mathbf{1}^T \mathbf{G}_0 |\mathbf{A}^T \mathbf{y}|^p - \lambda \mathbf{1}^T |\mathbf{y}|^q, \quad (6.36)$$

for some $\lambda > 0$. As in the node case, the solution $\mathbf{y}$ may be thresholded to produce a bipartitioning of the edge set. This energy takes a minimum for $p > 1$ when $\mathbf{y}$ satisfies

$$\mathbf{B}^T \mathbf{G}_2^{-1} |\mathbf{B} \mathbf{y}|^{p-1} + \mathbf{A} \mathbf{G}_0 |\mathbf{A}^T \mathbf{y}|^{p-1} = \lambda |\mathbf{y}|^{q-1}. \quad (6.37)$$

We briefly consider some combinations of the norm parameters p and q values with a focus on a contrast with the node clustering case.

The $p=2$ and $q=2$ case also yields an eigenvector problem

$$\left(\mathbf{B}^\mathsf{T}\mathbf{G}_2^{-1}\mathbf{B} + \mathbf{A}\mathbf{G}_0\mathbf{A}^\mathsf{T}\right)\mathbf{y} = \mathbf{L}_1\mathbf{y} = \lambda \mathbf{y}, \tag{6.38}$$

in the edge Laplacian matrix. The eigenvectors of the edge Laplacian matrix have been lightly treated in the literature to date. These early investigations [1, 80, 288] reveal that the eigenvectors corresponding to low eigenvalues of the edge Laplacian roughly correspond to *circulations* around the complex. When the complex has a higher genus (i.e., there is one or more "handles" in the domain), a number of eigenvalues equal to the genus reach zero exactly and the corresponding eigenvectors represent flows that encircle these handles. From the standpoint of clustering, such circulations may indeed be meaningful clusters, since each circulation is a set of flows with minimal curl and divergence. Figure 6.10 shows the result of this model for clustering the flows in a graph.

Similarly, the $p=2$ and $q=1$ case yields a minimum to (6.37) when

$$\left(\mathbf{B}^\mathsf{T}\mathbf{G}_2^{-1}\mathbf{B} + \mathbf{A}\mathbf{G}_0\mathbf{A}^\mathsf{T}\right)\mathbf{y} = \mathbf{L}_1\mathbf{y} = \mathbf{1}. \tag{6.39}$$

This expression represents the flow version of the isoperimetric algorithm of Ref. [169]. The isoperimetric algorithm on nodes required an additional step of "grounding" a node (in accordance with the electrical circuit analogy) at $x_g = 0$ to break the singularity of the node Laplacian. Although we assumed that we have a complete cycle basis (i.e., $\mathbf{L}_1$ is full rank), (6.39) still has a trivial solution unless some edge is chosen as a fixed "ground" where $y_g = 0$. Figure 6.10 gives an example of this model for clustering the flows on a graph.

6.5 Applications

Clustering has an enormous number of applications across many fields. In this section, we give several applications of these clustering techniques with the goal of demonstrating the variety of applications of these methods.

6.5.1 Image Segmentation

Clustering image data is a core problem in image processing in which the clustering procedure is known as *image segmentation*. Image segmentation is an essential part of many practical applications and graph-theoretic algorithms have historically played an important, and increasing, role in image segmentation [345, 407, 416]. Example applications include the quantification of objects in medical/scientific images, photo editing, and enhanced visualization of image contents. Additionally,

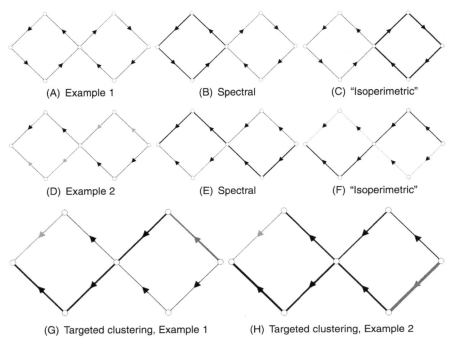

Fig. 6.10 Clustering flows. In all cases, the thickness of the line represents the strength of the **y** vector used to determine the partitioning (by thresholding). *Top row*: Example 1, showing two circulations around the cycles cluster easily by both the spectral edge clustering method and the "isoperimetric" edge clustering method. The clusters are obtained by thresholding the resulting flows. *Middle row*: Example 2, in which the orientation of four edges are flipped (marked in *blue*). A leftward flow is partitioned into two streams by both partitioning methods. *Bottom row*: The "random walker" model applied to targeted clustering. The labeled edges were the upper left edge (*green*) and the second labeled edge was in the lower right or upper right (*red*). In the first case, the two diagonal flows are clustered together while in the second case, the upper and lower flows are clustered together

image segmentation can be used as a preprocessing step before many other image analysis tasks, such as object recognition, image registration or image compression. Traditional two-dimensional image data consists of "picture elements" (pixels) typically arranged in a square grid and traditional three-dimensional image data consists of "volume elements" (voxels) arranged in a rectilinear grid. In our applications, we associate each pixel or voxel with a node and use a 4-connected lattice as the edge set (6-connected in three dimensions). In all of our examples, the edges were weighted using the Welsch function of the difference of image intensities (see Chap. 4).

Our goal in this example is to demonstrate the varieties of clustering algorithm that were discussed previously. Specifically, image processing applications permit both primal *and* dual segmentation algorithms, since the image content has a geometric arrangement that naturally admits description as a cell complex with a dual. Consequently, the segmentation problem may be cast as having the goal of labeling

6.5 Applications

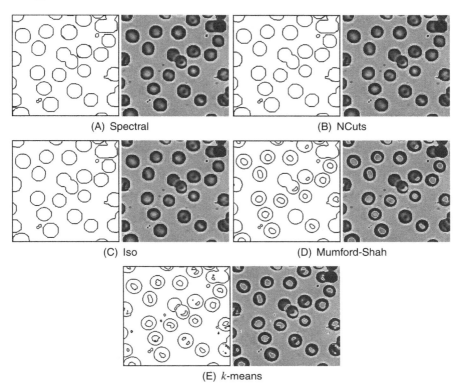

Fig. 6.11 Untargeted image segmentation of blood cells using spectral partitioning, Normalized Cuts (NCuts), isoperimetric clustering (Iso), k-means and the piecewise constant Mumford–Shah. The k-means algorithm employs image intensity and the Mumford–Shah algorithm employs both image intensity and spatial regularity causing them both to associate the bright center of each cell with the bright background. In contrast, the purely spatial isoperimetric, spectral and normalized cuts algorithms all associated this inner part with the cells themselves. For each algorithm, the outline of the computed cluster boundaries is presented on the *left*, and the same cluster boundaries are superimposed on the image on the *right*, indicating the final segmentation

each pixel (in the primal case) or of labeling the boundary of the pixels (in the dual case).

We begin by demonstrating the primal untargeted algorithms on the application of cell counting in an image of blood cells. Figure 6.11 displays the results of each algorithm on the task of separating the cells from the background. Note that no attempt was made to further subdivide the cell clusters. Due to the clarity of this image, all of the algorithms produced a reasonable clustering. Since the Mumford–Shah energy balances image intensity clustering with spatial clustering, and k-means removes the spatial aspect, these two algorithms associated the inner, brighter, part of the cells with the background. In contrast, the purely spatial isoperimetric, spectral and Normalized Cuts algorithms all associated this inner part with the cells themselves.

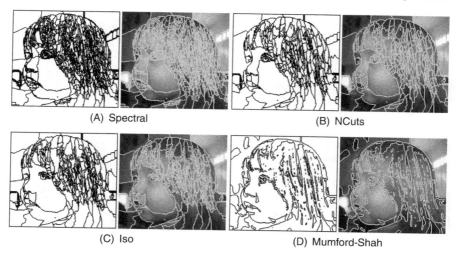

Fig. 6.12 Untargeted segmentation algorithms applied to a grayscale photograph of a girl. The algorithms used only image intensity to define edge weights and unary terms rather than more complicated models such as using texture boundaries to set weights. Consequently, the algorithms used in this way try to isolate each strand of hair as a separate object rather than as a unit

The blood cells image contained pixels belonging to two types, cell and background, which were generally separated by different intensity profiles. Images that contain more classes of objects are more difficult for an untargeted image segmentation method to parse, especially if the object classes have a differing intensity/texture profile. To illustrate this issue, the same set of untargeted algorithms were applied to find a segmentation of the photograph of a girl given in Fig. 6.12. As before, edge weights were derived from a Welsch weighting function of the image intensities of the neighboring pixels. The models could have been made more appropriate to the segmentation of this image by deriving weights from texture boundaries and using such information as features for the Mumford–Shah model. By basing the models on image intensity, the algorithms attempt to segment each strand of hair as a separate object, rather than segmenting all of the hair as a single unit. Since these algorithms all explicitly optimize a separation of two objects, the algorithms were used to generate multiple labels by applying the algorithms recursively and determining when to stop the recursion based on the energy obtained by the minimum produced by the segmentation.

An untargeted image segmentation algorithm tells us what the best clusters are. If we instead wanted to find a *particular* object, then we need to target that object through a targeted image segmentation approach. We may isolate the girl's head in this photograph by supplying some labeled pixels which identify the object to be isolated. In this example, these labeled pixels were supplied interactively by the authors and each of the targeted segmentation algorithms were run to produce a segmentation separating the image into foreground and background. Results of this experiment are displayed in Fig. 6.13. Although each of these algorithms also used edge weights derived from intensity differences, the targeting of a particular object

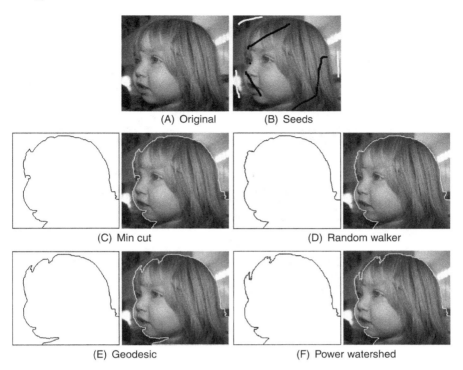

Fig. 6.13 Example of targeted image segmentation algorithms. These algorithms were used to isolate the girl's head by interactively supplying some labeled pixels. Although these algorithms also used the same intensity-based edge weighting as before, the act of targeting a particular object allows the algorithms to know which objects should be grouped together (e.g., each strand of hair belongs to the foreground label)

was sufficient for these algorithms to group together the parts of the objects (e.g., every strand of hair is grouped together in the foreground).

The previous example applied targeted segmentation algorithms to isolate the image foreground. However, all of these algorithms operated on the *primal* graph in the sense that the final output was a labeling of each pixel and the boundary was implied by the change in label. We discussed in the text how a *dual* algorithm could also be applied to image segmentation by finding a set of edges in the dual graph which separated the nodes into two pieces. Dual clustering algorithms are available for two-dimensional image processing because the planar graph defined by the 4-connected lattice has a well-defined 2-dual. The most popular dual segmentation algorithm in the computer vision literature is the intelligent scissors/live wire algorithm [129, 287], which inputs boundary points from a user which are then connected via the shortest path in the dual graph. Figure 6.14 gives an example of using this algorithm to target the same object in the photograph.

One advantage of a primal algorithm is that it easily extends to alternate embeddings (or no embedding). However, since the dual graph (or lack thereof) depends on the availability of an embedding, these algorithms are less portable between prob-

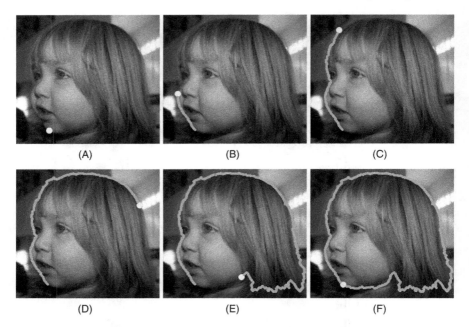

Fig. 6.14 Targeted segmentation using a dual algorithm. The dual algorithm operates on the dual graph to find a set of edges in the dual graph comprising the boundary of the nodes in the primal. We applied the intelligent scissors/live wire algorithm [129, 287] to segment the girl's head in this image. At each step of the algorithm the user determines a point on the boundary (shown by the *bright circle*) and a shortest path is calculated between the new point and the previous point (shown by the *thick line*). The process is continued until the boundary is completed. In this example, six points total were manually chosen

lems. An example of this change in embedding is the application of a segmentation algorithm in a two-dimensional image compared to a three-dimensional image. A primal segmentation algorithm can just as easily be applied in two or three dimensions, since the output is a labeling of the pixels in two dimensions or the voxels in three dimensions. However, a dual algorithm that finds a one-dimensional boundary in two dimensions that isolates two-dimensional regions of an image will not subdivide the three-dimensional image. Consequently, such an algorithm must be reworked to apply to find the boundary of regions in three dimensions, i.e., to find two-dimensional boundary surfaces. This topic was addressed earlier in Sect. 6.1.2.1 in which we showed how the intelligent scissors/live wire algorithm could be extended to three-dimensional segmentation by computing a minimal surface in the dual complex. Instead of inputing points and computing a shortest path between them, the three-dimensional algorithm inputs closed contours and computes a minimal surface between them. Figure 6.15 gives three examples from medical imaging of this dual algorithm in three dimensions. Note the bottom image of the aorta which splits into two pieces at the iliac branch. Although one contour was input above the split and two contours were input below the split, the resulting segmentation naturally merges these contours into a single surface.

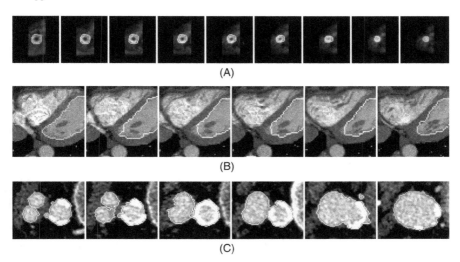

Fig. 6.15 Dual segmentation in three dimensions applied to the segmentation of three-dimensional medical data. Several slices of three-dimensional volumetric data are shown. In each example, the *green contours* were placed on the bounding slices and the intermediate *yellow contours* represent the minimum-weight surface between these contours. (**A**) SPECT cardiac data. (**B**) CT cardiac data. (**C**) CT aorta near iliac branch. Multiple closed contours may be placed and a single surface may be found that splits accordingly to accommodate the prescribed boundaries. Note that not all slices between the closed contours are displayed

6.5.2 Social Networks

Clustering is a common task in the study of social networks, where it is often known as *community finding*. In a social network, each node typically represents a person (or group) and the edges represent a social connection (e.g., friendship). If the network is weighted, the edge weight indicates the strength of the social connection. Since a larger weight typically represents a stronger social bond, we will treat these edge weights as *affinity weights* (similar to conductances in the circuit theory analogy).

In this application, we consider the social network studied by Zachary [415], which has appeared many times in the study of social networks (e.g., [148, 292]). Zachary observed a university karate club for two years which consisted of 34 members. The club split when the karate instructor "Mr. Hi" wanted more money and the club president "John A" fired him. Zachary's goal in studying this social network was to determine if it was possible to predict the faction joined by each member based purely on the social structure of the club. The social structure of the club was modeled as a network in which each member was associated with a node and an edge was assigned between two nodes if the two members met in some venue outside the club (e.g., the campus pub, common classes, outside karate tournaments, etc.). Each edge was assigned a weight equal to the number of outside venues that the two members had in common.

Zachary's method of predicting the split was to designate "Mr. Hi" (node 1) as a source node and "John A" (node 34) as a sink node and then determine the minimum cut (maximum flow) between these nodes in the network. This approach may be seen as an example of the *targeted* clustering problem in which the goal was to isolate a particular cluster (Mr. Hi's cluster) from the club (John A's cluster). In fact, the max-flow/min-cut algorithm employed by Zachary corresponds precisely to the targeted clustering approach defined by solving the Basic Energy Model (6.6) with norm $p = 1$.

The max-flow/min-cut approach taken by Zachary correctly predicted the faction split for every member except for node 9. Zachary's algorithm assigns person 9 to John A's faction when, in fact, this person joined Mr. Hi's club. Zachary explains this discrepancy by noting that person 9 was indeed aligned with John A's faction but he would have had to give up his high belt ranking if he had joined with John A (because the new club planned to teach a different style of karate). In addition to Zachary's method, we employed the random walker ($p = 2$) and geodesic ($p = \infty$) algorithms to this targeted clustering problem. Both of these algorithms also fail to correctly predict the faction joined by person 9 after the split. Additionally, the geodesic algorithm incorrectly predicts the faction joined by person 14. Although person 14 is clearly more strongly connected to Mr Hi's faction, person 14 is connected directly to Mr. Hi and John A with the same weight. In this case, the shortest path from person 14 to Mr. Hi takes the same weight as the shortest path from person 14 to John A and the tie was broken in the wrong direction. This situation illustrates the common problem with the geodesic algorithm that it is sensitive to the characteristics of the shortest path without considering the global structure of the network. However, the geodesic algorithm is still mostly correct (with only two incorrect assignments) and very fast compared to the other algorithms (especially on large networks). Figure 6.16 displays Zachary's karate club network with the true split and the targeted clustering obtained by the three algorithms.

6.5.3 Machine Learning and Classification

Machine learning is a vast topic. The goal of a machine learning algorithm is to reduce a phenomenon of interest to a series of quantities that may be used to identify the phenomenon and to generalize determinations made about the phenomenon to other objects with a similar set of quantities. Machine learning techniques are generally divided into *supervised* learning and *unsupervised* learning techniques. A supervised learning technique inputs a small set of labeled training data that is used to build a model that may be applied to label unseen data. In contrast, an unsupervised learning technique inputs unlabeled data with the goal of producing a division of this data into meaningful labels which may be applied to future unseen data. Between supervised and unsupervised techniques lies the body of *semi-supervised* and *transductive learning* techniques which input a small amount of labeled data *and* a larger amount of unlabeled data which is used together to build a model for the labeling of unseen data.

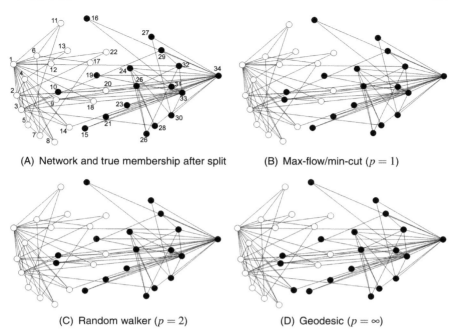

Fig. 6.16 Zachary's Karate Club network [415]. Zachary tried to predict the actual split of the karate club into two groups based on the social interactions of each member. This problem may be viewed as an instance of the targeted clustering problem in which the leaders of the two factions, "Mr. Hi" (person 1) and "John A" (person 34), are treated as known and Zachary's goal was to predict the membership of each other member

Learning and clustering are intimately related topics. The goal of both learning and clustering algorithms is to assign a labeling to data. Further, untargeted clustering algorithms are similar to unsupervised learning algorithms in the sense that both types of algorithm must find a good method for separating data into different labels. In contrast, the targeted clustering algorithms are similar to supervised learning algorithms (and particularly semi-supervised and transductive learning algorithms) in the sense that these algorithms must take a small amount of labeled training data and determine how to assign labels to a large number of unlabeled data points. The primary difference between learning and clustering is that a clustering algorithm is generally focused on labeling a particular set of data which is available, while a learning algorithm must be able to generalize the clustering to label new (unseen) data. However, transductive learning algorithms are generally defined to label only a certain set of unlabeled data points (provided during training) [73], and therefore the targeted clustering algorithms described in this chapter are more properly viewed as transductive learning algorithms in the context of machine learning. Additional work has been done to further blur the lines between machine learning and clustering by examining how to extend some clustering algorithms to unseen data (see, e.g., Refs. [28, 29]).

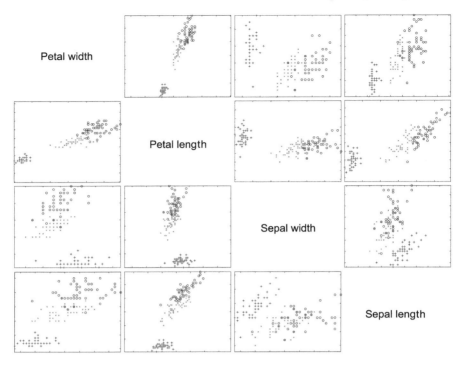

Fig. 6.17 A scatter plot of the Fisher iris data [138]. Each flower is described by four measurements: petal width, petal length, sepal width and sepal length. The *setosa* are indicated with the marker '+', *Iris versicolor* with the marker 'x', and *Iris virginica* with 'O'. Each row/column represents a variable and the off-diagonal image shows the row variable on the x-axis and the column variable on the y-axis. The measurements from the *setosa* are well-separated, while the *Iris versicolor* and *Iris virginica* measurements overlap substantially

In this application example, we do not address the issue of extending a clustering algorithm to unseen data, but rather to address the use of clustering to group data used to describe an object that we wish to model. To illustrate these clustering approaches, we use the classical Fisher iris data [138]. Fisher's paper was an early example of the standard approach used now in machine learning. Fisher wanted to determine whether or not it was possible to distinguish three types of irises from a set of measurements of each iris. Fisher's data consisted of 50 samples of each of three different species of iris, *setosa*, *Iris virginica* and *Iris versicolor*. Four measurements were taken from each flower: the petal width, petal length, sepal width and sepal length. Based on these measurements, the data for the *Iris virginica* and *Iris versicolor* are substantially overlapping while the data describing the *setosa* is well-separated from the other two. Figure 6.17 provides a scatter plot illustrating the distribution of the data for each type across measurement dimensions.

We first applied the untargeted clustering algorithms to determine if they could produce the three clusters. The initial step was to generate a graph from the data in which each data point is treated as a node and the nodes are connected with edges

6.5 Applications

Table 6.3 The Rand index, as defined in (6.40), for the clusters produced by the various untargeted clustering algorithms when applied to untargeted clustering of the Fisher iris dataset

	k-Means	Mumford–Shah	Isoperimetric	Spectral	Normalized cuts
Rand index	0.8623	0.8797	0.8797	0.8797	0.8797

via a K-nearest-neighbor approach (see Chap. 4) in which we set $K = 25$ and all edge weights to unity. Since the algorithms described above partition the graph into only two parts, the algorithms were run recursively to produce three partitions. In each case, the energy value of the partition was used to determine when to stop the recursion with the stop criterion set such that three partitions were produced. The k-means algorithm was also included (as a variant of the minimization of the Mumford–Shah energy), although this algorithm did not need to be run recursively in order to produce the prescribed three partitions.

The quality of the clustering produced by minimizing each objective function was evaluated against the true labeling by using the Rand index [315]. The Rand index between two labelings, x_1 and x_2, is defined as the ratio

$$\text{Rand}(x_1, x_2) = \frac{a+b}{\binom{n}{2}}, \qquad (6.40)$$

where a represents the number of pairs of nodes which share the same label in x_1 while also sharing the same label in x_2, and b represents the number of pairs of nodes which have different labels in x_1 while also having different labels in x_2. If x_1 and x_2 represent the same clustering then $\text{Rand}(x_1, x_2) = 1$, while if the two clusterings do not agree on any pair of points then $\text{Rand}(x_1, x_2) = 0$. Table 6.3 gives the Rand index for the clustering obtained by the various primal untargeted clustering algorithms described in the chapter. In this case, all of the algorithms produced a reasonable clustering in which the quality was equal for all algorithms except the k-means algorithm which performed slightly worse.

We next tested the primal targeted clustering algorithms on the iris data in a transductive learning context. Samples were chosen randomly from each of the three classes. The number of samples was increased from one to fifty in order to track the performance of each algorithm with respect to the number of training samples. At fifty samples, all of the data is used to train/test, so the labeling will be perfect (thus the Rand index will be equal to unity). For each number of samples, 100 trials were run with randomly generated samples for each algorithm and the Rand index of the resulting labelings were computed with respect to the ground truth. Figure 6.18 displays the results for the targeted clustering algorithms applied to this data using a K-nearest neighbor graph in which $K = 5$ and again with $K = 30$. In both cases, the edges were weighted using the Welsch function as a function of the Euclidean distance between data points (in normalized feature space). Figure 6.18 displays plots of the mean Rand index across 100 trials for each number of samples and the standard deviation of the Rand index across trials. These plots allow us to make several observations about the behavior of these algorithms in this scenario. First, all three algorithms behave roughly the same. However, the geodesic

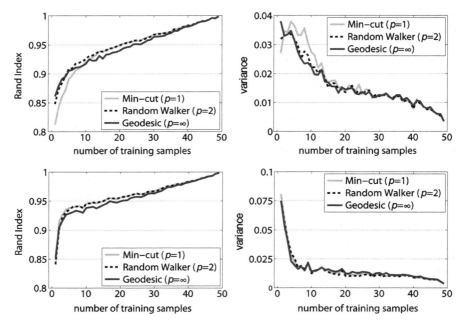

Fig. 6.18 Targeted clustering using randomly sampled seeds on the Fisher iris dataset. Each trial was run 100 times with an increasing number of samples. The x-axis represents the number of training samples and the y-axis represents the Rand index. The *left column* shows the mean Rand Index and the *right column* shows the variance of the Rand index. *Top row*: Run on a K-Nearest Neighbor graph with $K = 5$. *Bottom row*: Run on a K-Nearest Neighbor graph with $K = 30$. We see that the geodesic algorithm consistently has the worst performance for a mid-range number of labeled nodes, but generally the lowest variance in performance of the three algorithms

algorithm performs better than the other two with a very small number of samples but lags behind the others as the number of samples increases. This effect is more pronounced for the graph with lower connectivity. Secondly, all of the algorithms gave a better average performance with higher connectivity, but the variance of the quality was greater for a low number of samples.

6.5.4 Gene Expression

Grouping genes by their expression pattern can be useful to deduce the gene regulatory network, classify groups into particular phenotypes (e.g., cancerous or noncancerous) or to match gene expression with other macroscopic anatomical features such as a physiological atlas [41, 76, 198, 368, 408]. In this example, we follow Bohland et al. [41] to determine whether the grouping of gene expression profiles match the standard anatomical grouping.

Bohland et al. [41] studied the C57BL/6J mouse brain in the Allen Brain Atlas [255] and compared the clustering of locations based on their gene expression pat-

6.5 Applications

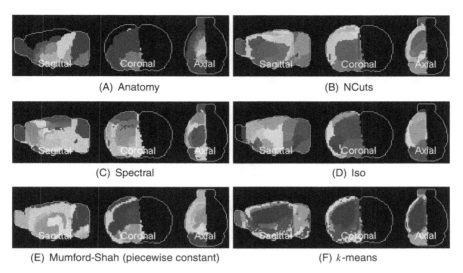

Fig. 6.19 Do the gene expression patterns match the accepted anatomy? In this example, we follow the Bohland et al. experiment [41] to demonstrate how well a clustering of gene expression patterns match the accepted anatomy. In each experiment, we chose parameters for each algorithm to generate a partitioning with the highest Rand index as compared to the anatomical atlas. The Rand index for each partitioning were, NCuts: 0.881, Spectral: 0.8783, Isoperimetric: 0.8767, Mumford–Shah: 0.8845, k-means: 0.8678. Consequently, it could be argued that there is a strong correspondence between anatomy and gene expression profile, but that the relationship is not perfect. Note how the k-means clustering is spatially fragmented due to the lack of any geometric regularization, as opposed to the Mumford–Shah result which is effectively k-means with geometric regularization

terns with a classically-defined anatomical reference atlas. The goal was to assess the level of correspondence between molecular level and higher-level information about brain organization. Each voxel in the sample was described by 3041 genes which were deemed to be consistent across experimental observations. Bohland et al. reduced these 3041 genes to produce a feature space in which each anatomical location is represented by a tuple consisting of 271-dimensions. They then applied a k-means clustering to this data. We replicated the methodology of Bohland et al. [41] using the untargeted clustering algorithms described in this chapter. Figure 6.19 displays views of the clusterings obtained by the various untargeted primal clustering methods. In each case, we weighted the edge weights using an ℓ_∞ norm to measure the distance between the (reduced) gene expression data tuples, which was then input into a Welsch function (see Chap. 4).

We can also use the tools of targeted clustering to determine how well a particular anatomical structure matches a gene expression pattern. In order to examine this question, we generated labeled seeds from the anatomical atlas by setting the central portion of the medial axis of each region as seeds. We then targeted the striatum by setting all of its seeds to foreground ('1') and seeds from all the other regions as background ('0'). Figure 6.20 displays the results obtained from applying the targeted clustering algorithms to this data. Each of the algorithms produce a seg-

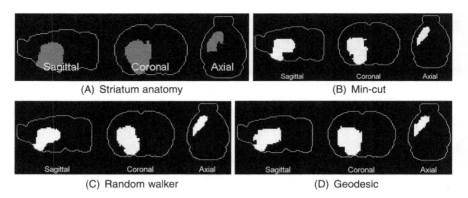

Fig. 6.20 Targeted clustering of the striatum. By using seeds derived from the anatomy of the striatum, we applied the various targeted clustering algorithms to determine the gene expression pattern associated with this anatomical structure

mentation that matches the anatomy well. Therefore, this analysis suggests that the gene expression profile of striatum is fairly well separated from the gene expression profile of surrounding structures.

6.6 Conclusion

In this chapter, we showed how the variational models of Chap. 5 could be applied to the targeted, untargeted and semi-targeted clustering problem on both a primal and dual complex. Essentially, the filtering algorithms modeled the denoised data as having low spatial frequency (possibly with discontinuities). These models could easily be applied to clustering by modeling the *cluster labels* as having a low spatial frequency (possibly with discontinuities). Clustering appears in many applications in image processing, machine learning and complex network analysis. We also showed how to apply these clustering models to the clustering of higher-order cells, such as edges to permit flow clustering.

Chapter 7
Manifold Learning and Ranking

Abstract A prominent theme of this book is the spatial analysis of networks and data *independent of an embedding* in an ambient space. The topology and metric of the network/complex have been sufficient to define the domain upon which we may perform data analysis. However, an intrinsic metric defined on a network may be interpreted as the metric that would have been obtained if the network had been embedded into an ambient space equipped with its own metric. Consequently, it is possible to calculate an embedding map for which the induced metric approximates the intrinsic metric defined on the network. The calculation of such embeddings by manifold learning techniques is one way in which the structure of the network may be examined and visualized. A different method of examining the structure of a network is to calculate an importance ranking for each node. In contrast to the majority of this book, the ranking algorithms are generally used to examine the structure of *directed* graphs.

In the context of network theory, **data discovery** refers to any procedure that condenses the vast amount of information contained in a network into a simpler form that can be used to summarize the network and thus simplify the representation. This simplification can be used either as an output to aid human understanding and information extraction or as an preprocessing step to be followed by subsequent analysis of the network.

We will address two data discovery problems using techniques that fit the overall theme of this work. The first problem we will address is dimensionality reduction. The problem of dimensionality reduction is to input a large number of data values that are assigned to each node and then reduce the number of values required to represent the data at each node. Due to the geometric viewpoint behind dimensionality reduction techniques, in this chapter we will interpret the data associated with each node as the node *coordinates*. This reduction is motivated by an assumption that the vertices in high-dimensional space lie on a lower dimensional *manifold* embedded in the high-dimensional space. Since the data is "really" low-dimensional, we may replace the high-dimensional coordinates with low-dimensional coordinates without losing information. An algorithm for producing low-dimensional coordinates for each vertex on the manifold is known as **manifold learning**. Manifold learning may be used for several purposes, such as: (i) visualization of the data for human understanding; (ii) providing an isotropic space where Euclidean distances are meaningful (e.g., allowing meaningful application of algorithms such as k-means); (iii) using the low-dimensional coordinates as inputs to machine learning algorithms; and (iv) generating an intrinsic set of coordinates to parameterize a dataset. The starting point for almost all manifold learning techniques is to apply some method for connecting data points at nodes via edges to form a network. Given this network, various algorithms have been derived for extracting the low-dimensional representation from the structure of the network.

The second data discovery problem that we address in this chapter is node ranking. Node ranking is a process that identifies the relative influence of nodes in a network. Knowing the relative influence of nodes is valuable in several areas, such as social network analysis (identifying the most important people), computer network analysis (identifying which critical machines to protect from attack), or in searching the World Wide Web. The application of web search motivated the development of both algorithms reviewed in this chapter: The PageRank algorithm and the HITS algorithm.

7.1 Manifold Learning

The classical approach to dimensionality reduction is Principal Components Analysis (PCA) [117]. The PCA method reduces a high-dimensional representation of a dataset into a lower-dimensional representation of the dataset such that the variability of the high-dimensional representation is optimally represented. However, PCA is a *linear* method for dimensionality reduction in the sense that PCA finds a suitable set of basis vectors and projects the high-dimensional data onto these vectors. In contrast, the manifold learning techniques may be viewed as *nonlinear* methods for dimensionality reduction. Many real-world datasets that may be described by a lower-dimensional representation do not admit a linear projection onto a lower-dimensional space. For example, Fig. 7.1 shows two examples of one-dimensional data embedded in two-dimensional space. In the first example, PCA will find a single basis vector which fully describes the variance of the data. However, no such basis vector may be found that describes the data when embedded in a spiral.

Formally, we may describe the manifold learning problem as follows: Given a set of points, v_i, each described by a tuple of coordinates $\tilde{s}_i \in \mathbb{R}^m$, find a mapping $f(\tilde{s}_i) = \tilde{x}_i$ such that $\tilde{x}_i \in \mathbb{R}^\ell$ for $\ell < m$. In general, we do not solve for the mapping f explicitly, but instead produce the set of coordinates $\tilde{x}_i$ from the given set of initial coordinates $\tilde{s}_i$. Note that the manifold learning problem is distinct from the classical surface reconstruction problem (e.g., [203]) in which the goal is to find a simplicial manifold which best fits a set of embedded points—in particular, in surface reconstruction the dimensionality is typically known *a priori* and the sampling of the points is often not dense enough for a unique solution.

7.1 Manifold Learning

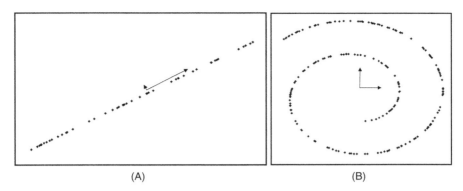

Fig. 7.1 Two examples of one-dimensional data embedded in two-dimensional space. The embedding in (**A**) is described well by the single basis vector found by PCA. However, PCA finds no useful basis-vector representation for the spiral in (**B**). The goal of manifold learning techniques is to find a one-dimensional description of the points that works in both situations

In this section, we review several influential manifold learning techniques which are related to the overall themes of this book: Isomap, Laplacian Eigenmaps and Locality Preserving Projections (LPP). Each of these algorithms solve the manifold learning problem described above. If the points lie on a manifold, then the neighborhood of each point (with a sufficiently dense sampling) is locally Euclidean. Consequently, the embedding space can be used to define *local* neighborhoods. Since the neighborhood of each node is defined by the incident edges, the edge set is typically chosen to connect nodes which are near each other as measured using Euclidean distances in the high-dimensional embedding space. Therefore, a preliminary step of all of these algorithms is to connect the input points locally via an edge set. The standard approach to creating an edge structure is to find the k-nearest neighbors of each node in the high-dimensional space and assign an edge between the node and each of these nearest neighbors (see Chap. 4 for other possibilities). These edges are then assigned distance weights by the Euclidean distance of the coordinates, i.e., $w_{ij} = \|\tilde{s}_i - \tilde{s}_j\|$. Having produced this initial weighted graph from the data, the various manifold learning algorithms proceed differently to produce the low-dimensional coordinates.

7.1.1 Multidimensional Scaling and Isomap

In order to understand the Isomap algorithm for manifold learning we must first consider the **Multidimensional Scaling** (MDS) method for dimensionality reduction. MDS was used for many years in the psychology and social sciences communities before being introduced in the context of manifold learning. MDS inputs a distance matrix between each pair of nodes (computed via any process), assumes that the distances describe (squared) *Euclidean* distances between every two points and, if it exists, finds a series of new coordinates for which the (squared) Euclidean distances between the new coordinates match the desired input distances. Formally, if our

m-dimensional input coordinates for each of the n nodes are represented by a set of n coordinate vectors $\tilde{c}$ of dimension m, and we are given a symmetric $n \times n$ distance matrix **S** defined as $S_{ij} = \mathcal{D}(v_i, v_j)$, then the goal of MDS is to compute coordinates in an ℓ-dimensional space (where $\ell < m$) in the form of a set of n coordinate vectors $\tilde{x}$ of length ℓ such that

$$\|\tilde{x}_i - \tilde{x}_j\|_2^2 = S_{ij}. \tag{7.1}$$

If there exists a set of $\tilde{x}$ such that $S_{ij} = \|\tilde{x}_i - \tilde{x}_j\|_2^2$ then we will call **S** a **Euclidean distance matrix**. When **S** is a Euclidean distance matrix then the minimal embedding may be found by employing the following theorem (see [65, 333] for proof):

Theorem 7.1 *A nonnegative symmetric matrix* $\mathbf{S} \in \mathbb{R}^{n \times n}$ *with zeros on the diagonal is a Euclidean distance matrix if and only if* $\mathbf{P} = -\frac{1}{2}\mathbf{HSH}$, *where* $\mathbf{H} = \mathbf{I} - \frac{1}{n}\mathbf{11}^\mathsf{T}$, *is positive semi-definite. Additionally, the minimum embedding dimension m is the rank of* **P**.

Since **P** is a symmetric positive semi-definite matrix, it admits a Cholesky decomposition $\mathbf{P} = \mathbf{X}^\mathsf{T}\mathbf{X}$. A convenient method for computing the Cholesky factors is to consider the eigenvector decomposition of $\mathbf{P} = \mathbf{Q}\mathbf{\Lambda}\mathbf{Q}^\mathsf{T}$, in which it is clear that the Cholesky factors $\mathbf{X} = \mathbf{Q}\mathbf{\Lambda}^{\frac{1}{2}}$. Computation of the Cholesky factors via an eigenvector decomposition is particularly useful. In particular, if m is the minimum number of dimensions necessary to embed the points with Euclidean distances represented by the distance matrix **S**, then all but m eigenvectors will correspond to eigenvalues equal to zero.

In practice, it may not be possible to embed the vertices in a low dimension such that Euclidean distances between the low-dimensional coordinates match the prescribed distances given by the matrix **S**. However, it may be possible to *approximately* embed the vertices into $\hat{m} < m$ dimensions (i.e., so that the embedded Euclidean distances in $\mathbb{R}^{\hat{m}}$ approximate **S**). Therefore, the standard approach to projecting the data is to define a threshold β such that **x** is comprised of $\mathbf{x} = \hat{\mathbf{Q}}\hat{\mathbf{\Lambda}}^{\frac{1}{2}}$ where $\hat{\mathbf{Q}}$ represents the $\hat{r}$ eigenvectors corresponding to the set of $\hat{\mathbf{\Lambda}}^{\frac{1}{2}}$ eigenvalues less than β.

The MDS algorithm may therefore be summarized by the following steps:

1. Input a distance matrix **S**.
2. Construct $\mathbf{P} = -\frac{1}{2}\mathbf{HSH}$, where $\mathbf{H} = \mathbf{I} - \frac{1}{n}\mathbf{11}^\mathsf{T}$.
3. Compute the eigenvector decomposition of $\mathbf{P} = \mathbf{Q}\mathbf{\Lambda}\mathbf{Q}^\mathsf{T}$.
4. Set the low-dimensional coordinates as $\mathbf{X} = \mathbf{Q}\mathbf{\Lambda}^{\frac{1}{2}}$. Each column i of matrix **X** is the output coordinate vector $\tilde{x}_i$ of low-dimensional coordinates for v_i.
5. Truncate $\tilde{x}_i$ by removing the coordinates corresponding to any eigenvectors with eigenvalues having a magnitude less than the parameter β.

One popular way for finding a distance matrix **S** was proposed in the Isomap algorithm of [373]. The Isomap algorithm begins by building a weighted graph from a set of points embedded in high-dimensional space (see above). Following this graph construction, the Isomap algorithm uses Floyd's algorithm to compute the shortest distance within the graph between every pair of points and assigns these distances to a matrix **S**, i.e., $S_{ij} = \mathcal{D}(v_i, v_j)$. This matrix is then used in the MDS procedure to produce a reduced set of low-dimensional coordinates $\tilde{x}$, such that the Euclidean distance between nodes approximates the distances prescribed by **S**.

The Isomap algorithm consists of the following steps:

1. Input a set of points with a high-dimensional coordinates $\tilde{c}_i$ for each v_i.
2. Construct a graph by connecting every node to its k-nearest neighbors (measured as Euclidean distances between the high-dimensional coordinates).
3. Weight each edge in this graph by the Euclidean distance between the coordinates of the incident nodes.
4. Compute the matrix of all-pairs shortest paths, and calculate the distance matrix **S**.
5. Use **S** as the distance matrix in MDS to compute the set of low-dimensional coordinates, $\tilde{x}_i$, for each v_i.

7.1.2 Laplacian Eigenmaps and Spectral Coordinates

A different approach to manifold learning is given by the (LE) algorithm of Belkin and Niyogi [23]. This algorithm substitutes the **spectral coordinates** of the graph Laplacian matrix as the low-dimensional coordinates. We define the spectral coordinates of node v_i with respect to the symmetric Laplacian matrix **L** under the eigenvector decomposition $\mathbf{L} = \mathbf{Q \Lambda Q}^\mathsf{T}$ to be the *spectra* or the columns of the eigenvector matrix **Q**, i.e., $\mathbf{x}_i = [q_{i1}, q_{i2}, \ldots, q_{in}]^\mathsf{T}$. The spectra therefore assigns a coordinate in a n-dimensional space to each of the n nodes in the graph. If the eigenvectors represented in **Q** are sorted by the magnitude of their corresponding eigenvalues λ_i, then lower-dimensional coordinates can be obtained either by assuming a threshold β as above or by choosing the $\ell < n$ eigenvectors corresponding to the ℓ smallest eigenvalues.

The spectral coordinates of the Laplacian matrix have been used for a long time as a predictor for the "closeness" of two nodes in a graph (see [71, 345]). Specifically, a Euclidean measure of the distance between the spectral coordinates of two points may be interpreted as the *commute time* between these points, when the spectral coordinates are normalized by the eigenvalues. The commute time $T(v_i, v_j)$ is defined as the expected number of steps required for a random walker to pass from node v_i to node v_j and then back again to node v_i (see Chap. 3 for more information). In general, the commute time between two nodes v_i and v_j can be calculated from the normalized eigenvectors of the Laplacian matrix with the expression

$$T(v_i, v_j) = \text{Vol} \sum_{k=2}^{n} \frac{1}{\lambda_k}(q_{ki} - q_{kj})^2, \tag{7.2}$$

where $\text{Vol} = \sum_i d_i$.

Several authors have approached the dimensionality reduction problem directly from the standpoint of commute times [313, 314, 327, 412]. A different view on why the spectrum of a graph can characterize its intrinsic shape is to consider that the spectrum of a membrane has long been linked to the membrane shape and geometry. Kac famously asked whether one can "hear" the shape of a drum (membrane) by listening to its spectral structure [220]. Although it has since been shown that two different drums can share a spectrum [156], there is still a strong correspondence between drum geometry and the spectrum of its Laplacian operator. Similarly, although two (co-spectral) graphs can have the same Laplacian spectrum [35, 151], the spectrum of the Laplacian encodes information about the graph structure. It is an open question to characterize which graphs have a unique spectrum [387].

With this interpretation of the commute time, the LE algorithm and the Isomap algorithm both appear to seek to preserve distances measured along the graph, however there is an important distinction between how these distances are quantified in the two methods. The Isomap algorithm can be viewed as embedding the vertices such that the resulting Euclidean distances between two nodes reflects the length of the shortest path between the nodes (within the graph), while the LE algorithm embeds the vertices such that the resulting Euclidean distances reflect a distance averaged across parallel paths between the nodes (as a result of reflecting the commute time between nodes in the graph). Since the Isomap algorithm treats edge weights as shortest-path distances and the LE algorithm employs the edge weights in the Laplacian matrix, it is important to carefully distinguish the role of the edge weights as they are used in the two algorithms. Specifically, Chap. 2 discussed the important distinction between *distance weights* and *affinity weights*. The affinity weights must be employed whenever the gradient or Laplacian operators are employed, which is the case in the LE method, whereas distance weights are more appropriate for the Isomap method.

Recall that the affinity weights are inversely related (reciprocal) to the distance weights. Furthermore, the convention in the implementation of LE is to employ a nonlinear weighting function such as the Welsch function [23] (see Chap. 4 for more details and possibilities).

To summarize, the LE algorithm computes low-dimensional coordinates **x** by the following steps:

1. Input a set of points with a high-dimensional tuple of coordinates, $\tilde{c}_i$ for each node v_i.
2. Construct a graph by connecting every node to its k-nearest neighbors (measured as Euclidean distances between the high-dimensional coordinates).
3. Weight each edge in this graph by an affinity weighting function (e.g., the Welsch function) from Chap. 4.
4. Construct the Laplacian matrix **L** for this weighted graph.
5. Calculate the generalized eigenvectors of **L**, satisfying **Lq** = λ**Dq** where **D** is the diagonal matrix of (weighted) node degrees.
6. Truncate the output low-dimensional coordinates as a subset of the generalized eigenvectors calculated above. The truncation may be established by either keeping only those eigenvectors corresponding to eigenvalues lower than some

threshold β or by simply keeping a predefined number of eigenvectors (corresponding to the ℓ smallest eigenvalues). The final coordinates for each node are then given by $\tilde{x}_i = [q_{i2}, q_{i3}, \ldots, q_{i\ell}]^\mathsf{T}$.

7.1.3 Locality Preserving Projections

Both the Isomap and LE algorithms require the solution of an eigenvector problem for a matrix of size $n \times n$ where n is the number of nodes. When n is large, these algorithms can be computationally demanding. Furthermore, another weakness shared by both the Isomap and LE algorithms is that after the dimensionality is reduced for a point set, it is unclear how to efficiently reduce the dimensionality of new points that arrive after the initial computation. The Locality Preserving Projections (LPP) algorithm [193] was developed primarily to reduce the computational burden, but was also designed to permit fast dimensionality reduction of new points.

The LPP algorithm is essentially the LE algorithm with the extra assumption that a low-dimensional coordinate for each node is a linear combination of the original high-dimensional coordinates of that node. Additionally, this linear combination of high-dimensional coordinates to produce a low-dimensional coordinate is the same for all nodes. If we let **C** represent the $m \times n$ matrix where each column consists of the original m-dimensional coordinates of a node, and **r** represents the m coefficients of the linear combination used to compute one of the new coordinates for all n nodes, then $\mathbf{z} = \mathbf{C}^\mathsf{T}\mathbf{r}$. Substituting this combination into the LE algorithm yields the coefficients **r** as satisfying

$$\mathbf{CLC}^\mathsf{T}\mathbf{r} = \lambda \mathbf{CDC}^\mathsf{T}\mathbf{r}. \tag{7.3}$$

This relation may be used to compute all $\ell < m$ coordinates for each of the n nodes via

$$\mathbf{X}^\mathsf{T} = \mathbf{C}^\mathsf{T}\mathbf{R}, \tag{7.4}$$

where **X** is the $\ell \times n$ matrix for which the columns contain the ℓ-dimensional tuple of coordinates for each point and **R** is an $m \times \ell$ matrix whose columns are composed of a predetermined number of eigenvectors computed in (7.3). Once **R** is calculated, any new high-dimensional point, **c**, may be projected onto low-dimensional coordinates via the operation $\mathbf{x}^\mathsf{T} = \mathbf{c}^\mathsf{T}\mathbf{R}$.

As in the LE algorithm, the presence of the Laplacian means that all edge weights should be *affinity weights*, for the reasons given in Chap. 4. Following the convention for LE, we may set edge weights using the Welsch function [23, 193] (see Chap. 4 for more details and possibilities).

The LPP algorithm computes low-dimensional coordinates **x** by the following steps:

1. Input a set of points with a high-dimensional tuple of coordinates, $\tilde{c}_i$ for each node v_i.

2. Construct an undirected graph by connecting every node to its k-nearest neighbors (measured as Euclidean distances between the high-dimensional coordinates).
3. Weight each edge in this graph by an affinity weighting function (e.g., the Welsch function), see Chap. 4.
4. Construct the Laplacian matrix **L** for this weighted graph.
5. Solve the generalized eigenvector problem in (7.3) for a predetermined number of the eigenvectors corresponding to the smallest generalized eigenvalues.
6. Output low-dimensional coordinates, **X**, via (7.4).

Figure 7.2 shows the application of these three manifold learning algorithms to flatten to two dimensions the three-dimensional coordinates of the "Swiss Roll"

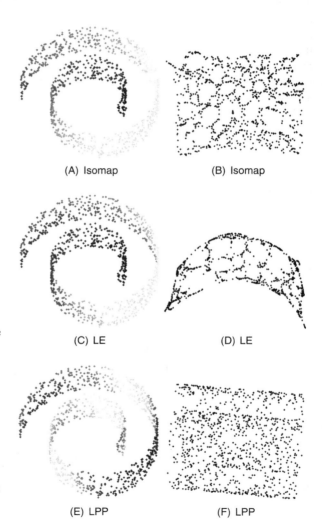

Fig. 7.2 Flattening of the "Swiss Roll" data from Tenenbaum et al. [373]. These points lie on a two-dimensional manifold with a complicated embedding in three-dimensional space. The goal of a manifold learning algorithm is to uncover the "intrinsic" two-dimensional structure and assign meaningful two-dimensional coordinates to each points. The *right column* shows the two-dimensional coordinates assigned to each node by the three algorithms, while the *left column* shows a shading of the three-dimensional data by the horizontal coordinate in the right column

dataset from [373]. Note that the Swiss Roll is a three-dimensional version of the spiral example in Fig. 7.1 which causes difficulties for a PCA-based dimensionality reduction. We can see that the Isomap and LE algorithms produce a meaningful two-dimensional set of coordinates for the Swiss Roll data, while the LPP algorithm produces a projection of the data onto the plane.

7.1.4 Relationship to Clustering

One interpretation of the dimensionality reduction problem is that the goal is to produce low-dimensional coordinates such that the Euclidean distance between the new coordinates gives some measure of the similarity between data points. Consequently, similar data points should be mapped close together and dissimilar data points should be mapped further apart. The clustering problem has a similar goal: Assign a value to each node such that similar nodes (i.e., nodes in the same cluster) have similar values. Once all of the nodes have been assigned a meaningful value, then they may be clustered by thresholding or by using an algorithm such as k-means. From the standpoint of manifold learning, the clustering problem is therefore a reduction to the ultimate low-dimensionality of one dimension.

Figure 7.3 illustrates the concept of using manifold learning techniques for clustering through the application of all three manifold learning techniques to two point clusters. Each point cluster is initially embedded in the "high-dimensional" space of two dimensions and then reduced to one dimension via the manifold learning technique. Each point is shaded with the one-dimensional coordinate that it is mapped to. Using manifold learning techniques in this way effectively removes any distinction between LE and the Normalized Cuts clustering algorithm reviewed in Chap. 6. Since Normalized Cuts is just one specific case of the more general Untargeted Basic Energy Model algorithm presented in Chap. 6, it may be possible to generate additional manifold learning techniques from this same Untargeted Basic Energy Model clustering algorithm for other choices of the p and q parameters.

The most natural connection between manifold learning and clustering is via a reduction of the nodes to one-dimensional coordinates which may be thresholded to produce a hard clustering. However, a different approach to using manifold learning techniques for clustering is to map the nodes to ℓ dimensions as an initial clustering step and then employ an additional clustering method in the dimensionality-reduced space to produce a final clustering. This approach was also taken in the application of Normalized Cuts to clustering [345] and in other studies [71].

7.1.5 Manifold Learning on Edge Data

Instead of high-dimensional data associated with each node, we could also seek a dimensionality reduction of high-dimensional flows associated with each edge. In

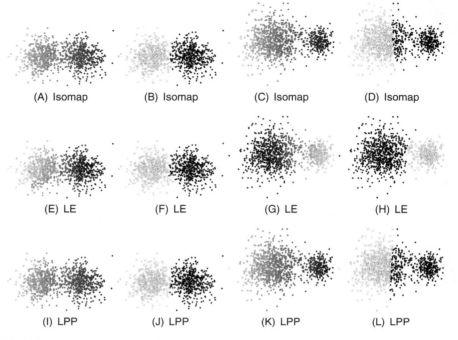

Fig. 7.3 Manifold learning as a clustering algorithm. Each algorithm is applied to two point clusters to reduce the two-dimensional coordinates into a one-dimensional coordinate. The images in the *first* and *third column* display the mapped one-dimensional coordinate as a shading of each point. The images in the *second* and *fourth column* display a labeling of each point which was obtained by thresholding the one-dimensional coordinate to produce two partitions of equal size (other thresholding strategies are possible, see Chap. 6). All manifold learning techniques produced a suitable clustering when the clusters were symmetric, but only LE produced two distinct clusters using this method when the clusters were not symmetric

this case, each edge of the graph has associated with it multiple flow values. Given this circumstance, we might ask whether or not it would be possible to replace these high-dimensional input flows with a representative set of low-dimensional flows.

One motivation given for applying manifold learning techniques was to improve the visualization of the data by reducing the data to a lower, more manageable, dimension for display. However, visualization of edge data is a less important application since visualization of edge data of any dimension is more difficult than the visualization of node data.

A different motivation for applying manifold learning techniques was to reduce the coordinates to a smaller, more meaningful set of coordinates which could be used as inputs to a classification algorithm. This motivation can also be applied to high-dimensional flow data that we wish to "distill" to more meaningful low-dimensional data for purposes of classification. Consequently, it is reasonable to ask whether any of the manifold learning techniques could be applied to high-dimensional edge flow data.

The MDS approach may be used for calculating low-dimensional edge coordinates **y** if a distance metric can be defined for each pair of edges. However, we are unaware of any method in the literature for computing a meaningful distance between a pair of edges on a connected 1-complex that could be used in multidimensional scaling.

The Laplacian Eigenmaps algorithm for manifold learning could be extended to edge data by employing the *edge Laplacian* matrix. Since the degree matrix **D** was interpreted in Chap. 2 as a weighting for each node, then a natural extension of the Laplacian Eigenmaps algorithm to manifold learning on edge data would be to solve the eigenvector problem

$$\mathbf{L}_1 \mathbf{y} = \lambda \mathbf{y}, \tag{7.5}$$

where $\mathbf{L}_1$ represents the edge Laplacian matrix. Similar to the node-based Laplacian Eigenmaps, we could solve (7.5) for a predetermined number of eigenvectors corresponding to the smallest generalized eigenvalues. To our knowledge, manifold learning has not been pursued to reduce the dimension of edge flow data.

7.2 Ranking

One task of data discovery is to determine the most significant or influential nodes in a network. The significance of a node is represented numerically by a variable assigned to each node which is called a node *rank*. The study of algorithms to determine node rank were substantially motivated by the problem of searching the World Wide Web [318]. The goal of the web search problem is to rank the significance of nodes (webpages) that are associated with the search terms. In this section we review the PageRank algorithm underlying the Google search engine [60] and the Hyperlink-Induced Topic Search (HITS) algorithm underlying the Teoma search engine [318]. In contrast to the majority of the rest of the book, the algorithms in this section will apply exclusively to *directed* graphs. The recent paper by Jiang et al. [219] gives an analysis of ranking algorithms from the standpoint of discrete calculus operators (notably the edge Laplacian and curl operators).

7.2.1 PageRank

In node ranking, the goal is to calculate a value x_i, assigned to each node, v_i, (a 0-cochain) that represents the rank of node v_i. As before, $x_i \in \mathbb{R}$. The magnitude of x_i does not matter as much as the relative values of x_i between different nodes, which define the node ranking.

The first principle underpinning the **PageRank** algorithm [60] is that the rank of a node should be high if (i) the rank of nodes linking to that node are high and (ii) should be low if the node has few incoming edges or if the nodes linking to the node have low rank. By itself, this insight has been in use for a long time [223]. Specifically, we may calculate the rank iteratively via the equation

$$x_i^{[k+1]} = \sum_{v_j \in \mathcal{N}_{\text{in}}} \frac{1}{d_j} x_j^{[k]}, \tag{7.6}$$

where $\mathcal{N}_{\text{in}}$ is the set of nodes with an outgoing edge connected to v_i and d_j is the outgoing degree of node v_j. When the directed graph is strongly connected and aperiodic, the iteration in (7.6) converges to the solution

$$x_i = \sum_{v_j \in \mathcal{N}_{\text{in}}} \frac{x_j}{d_j}. \tag{7.7}$$

The steady-state solution of (7.7) may be represented in matrix form using the notation from before,

$$\mathbf{WD}^{-1}\mathbf{x} = \mathbf{x}, \tag{7.8}$$

where we can view **W** as the adjacency matrix of the *directed* graph defined by

$$W_{ij} = \begin{cases} 1 & \text{if } e_{ji} \in E, \\ 0 & \text{else.} \end{cases} \tag{7.9}$$

Note that in the case of directed graphs, e_{ij} represents an oriented and directed edge from node v_i to node v_j. Therefore, the steady-state solution of the PageRank algorithm is given by the right eigenvector of $\mathbf{WD}^{-1}$ corresponding to the eigenvalue equal to unity (which is guaranteed to exist if the graph is strongly connected [60]). The power method is a common method for finding this eigenvector (see Appendix B).

Viewed iteratively, the matrix form of (7.8) has been interpreted as a random walk across the nodes of the network in which a walker located at a node has equal probability of transitioning to each node connected with an outgoing edge. Under the interpretation of a random walker model, the node rank of a given node may be interpreted as the probability that a random walker dropped at any location in the network can be found at the node after a substantial amount of time has passed (i.e., the system has equilibrated into a steady-state). In the context of ranking webpages on the Internet, the random walker model is sometimes called the **random surfer model**.

An important innovation of PageRank over older literature (see [149]) is that the graph is modified such that, in addition to the original edges, *every* node is connected to *every* other node with an outgoing edge. Any nodes that were previously connected before this addition are now connected with two edges. The original edges are given weight α (sometimes called the **teleportation parameter**) and the edges of these additional connections are given weight $(1-\alpha)$. Empirically, the value of $\alpha \approx 0.85$ appears to work best in practice. These additional edges are also referred to as a **dampening factor** in the literature on PageRank.

7.2.1.1 PageRank as Advection

The random walker model is a common interpretation of the PageRank algorithm. In this section, we additionally show that the PageRank algorithm may be interpreted as an *advection* equation in the same form as seen in Chap. 2. To see this connection between PageRank and advection, we begin by letting $\tilde{\mathbf{x}} = \mathbf{D}^{-1}\mathbf{x}$. We may then rewrite our matrix equation for the node rank in terms of a Laplacian matrix for a directed graph,[1] defined as

$$L_{ij} = \begin{cases} d_i & \text{if } i = j, \\ -1 & \text{if } e_{ji} \in \mathcal{E}, \\ 0 & \text{else}. \end{cases} \quad (7.10)$$

Rewriting our matrix equation for PageRank in terms of this Laplacian matrix consists of the following operations:

$$\mathbf{W}\tilde{\mathbf{x}} = \mathbf{D}\tilde{\mathbf{x}}, \quad (7.11\text{a})$$

$$\mathbf{L}\tilde{\mathbf{x}} = \mathbf{0}, \quad (7.11\text{b})$$

$$\mathbf{L}\mathbf{D}^{-1}\mathbf{x} = \mathbf{0}. \quad (7.11\text{c})$$

Therefore, the ranks in the PageRank algorithm are computed by finding the first left eigenvector of this Laplacian matrix representing the directed graph. We note that it is easy to see from (7.11c) that if the (directed) graph is symmetric then $\mathbf{L}^\mathsf{T} = \mathbf{L}$ and the solution to (7.11c) trivially ranks each node as proportional to its degree, since the zero eigenvector of the Laplacian matrix is constant.

Recall from Chap. 2 that we can decompose this directed Laplacian matrix into the components $\mathbf{L} = \mathbf{A}^\mathsf{T}\mathbf{A}^+$, where $\mathbf{A}^+$ indicates the incidence operator in which all non-positive values are set to zero (i.e., $A^+_{ij} = A_{ij}$ if $A_{ij} > 0$ and $A^+_{ij} = 0$ otherwise). Consequently, we can rewrite the eigenvalue equation of (7.11c) as

$$\mathbf{L}\mathbf{D}^{-1}\mathbf{x} = \mathbf{A}^\mathsf{T}\mathbf{A}^+\mathbf{D}^{-1}\mathbf{x} = \mathbf{A}^\mathsf{T}\mathbf{G}^{-1}\mathbf{A}^+\mathbf{x} = \mathbf{0}, \quad (7.12)$$

where $\mathbf{G}^{-1} = \text{diag}(\mathbf{g}^{-1})$ and $g_{e_{ij}} = 1/d_i$. Recall that in the context of the directed graphs of this section, d_i represents the degree of *outgoing* edges at node v_i.

Equation (7.12) represents an advection process on a graph described in Chap. 2. If we interpret PageRank as an advection process, then (7.12) describes a graph having an advection flow field with direction equal to the edge directions and flow field for each outgoing edge from node v_i having magnitude of $1/d_i$. Viewed as an advection process, PageRank finds the unique distribution of concentrations (ranks)

---

[1] Note that Chung defined a symmetric conception of the Laplacian operator on a directed graph [79]. See Chap. 2 for more information on this advection process and the corresponding Laplacian matrix used here.

whose transport under this advection process has zero divergence. We may write the divergence for each node in summation form

$$\mathrm{div}(v_i) = \sum_{v_j \in \mathcal{N}_{\mathrm{in}}} \frac{1}{d_j} x_j - \sum_{v_j \in \mathcal{N}_{\mathrm{out}}} \frac{1}{d_i} x_i = \sum_{v_j \in \mathcal{N}_{\mathrm{in}}} \frac{1}{d_j} x_j - x_i, \qquad (7.13)$$

where $\mathcal{N}_{\mathrm{out}}$ indicates the set of nodes receiving an outgoing edge from v_i. This equation for the divergence returns us to the original summation formulation of PageRank given previously in (7.7).

One value of this interpretation for the PageRank algorithm would be to form a basis for the development of a PageRank algorithm for ranking edges. Specifically, the advection interpretation suggests that the natural extension of PageRank to edges would require a definition of an *edge* (vector) advection process which could be brought to steady-state. At present, such an extension would be the subject of future work.

7.2.2 HITS

The PageRank algorithm determines the influence of a node based on the number of incoming edges from other influential nodes. However, real-world networks are often built organically over time in which the links are primarily directed backward in time. This phenomenon is exhibited in networks such as the World Wide Web or the network of judicial citations (see Sect. 7.3.4). Consequently, the PageRank algorithm favors older, established nodes even if more recent nodes are very important. One approach to overcoming this bias toward older nodes is to identify some nodes as *hubs*, which are nodes with a good track record of linking to important nodes. If we have identified several good hubs, a new important node can immediately be identified if it is linked to by the hubs. This approach to the ranking problem is taken by the **Hyperlink-Induced Topic Search** (HITS) of Kleinberg [236]. The HITS algorithms assigns each node two scores, a *hub score* and an *authority score*. The authority score corresponds to the rank value of the PageRank algorithm, while the hub score identifies a node which is good at linking to nodes with a high authority score.

Specifically, the HITS algorithm assigns the authority score, x_i, and the hub score, $\bar{x}_i$, for v_i via the iteration

$$x_i^{[k+1]} = \sum_{v_j \text{ if } \exists e_{ji} \in \mathcal{E}} \bar{x}_j^{[k]}, \qquad (7.14)$$

$$\bar{x}_i^{[k+1]} = \sum_{v_j \text{ if } \exists e_{ij} \in \mathcal{E}} x_j^{[k]}, \qquad (7.15)$$

which is shown to converge [236]. Note that the authority score looks backward by searching over edges e_{ji} and the hub score looks forward by searching over edges e_{ij}. These equations may be written in matrix form using the directed adjacency matrix as

$$\mathbf{x}^{[k+1]} = \mathbf{W}^T\bar{\mathbf{x}}^{[k]}, \quad (7.16)$$

$$\bar{\mathbf{x}}^{[k+1]} = \mathbf{W}\mathbf{x}^{[k]}. \quad (7.17)$$

Consequently,

$$\mathbf{x}^{[k+1]} = \mathbf{W}^T\mathbf{W}\mathbf{x}^{[k]}, \quad (7.18)$$

$$\bar{\mathbf{x}}^{[k+1]} = \mathbf{W}\mathbf{W}^T\bar{\mathbf{x}}^{[k]}. \quad (7.19)$$

Therefore, HITS computes the node rank $\mathbf{x}$ as the eigenvector of $\mathbf{W}^T\mathbf{W}$ corresponding to the eigenvalue equal to unity. Similarly, the node hub score $\bar{\mathbf{x}}$ is the eigenvector of $\mathbf{W}\mathbf{W}^T$ corresponding to the eigenvalue equal to unity. These eigenvectors are often calculated via the power method (see Appendix B).

The HITS algorithm has several important differences with the PageRank algorithm for computing node rank. First, the HITS algorithm computes not only a rank for each node, but also a hub score. Second, the PageRank algorithm contains a free parameter α which is not typically included with the HITS algorithm. Third, the PageRank algorithm tends to favor older, downstream nodes, while the HITS algorithm has less bias in this regard. Fourth, the PageRank algorithm finds a unique solution while HITS may not [130]. Finally, the PageRank algorithm has two natural interpretations as a PDE on a graph (diffusion and advection), while the HITS algorithm is not often interpreted in these terms. As a result of these differences, the HITS algorithm and the PageRank algorithm often compute quite different node ranks for a directed network.

7.3 Applications

We examine several applications of the manifold learning and ranking algorithms presented in this chapter. Manifold learning is applied in a conventional manner to study the shape deformations of the human liver and in an unconventional manner to generate correspondences between vertices of a mesh. The ranking algorithms are then applied to the Internet search problem for which they were designed and then to analyze the network of US Supreme Court citations.

7.3.1 Shape Characterization

The most common use of manifold learning techniques is to reduce a high-dimensional signal to a low-dimensional representation. This low-dimensional representation may be used to accomplish several tasks, such as

1. Data discovery: Generating an intuitive understanding and visualization of the data.
2. Extracting a small set of numbers for each object (the low-dimensional coordinates) which may be used with a machine learning algorithm to classify objects.
3. Interpolating between data points, e.g., to create an "average" object within the space.

One area in which manifold learning can help with these tasks is in image processing. Given a set of images to classify, the intensity of each pixel in an image may be considered as one of the image dimensions. Therefore, an image with m pixels can be viewed as an m-dimensional signal which may be described by manifold learning techniques as an ℓ-dimensional signal in which $\ell \ll m$. For example, in the original Isomap study [373] the authors used manifold learning to characterize the space of binarized handwritten digits of the number '2'. Each image with m pixels was first rearranged into a column vector and then reduced to two dimensions with the manifold learning technique. The authors observed that each dimension was semantically meaningful, capturing the "bottom loop articulation" and the "top arch articulation".

To illustrate this type of manifold learning application, we have applied our techniques to characterize the shape of the human liver, obtained from segmentations of CT scans. Our data was comprised of 108 segmented (binarized) livers in three dimensions that were registered to a common space with $128 \times 128 \times 128$ voxels (a space with 2,097,152 dimensions) [238]. Treating each segmented liver as a node, we used the Hamming distance to compare two nodes and used this distance to establish an edge between each liver and its fourteen nearest neighbors. Each edge was then weighted with the Hamming distance between the two nodes and the Isomap and Laplacian Eigenmaps algorithms were used to generate a two-dimensional characterization of the space of liver shapes.

Figure 7.4 displays the results of these two-dimensional embeddings. Selected nodes (indicated with a circle) are illustrated with a small image displaying a three-dimensional rendering of the liver corresponding to the node. These renderings of the livers were intended to be displayed from a common viewing angle. The (normalized) two-dimensional coordinates of each liver obtained from Isomap may be interpreted semantically as characterizing the functional lobes of the liver. Each liver is comprised of two functional lobes, the larger right lobe (pictured toward the left and bottom) and the smaller left lobe (pictured toward the upper right). The horizontal coordinate assigned to each liver by Isomap measures the distinctiveness of these two functional lobes such that a liver mapped to a small horizontal coordinate has distinct right and left lobes, while those livers mapped to a large horizontal coordinate have a much less distinct right and left lobe. Similarly, the vertical coordinate assigned by Isomap may be interpreted as describing the relative prominence of the right or left lobe. A liver mapped to a smaller vertical coordinate has a more prominent right lobe while a liver mapped to a larger vertical coordinate has a more prominent left lobe. Although the coordinates produced by the Isomap embedding had a natural semantic interpretation, the semantic interpretation for the embedding obtained via Laplacian Eigenmaps is much less clear. However, we saw in the Swiss

7.3 Applications

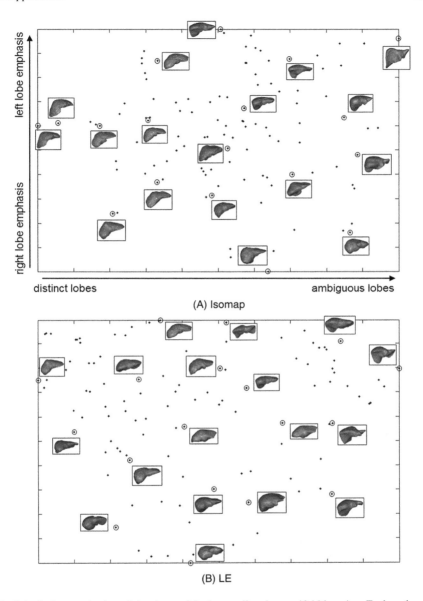

Fig. 7.4 A characterization of the shape of the human liver by manifold learning. Each node represents one liver segmented from a database of CT images. Selected nodes (indicated with a *circle*) are illustrated with a small image displaying a three-dimensional rendering of the liver corresponding to the node. The embedding obtained from Isomap provides a semantic interpretation: The *horizontal coordinate* of a liver describes how distinguishable the right and left lobes are, while the *vertical coordinate* describes the relative prominence of the right or left lobes. These coordinates could be used to train a classifier to diagnose hepatic pathologies, to cluster livers into different types, to work in conjunction with an image segmentation algorithm to constrain the segmentation to lie within the liver manifold, or simply as a visualization device to explore the dataset of livers

Roll example of Fig. 7.2 that the two algorithms could produce a very different embedding of the same graph.

7.3.2 Point Correspondence

The standard use of manifold learning is to reduce a high-dimensional dataset to a lower number of dimensions. Indeed, a manifold learning technique should be able to flatten a low-dimensional dataset embedded into a high-dimensional space regardless of how the dataset is embedded. Therefore, a two-dimensional sheet embedded in three dimensions should flatten back to two dimensions, regardless of whether the sheet is embedded in three dimensions as a Swiss Roll or a crumpled plane. This principle can be exploited to turn any manifold learning technique into a point correspondence method between two high-dimensional objects [258, 274, 421]. Specifically, two manifolds which are fundamentally the same should flatten to the same locations (up to a rigid transformation and sign ambiguity) regardless of the initial embedding in higher dimension. Using this principle, the points of a dataset with two different embeddings may be matched by "flattening" each object and using the flattened coordinates assigned to each point as a signature which may be used to produce point correspondences.

In this example, we show how to produce a point correspondence between the three-dimensional meshes of two horses taken with a different pose [365]. Each horse mesh is composed of the same number and topology of nodes, edge and faces. Each edge was weighted by the Euclidean length of the edge in three dimensions. Isomap, Laplacian Eigenmaps and LPP were used to "flatten" these meshes to three dimensions, meaning that each three-dimensional coordinate of the mesh was replaced by a different set of canonical coordinates. These new three-dimensional coordinates assigned to each vertex can then be used to find a point correspondence between all of the points in each pose.

Previous studies in the literature of this point correspondence technique have focused on Laplacian Eigenmaps, after which the two coordinate sets are aligned via a rigid transformation. In this example, the rigid transformation was ignored. Figure 7.5 (contained in the color plate section at the end of the book) displays the results of the point correspondence where the horses are rendered with the original three-dimensional coordinates of each vertex which are colored with an RGB value that represents the "flattened" three-dimensional coordinates produced by each manifold learning algorithm. If the algorithm works well, then the location of each color on the manifold should be unique and vertices with the same colors in the two meshes may be considered to be in correspondence. In this example, Laplacian Eigenmaps appears to produce the best correspondence by successfully matching the legs, head, and tail between the two poses, whereas the LPP method performs slightly less well in matching the front right leg, and the Isomap method mis-registers the front legs as well as the tail.

Point correspondences are often computed for surface matching in medical imaging data where two anatomical regions are to be compared across individuals. For

7.3 Applications

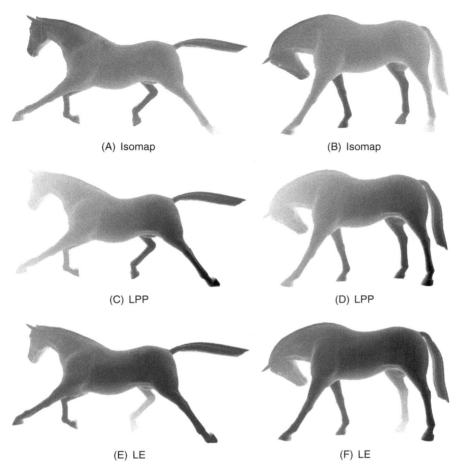

Fig. 7.5 Two three-dimensional meshes of a horse in different poses. Each mesh was mapped to three dimensions using the various manifold learning techniques. The mapped three-dimensional coordinates of each point were mapped to RGB space for display. Two vertices having similar colors should therefore be identified as the same coordinate location in the two meshes

example, the surface of the human brain exhibits a stereotyped folding pattern and many specific brain functions are localized to specific folds, yet the overall size shape of the brain and its folds—or, in other words, the *embedding* of the brain surface—varies across individuals. Although powerful techniques exist for computing correspondences between brain surfaces generated from anatomical Magnetic Resonance Imaging (MRI) data that align corresponding folds across individuals (e.g., [137]), brain surface models contain up to 250,000 faces and therefore many of these surface matching techniques are quite computationally demanding. The Laplacian Eigenmaps method was applied to match two mesh representations of the surface of the left cerebral hemisphere of the human brain generated from anatomical MRI data collected in two patients, where the goal in this context is to align

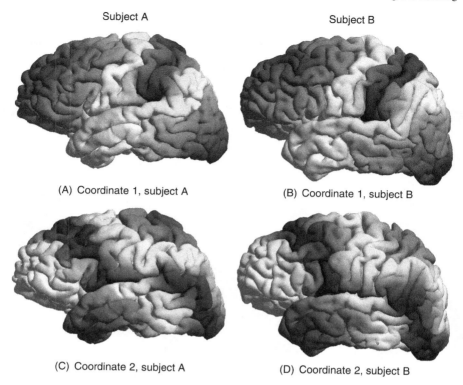

Fig. 7.6 Brain surface matching via the Laplacian Eigenmaps method. The first two eigenvectors of the graph Laplacian are used to establish corresponding two-dimensional coordinate parameterizations of the surfaces which can be utilized as a correspondence map providing the matching. The surfaces represent the outer surface of the cortical gray matter (i.e., the "pial surface") and each surface was generated automatically with the FreeSurfer software environment [96, 136]. The surface meshes for subjects A and B contained 248,868 and 259,792 triangles, respectively

the cortical folds of the two individuals. As before, the Laplacian matrices were weighted by the edge length under the input embedding in three dimensions measured with the Euclidean metric, but only the computation of the largest two eigenvectors is required, making the method extremely fast. The results are presented in Fig. 7.6 (contained in the color plate section at the end of the book). Corresponding folds in the two surface models were assigned similar coordinates, indicating that the Laplacian Eigenmaps performed well at matching the folding patterns.

7.3.3 Web Search

The ranking algorithms in the text were originally devised in the context of searching the World Wide Web, which is treated as a directed network in which each node is identified with a web page and each directed edge is identified with a link from

7.3 Applications

Table 7.1 The top ranked web pages in a snapshot of the English language Wikipedia in 2005. The PageRank and HITS algorithms emphasize different qualities, such that the 'Year 2000' is the only page with a high authority ranking in both algorithms. In addition to the authority ranking, the HITS algorithm provides a ranking of pages which link to many pages with a high authority score, which are called hubs

Rank	PageRank top pages	HITS top authority pages	HITS top hub pages
1	United States	Year 2000	Taunton, MA
2	Race (U.S. Census)	Population density	Minneapolis, MN
3	United Kingdom	Square kilometer	Syracuse, NY
4	Year 2000	Census	Philadelphia, PA
5	France	Square mile	Savannah, GA

one page to another. A ranking algorithm is then applied to this network (or a subset). Ultimately, the pages with highest rank that are deemed relevant to the search terms are returned. However, the details of how the search terms are used vary from one search engine to another. One example is the approach taken by Kleinberg [236] in which a subgraph of the web may be built by including (a) every web page containing a keyword, (b) every page pointed to by those pages, and (c) some fraction of the pages that point to a page containing the keyword. The ranking algorithm may then be applied to this subgraph to produce a ranking which is returned to the user.

In this example, we sidestep the issues of properly utilizing the search terms and simply offer an example of the rankings for a subgraph of the World Wide Web taken at a point in time. We obtained the (English language) Wikipedia graph taken from a snapshot on Nov. 11th, 2005 [86]. Running the PageRank (with $\alpha = 0.85$) and HITS algorithms on this subgraph of the entire Wikipedia in 2005 produced the rankings displayed in Table 7.1. The PageRank algorithm assigned its highest ranking to countries which are presumably linked by any page (influential or not) which has any relationship with the country. In particular, the United States and United Kingdom featured prominently in the rankings produced by PageRank, presumably because it was the English language Wikipedia pages that were included in our experiment and because Wikipedia originated in the United States. In both the PageRank and HITS rankings, the 'Year 2000' appeared with high rank. One explanation for this high rank is that many pages in Wikipedia include a progressive timeline with several dates, often with a tilt in the information toward the present time. In contrast to the "authoritative" concepts in the PageRank and HITS authority rankings, the HITS algorithm also provides a set of *hubs*, which are pages that link to pages with high authority. All of the pages with high hub scores in 2005 were cities in the United States. If we consider that these hubs link to pages with high authority ranking, then it is not so surprising that cities provide good hubs—all of the concepts with high authority shown to have the highest authority ranking would appear in an overview page of one of these cities.

7.3.4 Judicial Citation

The ranking algorithms described in this chapter were originally designed for web search, but may be applied to any directed graph (or subgraph) to produce a list of the most influential or important nodes. Instead of using the rankings to search a network, the ranking algorithms may also be used to study a network and test hypotheses about its structure. An example of this usage for the ranking algorithms is given by the network of US Supreme Court judicial citations compiled and studied by Fowler et al. [140, 141]. The network consists of 30,288 nodes where each node is identified with a case decided before the US Supreme Court and a directed edge between two nodes is identified with a case's citation of a previous court case.

The judicial citation network allows for the analysis of many interesting questions. What were the most influential cases within a subgroup (e.g., criminal, civil)? How did the influence of a particular case change over time? How did the judicial priorities of the court change over time? Is there a correlation between the influence of a case and other factors (e.g., whether the case was overturned)? Several of these questions were addressed in the work by Fowler et al. In our experiments we simply computed the PageRank and HITS rankings for the entire network in order to identify the level of influence of each case (using $\alpha = 0.85$ for PageRank). Identifying the most influential cases is a standard feature of several legal references, which is intended to help focus the reader's attention on which cases they should give particular attention. However, these lists of influential cases are generated in a somewhat subjective manner which causes disagreement between different legal sources. For example, Fowler et al. include the lists of influential cases provided by the Oxford Guide's list of salient cases and the Legal Information Institute's (LII) list of important cases. The Oxford Guide and the LII includes about 1–2% of the cases as influential. When one list names a case as influential, there is only about a 50% chance that the other list also names the case as influential, indicating a high level of subjectivity in the production of these lists. One appeal of a computational analysis of the history of court citations is the ability to numerically assign an objective score of the influence of each case.

Figure 7.7 shows the computed influence for each court case using the two algorithms. Some cases are highlighted which have a high rank relative to the average rank during that historical period. The PageRank algorithm favored older cases over newer ones, while the HITS algorithm preferred cases which were more recent. PageRank found that the most influential cases were *Brown v. Maryland* and *Gibbons v. Ogden*. The Oxford guide listed both of these cases as influential, while the LII listed only *Gibbons* as influential. These two cases represent landmark decisions in the early court of Chief Justice John Marshall who used these cases to establish the court's interpretation of the commerce clause in the US Constitution. The commerce clause is an extremely important aspect of the US Constitution because much of the ability of the federal government to regulate states is derived from its ability to regulate commerce. Consequently, the commerce clause of the US constitution is often a flashpoint of contention between the state and federal governments. In contrast, the HITS algorithm identified *Cantwell v. Connecticut* and *Schneider v. State*

7.3 Applications

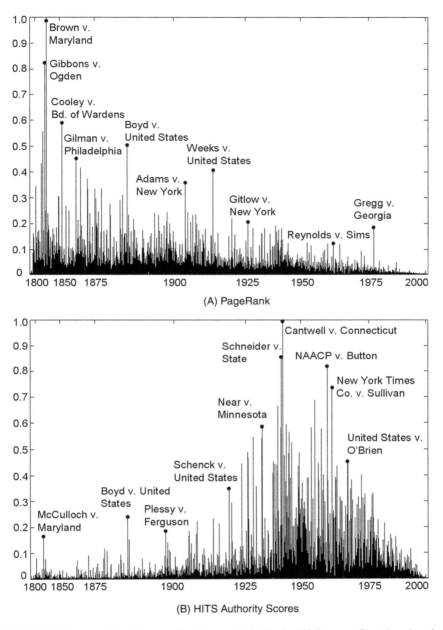

Fig. 7.7 Calculation of the influence of each case decided by the US Supreme Court based on the citation network compiled by Fowler et al. [140, 141]. The PageRank and HITS algorithms were both applied and the rank scores for each algorithm were normalized. Cases with a high influence score relative to the contemporary cases are labeled with a *solid dot* and the name of the case

of New Jersey as the most influential cases. These two cases helped establish the court's interpretation of freedom of religion and specifically the right of a state to restrict the practice and preaching of a religion. Although *Cantwell* was recognized by both the Oxford Guide and the LII as an influential case, neither of these sources listed *Scheider* as an influential case. One of the only court cases that was given a high ranking by both PageRank and HITS relative to its contemporaries was *Boyd v. United States*. The *Boyd* case was a landmark civil liberties case which developed the protections of an individual's right to privacy. The Oxford Guide lists *Boyd* as an influential case, while the LII does not.

7.4 Conclusion

In this chapter we reviewed various methods for summarizing data and graph structure present in a network. The manifold learning methods all proceeded by first generating a graph from a set of points and then computing the eigenvectors of a matrix associated with the graph. We showed that manifold learning methods can be useful for summarizing data and finding intrinsic coordinates to describe the data. Additionally, we showed that manifold learning methods are strongly related to node clustering and considered the possibility of extending these methods to edge flow data. These manifold learning algorithms were seen to fit directly into the theme of this work by the central appearance of the spectrum of the Laplacian operator.

The ranking algorithms presented in the chapter also provided a mechanism for analyzing the importance of nodes in directed graphs. Due to the formulation on directed graphs, the ranking algorithms could be interpreted as an advection equation on a graph (reviewed in Chap. 2), as well as a directional diffusion operation.

Chapter 8
Measuring Networks

Abstract We have adopted the view of graphs and, more generally, cell complexes as a domain upon which we may apply the tools of calculus to formulate differential equations and to analyze data. An important aspect of the discrete differential operators is that *the operators are defined by the topology of the domain itself*. Therefore, in an effort to provide a complete treatment of these differential operators, we examine in this chapter the properties of the network which may be extracted from the structure of these operators. In addition to the network properties extracted directly from the differential operators, we also review other methods for measuring the structural properties of a network. Specifically, the properties of the network that we consider are based on distances, partitioning, geometry, and topology. Our particular focus will be on the *measurement* of these properties from the graph structure. Applications will illustrate the use of these measures to predict the importance of nodes and to relate these measures to other properties of the subject being modeled by the network.

Measures of network properties have been important in several applications. Graph measures may be used to summarize node or network properties to help understand the network. These measures may also be used to make predictions about the separability of the network or the importance of individual nodes or edges. A different use of graph measures has been to predict properties of the subject which is being modeled computationally as a network. For example, a series of network measures have been used to determine whether the configuration of a network of cells inside tissue imaged in a histological section can predict cancer [101, 180]. These network measures have also been widely used in chemical graph theory to predict various structural and behavioral properties of molecules. In this chapter, we review several types of measures that describe the connectedness, topology, or geometry of a network and then give applications of these measures in social networking and chemical graph theory. Readers wishing to explore the network measurement topic further are referred to the excellent review article of Costa et al. [89].

8.1 Measures of Graph Connectedness

In 1947 the chemist H. Wiener observed that a particular measure on the graph that represents an alkane molecule allows one to predict the boiling point of the compound. Pursuing this line of research led Wiener to show that this same distance measure could be used to calculate a series of chemical properties that had previously escaped prediction. When interest in this work was reawakened in the 1970s, it triggered a search within chemical graph theory for other distance-based "topological indices" that could be used to predict various properties of molecules (see Sect. 8.5.2 for a detailed example of this line of work). More recently, a set of similar measures has been used to describe the characteristics of complex networks. Almost all of these quantities are derived from measures of distance on a graph, which we now review.

8.1.1 Graph Distance

We first review the distance between two nodes on a weighted graph that was defined in Chap. 4. (If there are no weights specified and the graph is embedded in $\mathbb{R}^N$, then the edge weights can be set to reflect the Euclidean length of the edge.) If we consider a graph with any set of positive weights associated to the edge set, then given these edge weights we may define the distance between any pair of nodes using the distance operator $\mathcal{D}(\cdot, \cdot)$ as

$$\mathcal{D}(v_i, v_j) = \min_{\Pi_{i,j}} \sum_{e_{ij} \in \Pi_{i,j}} w(e_{ij}), \quad (8.1)$$

where $w(e_{ij})$ is the distance weight of edge e_{ij}, $\Pi_{i,j}$ is a set of edges representing a path between v_i and v_j, and we define $\mathcal{D}(v_i, v_i) = 0$. The optimal path connecting the pair of nodes is called the **shortest path**, and thus the distance is defined as the length of the shortest path between the pair of nodes along the edges of the graph. If no path connects v_i and v_j (the graph is disconnected), then we define the corresponding distance as $\mathcal{D}(v_i, v_j) = \infty$. When the weights are positive, the distance defined in this way establishes a formal *metric* between nodes on the graph, since the distance is nonnegative, symmetric, discernible and satisfies the triangle inequality (see Sect. 4.4).

> A **metric** can be defined for any graph with positive edge weights by defining the distance between any pair of nodes by the length of the shortest path connecting them.

There are many fast algorithms available to compute the distance between two nodes. The most common algorithm for computing distance is Dijkstra's algorithm [108] which may be applied to any connected graph with nonnegative weights.

8.1.2 Node Centrality

Armed with the notion of distance between two nodes given by (8.1), we may examine the importance of a particular node by considering the distance between the node and the rest of the network. Measures of node importance are called **node centrality** measures due to the use of distances to determine how "central" a node is within a network (for a more extensive review of node centrality, see [247]). The most natural method for using distance to measure the centrality of a particular node is to examine the distance from the node to all other nodes in the graph. This measure of node centrality is known as the **total distance** of a node v_i, which is defined as

$$\text{TD}(v_i) = \sum_{v_j} \mathcal{D}(v_i, v_j). \tag{8.2}$$

The total distance is proportional to the **closeness** measure of a node, which is defined as the average distance from the node to all other nodes in the graph, i.e.,

$$\text{Closeness}(v_i) = \frac{1}{n-1} \text{TD}(v_i), \tag{8.3}$$

where n is the number of nodes in the graph, $n = |\mathcal{V}|$. Total distance for a single node may be computed efficiently by Dijkstra's algorithm.

The total distance and closeness both measure node centrality by examining the distance between a single node and the entire network. Therefore, each quantity provides an aggregated measure of node centrality. Instead of examining an aggregated measure of node centrality, we could adopt another view of node centrality by considering only the worst-case distance between a node and the remaining network. A measurement of worst-case node centrality is provided by the node **eccentricity**, which measures the maximum distance between the node and any other node in the network. Eccentricity is defined for node v_i as

$$\text{Eccentricity}(v_i) = \max_{v_j} \mathcal{D}(v_i, v_j). \tag{8.4}$$

A different approach to measuring node centrality is to examine the importance of a node as a link between other pairs of nodes. This **betweenness** has been characterized numerically as

$$\text{Betweenness}(v_i) = \sum_{\substack{v_j, v_k \\ v_i \neq v_j \neq v_k}} \frac{\sigma_{v_j, v_k}(v_i)}{\sigma_{v_j, v_k}}, \tag{8.5}$$

where $\sigma_{v_j, v_k}(v_i)$ indicates the number of shortest paths between v_j and v_k that pass through v_i, and σ_{v_j, v_k} indicates the total number of shortest paths joining v_j and v_k. Therefore, the betweenness measure represents the fraction of optimal paths between every pair of nodes that cross through v_i. Calculation of node betweenness is more expensive than the previous measures due to the fact that it is necessary to know the optimal paths between all pairs of nodes. The classic algorithm for computing all-pairs shortest paths is the Floyd–Warshall algorithm [87], although Johnson's algorithm or repeated applications of Dijkstra's algorithm are considered more efficient for sparse graphs [87]. Due to the usefulness of the betweenness measure in practice, specialty algorithms to compute betweenness have been developed that are faster than computation of the all-pairs shortest paths (the most efficient specialty algorithm is by Brandes [57]). However, even these specialty algorithms for computing betweenness are too expensive for large graphs, causing continued work on fast algorithms to approximate betweenness [16].

Not every measure of node centrality is based on shortest-path distance. Another common measure of node centrality is simply the number of neighbors of the node (the node degree). Degree may be a useful measure of node centrality for networks in which there is a large range of different node degrees, but may be less useful to describe the importance of nodes in networks for which the degree distribution has low variance.

8.1.3 Distance-Based Properties of a Graph

Having explored methods for measuring the centrality of nodes from distance calculations, we may use these methods to define a series of distance-based measures to describe the "connectedness" of the entire graph. The first of these global measures is the most successful measure used in chemical graph theory, and was defined by Wiener. Although many topological indices have since been proposed in the chemical graph theory literature, the measure used by Wiener has had the most success and influence. This measure is known as the **Wiener index** or **Wiener number** which is defined for a graph $\mathcal{G}$ as

$$W(\mathcal{G}) = \frac{1}{2} \sum_{v_i} \sum_{v_j} \mathcal{D}(v_i, v_j). \tag{8.6}$$

The Wiener index represents the sum of the shortest path lengths between all pairs of nodes in the graph. Therefore, a graph with a small Wiener index is more well-connected than a graph (with the same number of nodes) having a large Wiener index. In the network theory literature, it is more common to use the **average path length** which normalizes the distances comprising the Wiener index by the number of node pairs. The average path length is defined as

$$\text{AveragePathLength}(\mathcal{G}) = \frac{2W(\mathcal{G})}{n^2 - n} = \frac{1}{n^2 - n} \sum_{v_i} \sum_{v_j} \mathcal{D}(v_i, v_j). \tag{8.7}$$

Calculation of the Wiener index and average path length requires knowledge of the shortest paths between all vertex pairs in a network. Consequently, the Floyd–Warshall algorithm or the Johnson algorithms [87] are the most common methods for calculating the Wiener index and average path length.

The average path length and the Wiener index are measures of the connection strength between all pairs of nodes in the graph. However, both of these measures define the connection strength between nodes by the length of the *optimal* path. Since these measures are based purely on the length of a single optimal path, they may not reflect a more global measure of connectedness in the graph defined by multiple paths. Therefore, a different view of node connectivity is to define the strength of a connection between two nodes by the number of parallel paths. One way of measuring the strength of parallel connections is to use the resistance distance [235] between two nodes when the graph is viewed as an electrical circuit (see Chap. 3 for more discussion of the effective resistance and circuit analogy). Specifically, the resistance distance has been used to measure the strength of parallel paths connecting all pairs of nodes by defining the **Kirchhoff index** [43, 109]

$$\text{KI}(\mathcal{G}) = \frac{1}{2} \sum_{v_i} \sum_{v_j} R_{\text{eff}}(v_i, v_j) = n \, \text{trace}\{\mathbf{L}^\dagger\}, \tag{8.8}$$

where $\mathbf{L}^\dagger$ indicates the pseudoinverse of the Laplacian matrix and $R_{\text{eff}}(v_i, v_j)$ is the effective resistance between nodes v_i and v_j, as defined in Sect. 4.4. The Kirchhoff index of any graph is always smaller than the Wiener index, since $\mathcal{D}(v_i, v_j) \geq R_{\text{eff}}(v_i, v_j)$ implies that $W(\mathcal{G}) \geq \text{KI}(\mathcal{G})$. This inequality becomes an equality when there exists only a single path between all pairs of nodes, i.e., the graph is a tree.

When computing the Wiener index or Kirchhoff index for a weighted graph, care must be taken in the interpretation of the weights. Weights are incorporated in the definition of the distance measure (8.1) required by the Wiener index, and weights are also incorporated in the definition of the Laplacian matrix (2.96) required by the Kirchhoff index. However, in Chap. 2 we saw that the roles of the prescribed weights are *not the same* in these two cases. Given a weighted graph, one must take care to choose, depending on the problem and on the origin of the weights, whether the weights are "distance weights" or "affinity weights". Chapter 2 contains a discussion of these two interpretations of weights, and there it is shown that distance weights correspond to the primal metric tensor and affinity weights to the dual metric tensor. If the measures of distance are to be equivalent when quantified using the distance operator and using the Laplacian operator, distance weights must be used in the definition of the distance measure, and affinity weights in the definition of the Laplacian.[1] Specifically, for the two measures to operate using the same underlying metric on a single graph, i.e., for the two measures to be compatible, the weights used in the Laplacian matrix must be the *reciprocal* of the prescribed edge weights used in the distance measure. For example, in the case of a tree the two indices should be equal. However, improper interpretation of the weights will cause the two calculated indices to be unequal (i.e., using distance weights for both measures or

---

[1] In the circuit theory analogy, for the two indices to be compatible the prescribed edge weights of a weighted graph are interpreted as *resistances* when measuring the Wiener index, and the same prescribed edge weights on the weighted graph are interpreted as *conductances* when measuring the Kirchhoff index.

affinity weights for both measures will not yield the same value). Therefore, the comparison of the Wiener index and the Kirchhoff index provides a good example of the importance of distinguishing these two interpretations of the prescribed edge weights.

The **quasi-Wiener index** was also introduced [109, 279] to measure the strength of parallel paths connecting all pairs of nodes by calculating the eigenvalues of the Laplacian matrix. However, Gutman and Mohar proved that the quasi-Wiener index and the Kirchhoff index are equal for all graphs [182] and therefore we do not discuss the quasi-Wiener index any further.

Although the authors are not aware of any attempts to use topological indices defined on the *edge Laplacian* to measure other aspects of the graph (or the chemical properties of the molecule), the measures above suggest an easy extension to global edge–face (edge–cycle) relationships. Specifically, a higher-order Kirchhoff index could be defined as

$$\text{Higher-OrderKI}(\mathcal{G}) = m \operatorname{trace}\{\mathbf{L}_1^\dagger\}, \tag{8.9}$$

where $m = |\mathcal{E}|$ and $\mathbf{L}_1$ is the edge Laplacian (see Chap. 2). If the graph cycle set constitutes a basis (see Chap. 4), then the edge Laplacian matrix has full rank and the pseudoinverse is replaced by the true inverse of the edge Laplacian matrix. Recall from Chap. 4 that a set of $|\mathcal{E}| - |\mathcal{V}| + 1$ independent cycles will form a basis. This measure of the higher-order Kirchhoff index does not retain its original interpretation in terms of the effective resistance, since the effective resistance between two edges does not have a conventional definition. Despite losing this interpretation, this measure on the edge Laplacian matrix provides a value that indicates the relative "connectedness" of the edges via their incident nodes and cycles.

The Wiener index, average path length and Kirchhoff index all provide an aggregate measure of distance between all nodes in the graph. As before, we may use the definition of node eccentricity to define a measure of worst-case distance between all nodes. Specifically, the definition of node eccentricity allows for the definition of the **graph radius** and **graph diameter**

$$\text{Radius}(\mathcal{G}) = \min_{v_i} \text{Eccentricity}(v_i), \tag{8.10a}$$

$$\text{Diameter}(\mathcal{G}) = \max_{v_i} \text{Eccentricity}(v_i). \tag{8.10b}$$

A node for which $\text{Eccentricity}(v_i) = \text{Radius}(\mathcal{G})$ is called a graph **center** and a node for which $\text{Eccentricity}(v_i) = \text{Diameter}(\mathcal{G})$ is called a **peripheral node**.

Figure 8.1 gives an example of these distance-based measures used to describe a tree and a small lattice. The distance-based measures considered in this section have all been applied to measure some aspect of connectedness in a graph. In the next section we consider measures of graph *separability*.

8.1 Measures of Graph Connectedness

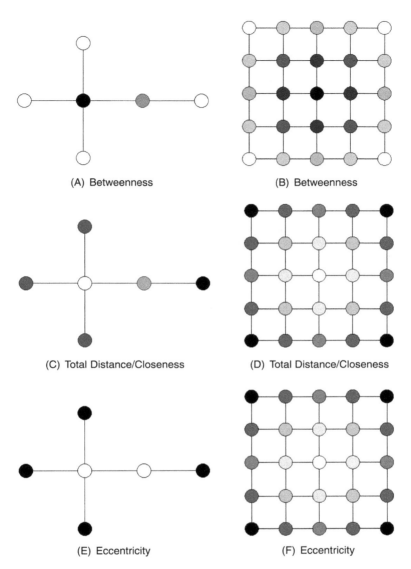

Fig. 8.1 Distance measures on two example graphs, displaying betweenness, TD and eccentricity for each node. Greater values of these quantities are represented by *darker shading*. Each example uses normalized values for the node quantities and unit weights for all edges. (*Left*) Measures for tree example: Radius = 2, Diameter = 3, Wiener Index = 28, Average Path Length = 1.8667, Kirchhoff Index = 28, Higher-Order Kirchhoff Index = 0. (*Right*) Measures for lattice example: Radius = 4, Diameter = 8, Wiener Index = 1000, Average Path Length = 3.333, Kirchhoff Index = 338.03, Higher-Order Kirchhoff Index = 814.42. From the node eccentricity examples we can determine which nodes are centers and which nodes are peripheral. *Tree graph*: all nodes with one neighbor are peripheral and the other two nodes are centers. *Lattice graph*: the corner nodes are peripheral and the middle node is the only center

8.2 Measures of Graph Separability

Similar to the distance-based measures of graph connectedness presented above, measures of graph separability are built upon metric properties of the graph. We first consider measures based on the volume of sets in a graph, then consider an example of separability based on distances.

8.2.1 Clustering Measures

Clustering measures use partitions of a graph to determine the separability of the graph. One of the most ancient measures of separability of a space arises from the **isoperimetric problem**, which seeks the shape with largest area/volume from the set of all shapes with the same perimeter/surface area. In the continuous, Euclidean $\mathbb{R}^3$ domain, the solution of this problem is known to be the sphere, and in $\mathbb{R}^2$ the circle. In a finite space (e.g., the surface of a closed object), one may define the **isoperimetric ratio** of an N-dimensional closed Riemannian manifold, $\mathcal{M}$, as [74]

$$h(\mathcal{M}) = \frac{\partial S}{\min(\text{Vol}(S), \text{Vol}(\bar{S}))}, \tag{8.11}$$

where S represents an N-dimensional submanifold, $\bar{S}$ represents its complement and ∂S represents the boundary length (surface area) of S. Instead of fixing a perimeter, ∂S, and seeking the submanifold S with greatest volume, we may generalize the isoperimetric problem to seek the node set $S \subset V$ that minimizes the isoperimetric ratio. Such a solution and its complement are called the **isoperimetric sets** of the domain. The value of the minimum isoperimetric ratio gives a notion of separability of the space. For example, if the domain were disconnected into two pieces, then the isoperimetric sets would consist of each piece individually and the isoperimetric ratio would be zero. Similarly, the solution to the isoperimetric problem on the surface of a "dumbbell" is also the two balls of the dumbbell with a small neck separating them [74]. Since the boundary of S would be measured on the neck, and the surface area on the two balls, the isoperimetric ratio of a dumbbell is small, meaning that a dumbbell is *nearly disconnected*. As a measure of separability for a domain, we may also consider the isoperimetric ratio of a graph. To do so, we must define the analogous concepts used above for a graph. Specifically, we let the set S refer to a set of nodes such that $S \subset V$, $S \neq \emptyset$, $|S| \leq \frac{1}{2}|V|$ with boundary defined as the sum of the weight of edges spanning the complementary subsets S and $\bar{S}$. Two definitions of the **volume** of a set of nodes in a graph are defined as [81, 285]

$$\text{Vol}_1(S) = |S|, \tag{8.12}$$

or

$$\text{Vol}_2(S) = \sum_{v_i \in S} d_i, \tag{8.13}$$

8.2 Measures of Graph Separability

where d_i represents the degree of node v_i. These two definitions of volume have led to two different definitions of the isoperimetric ratio of a graph.

We may write the isoperimetric ratio of the set S using an indicator vector, i.e., the vector $\mathbf{x}$ such that $x_i = 1$ if $v_i \in S$ and $x_i = 0$ otherwise. Using the indicator vector allows us to write the isoperimetric ratio as either

$$h_1(S) = \frac{\partial S}{\text{Vol}_1(S)} = \frac{\mathbf{x}^T \mathbf{L} \mathbf{x}}{\mathbf{x}^T \mathbf{x}} \tag{8.14}$$

or

$$h_2(S) = \frac{\partial S}{\text{Vol}_2(S)} = \frac{\mathbf{x}^T \mathbf{L} \mathbf{x}}{\mathbf{x}^T \mathbf{D} \mathbf{x}}. \tag{8.15}$$

As in the continuous case, the minimum isoperimetric ratio over all possible sets S gives a measure of the separability of the graph. This minimum of $h(S)$ over all possible S is called the **isoperimetric number**, **isoperimetric constant** or **Cheeger constant**. We note that there is some disagreement in the literature about these definitions, since all of these terms have been applied by various authors to either $h_1(\mathcal{G})$ or $h_2(\mathcal{G})$. We will use the term *isoperimetric constant*, which we denote as $h_1(\mathcal{G})$ or $h_2(\mathcal{G})$, depending on the definition of volume. The quantity $h_1(\mathcal{G})$ for a graph is additionally known as the **edge expansion** and the quantity $h_2(\mathcal{G})$ is known as the **graph conductance**.

The isoperimetric constant of a graph appears frequently in the literature. For example, the isoperimetric constant has been used to characterize **expander graphs** [6, 7]. Additionally, the **graph partitioning problem** is often formulated explicitly with the goal of finding the isoperimetric sets [169]. Unfortunately, calculation of the minimum isoperimetric ratio of an arbitrary graph is NP-Hard [285, 286]. Therefore, one approach to estimating the isoperimetric number of a graph is to apply several different graph partitioning algorithms and use the smallest isoperimetric ratio of these partitions as an estimate of the isoperimetric constant. Figure 8.2 shows examples of isoperimetric sets on two graphs, one of which exhibits a natural clustering and the other which does not.

A different approach to estimating the minimum isoperimetric constant of a graph is to use known bounds for the constant. The standard method for bounding the isoperimetric constant is through use of the **Fiedler value**, which is defined as the smallest nonzero eigenvalue of the Laplacian matrix. Fiedler observed that this eigenvalue was a good predictor of graph separability. Based on this observation, Fiedler named the smallest nonzero eigenvalue of the Laplacian matrix the **algebraic connectivity** [133] (although it is now known as the Fiedler value). The algebraic connectivity is a meaningful measurement of the graph separability, but it may also be used to bound the minimum isoperimetric constant $h_1(\mathcal{G})$ from both above and below. Specifically, the following expression combines **Cheeger's inequality** (upper bound) and **Buser's inequality** (lower bound) [74, 81] to provide

$$\sqrt{2 d_{\max} \lambda_2} \geq h_1(\mathcal{G}) \geq \tfrac{1}{2} \lambda_2, \tag{8.16}$$

where $d_{\max}$ is the maximum degree in the graph and λ_2 is the Fiedler value. Note that the these inequalities also hold for graphs with any set of positive edge weights. The right side of this equality may be seen easily since a real-valued relaxation of **x** causes (8.14) to be an expression for the Rayleigh quotient of **L**, which is minimized by λ_2 for all solutions of **x** orthogonal to the constant vector. Similarly, for $h_2(\mathcal{G})$ we have

$$\sqrt{2\lambda_2^*} > h_2(\mathcal{G}) \geq \tfrac{1}{2}\lambda_2^*, \qquad (8.17)$$

where λ_2^* indicates the smallest nonzero eigenvalue of the *normalized* Laplacian matrix. The appearance of the normalized Laplacian matrix is not surprising if we view the two definitions of volume in (8.12) and (8.13) as different definitions of *node weights*. Defining a node weight as unity (8.12) or defining a node weight as the node degree (8.13) lead naturally to the standard unweighted Laplacian matrix or the normalized Laplacian matrix, respectively (see Chap. 2 and [111, 112] for more details).

In addition to the node Fiedler value, we may define a *higher-order* Fiedler value defined on edges via the edge Laplacian. The higher-order Fiedler value is therefore defined as the smallest eigenvalue of the edge Laplacian. If the complex is simply connected (i.e., the number of faces included in the complex is greater than zero and equals $m - n + 1$), then this higher-order Fiedler value is positive. However, if there are fewer than $m - n + 1$ cycles in the complex, then the higher-order Fiedler value would correspond to the first nonzero eigenvalue (in accordance with the node case).

Another approach to measuring separability of a graph is the **clustering coefficient**, which looks locally for evidence of separability. The clustering coefficient is computed for a node by counting the number of pairs of neighbors of a node that are also neighbors of each other. This concept is motivated by social networks in which it has been observed that if one individual (node) knows two other people (has edges connecting them) then these people are likely to know each other (be connected via an edge). Since this set of connections close a triangle in the graph, this concept is sometimes known as **triadic closure**. Therefore, a pair of neighbors is considered *closed* if these neighbors are connected. With these definitions, the clustering coefficient at node v_i is computed as

$$\mathrm{CC}(v_i) = \frac{\text{Number of closed pairs of neighbors of } v_i}{\text{Total number of pairs of neighbors of } v_i}. \qquad (8.18)$$

Since $\mathrm{CC}(v_i)$ is undefined if $d_i < 2$, we adopt the convention that $\mathrm{CC}(v_i) = 0$ for these nodes. The clustering coefficient for the entire graph is given by

$$\mathrm{CC}(\mathcal{G}) = \frac{1}{n}\sum_i \mathrm{CC}(v_i). \qquad (8.19)$$

Therefore, this value takes a maximum at unity when $\mathcal{G}$ is fully connected and a minimum of zero when $\mathcal{G}$ represents a tree. Unfortunately, this definition of the clustering coefficient does not account for the graph weights. Several possibilities for extending this definition to include graph weights were proposed in [298] and an extension of this measure beyond triangles was given in [230]. Examples of the clustering coefficient and its comparison to the isoperimetric sets are provided in Fig. 8.2.

8.2.2 Small-World Graphs

The intuitive concepts of distance-based *connectedness* and *separability* seem to be opposite, i.e., that a separable network is not well-connected and *vice versa*. However, it is known that social networks contain clear groups of tightly coupled people but, despite this grouping, that the entire network is connected by short optimal paths. Watts and Strogatz explained this phenomenon with a model known as a **small-world network** [362, 396]. A small-world network was defined as any graph that has both a small average path length as defined in (8.7) and a large clustering coefficient as defined in (8.19). The name "small-world" network is derived from the social experiments of Milgram [283] who found that it was possible to send messages through a large, locally clustered, social network with approximately six steps. This result brought the concept of *six degrees of separation* to popular culture. Since the introduction of small-world networks, these models have been widely used to produce an understanding of many phenomena in physics, sociology and biology [397].

A key observation of Watts and Strogatz was that the average path length of a *random graph* is small when the graph is connected. Random graphs were initially (and comprehensively) studied by Erdős and Rényi [124, 125]. A random graph is defined as a graph which starts as a set of disconnected nodes to which k edges are progressively added. The added edges are chosen randomly from the set of all possible edges (node pairs), excluding multiple and self connections. Watts and Strogatz used the results on average path length to show that any locally connected graph (i.e., a graph with a large clustering coefficient) could be converted into a small world graph by randomly rewiring a small number edges. This randomization of a small number of edges has the effect of drastically lowering the average path length. This technique for producing a graph with small average path length has been exploited in the context of improving the speed of graph-based computer vision algorithms [167].

Following the publication of the initial work on small-world networks, other types of small-world networks have also been studied. A motivation for studying additional classes of small-world networks was the fact that the initial Watts–Strogatz model did not resemble many real networks in the sense that most nodes in the Watts–Strogatz model have the same degree. In fact, the **degree distribution** (the histogram of node degrees) of many real networks follows a power law [20] (e.g., the world wide web [3]). A network with a degree distribution following a power law is called a **scale-free network**. Recent interest in scale-free networks was generated by the work of Barabási and Albert [20], who suggested a mechanism for

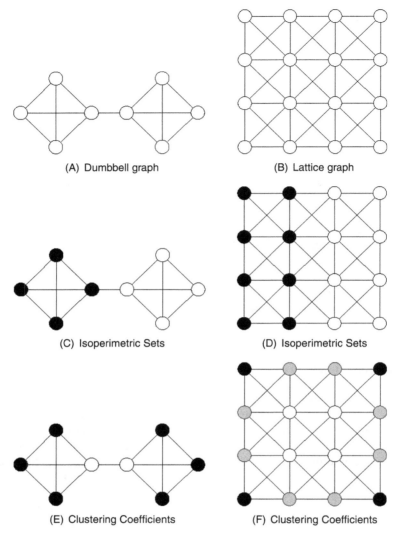

Fig. 8.2 Two examples of the clustering measures. The dumbbell graph on the *left* clusters well while the 8-connected lattice on the *right* does not. Node membership in the isoperimetric sets are indicated by coloring the nodes as *white* or *black*. Clustering coefficients are displayed for each node via *shading*, where *darker shading* represents nodes with a higher clustering coefficient. A node with a higher clustering coefficient may be interpreted as indicating that the node is more "interior". Dumbbell global clustering measures: $Vol_1 = 8$, $Vol_2 = 26$, $h_1 = 0.25$, $h_2 = 0.0769$, Fiedler Value $= 0.3542$, Graph Clustering Coefficient $= 0.8750$, Higher-Order Fiedler Value $= 0.3542$. Lattice global clustering measures: $Vol_1 = 16$, $Vol_2 = 84$, $h_1 = 1.25$, $h_2 = 0.2381$, Fiedler Value $= 1.4364$, Graph Clustering Coefficient $= 0.6571$, Higher-order Fiedler Value $= 1.0$. These measures tell us that, compared to the dumbbell, the lattice graph is larger (greater volume), harder to separate (larger isoperimetric constants and Fiedler value), and fewer of the neighbors of each node are connected (smaller clustering coefficient)

producing a scale-free network. This mechanism is known as **preferential attachment** and roughly states that as edges are added to a graph, the nodes with larger degree have an increased probability of being linked. This process therefore produces some nodes with very large degree (called "hubs") and the remaining degree distribution follows a power law [20]. An example of this type of process is in the author citation network, in which a paper with many citations is more likely to continue to be cited in the future. Although scale-free networks are commonly used as examples of small-world networks, the clustering coefficient is not always large (although the diameter is small), meaning that scale-free networks are not necessarily small-world graphs.

The measures of graph connectedness and graph separability considered thus far are measures that naturally pertain to 1-complexes or graphs. We now consider global topological measures for general p-complexes, and afterwards consider geometric measures defined specifically for surfaces or 2-complexes.

8.3 Topological Measures

The distance-based measures that we have studied so far are often called "topological indices" in the chemical graph theory literature. Although these methods do measure aspects of the network topology, they do not measure the usual topological invariants such as the Euler characteristic, genus, Betti numbers, torsion coefficients and orientability. In this section, we will describe how to calculate these invariants for a cell complex. Here we assume that we are measuring the topological properties of a p-complex. Recall from Chap. 2 that a p-complex is defined by sets of p-dimensional cells, S_p. A standard graph is therefore a 1-complex, which contains sets of nodes and edges only. The subject of computational topology on a complex is treated extensively in the literature [119, 254, 425].

We begin our treatment of topological measures with a discussion of the Betti numbers for a complex. The pth Betti number of an n-complex is defined as the rank of the pth homology group [254]. Informally, the Betti number may be viewed as the number of cuts that may be made without dividing a surface into two parts. Therefore, when $p = 0$, the Betti number represents the number of connected components, when $p = 1$, the Betti number represents twice the number of handles, and for $p = 2$, the Betti number represents the number of voids. Computation of the pth Betti number is possible from the incidence matrices via the formula [254]

$$\text{Betti}_p = |S_p| - \text{Rank}(\mathbf{N}_{p+1}) - \text{Rank}(\mathbf{N}_p), \tag{8.20}$$

where we may recall that $\mathbf{N}_p$ represents the pth incidence matrix of the complex. To make this definition hold for all p on a p-complex, we define $\text{Rank}(\mathbf{N}_p) = 0$ for $p \leq 0$ or for $p \geq n$. As an example, we may give the calculation of the Betti numbers for $p = 0$ and $p = 1$ using our conventional notation as

$$\text{Betti}_0 = |\mathcal{V}| - \text{Rank}(\mathbf{A}) - 0, \tag{8.21}$$

$$\text{Betti}_1 = |\mathcal{E}| - \text{Rank}(\mathbf{B}) - \text{Rank}(\mathbf{A}). \tag{8.22}$$

Recall that Rank($\mathbf{A}$) equals $|\mathcal{V}| - c$ where c represents the number of connected components (see Chap. 2). Therefore, $\text{Betti}_0 = c$. Similarly, when a complex is closed and simply connected, then we saw in Chap. 2 that Rank($\mathbf{B}$) = $|\mathcal{E}| - |\mathcal{V}| + 1$. Therefore, when $c = 1$ and the complex is closed and simply connected, $\text{Betti}_1 = 0$.

A simpler method for calculating the Betti numbers was given by Friedman [143] who noted that, by the definition of the pth-order Laplacian, it was possible to calculate the Betti numbers as

$$\text{Betti}_p = \text{Dimension}(\text{Nullspace}\{\mathbf{L}_p\}). \tag{8.23}$$

This approach to calculating the Betti numbers is often more straightforward since it may be computed by counting the number of zero eigenvalues of $\mathbf{L}_p$. This second expression for the Betti numbers in (8.23) also recovers the fact that $\text{Betti}_0 = c$, since it is well-known [34] that Rank(Nullspace$\{\mathbf{L}_0\}$) = c.

Multiplying the general expression for the Betti numbers in terms of the ranks of incidence matrices given in (8.20) on both sides by $(-1)^p$, results in the Euler–Poincaré theorem which states that

$$\sum_{i=0}^{p}(-1)^p|\mathcal{S}_i| = \sum_{i=0}^{p}(-1)^p \text{Betti}_i. \tag{8.24}$$

The value of this sum is known as the **Euler characteristic** of a p-complex, $\chi(\mathcal{G})$, i.e.,

$$\chi(\mathcal{G}) = \sum_{i=0}^{p}(-1)^p|\mathcal{S}_i| = \sum_{i=0}^{p}(-1)^p \text{Betti}_i. \tag{8.25}$$

When $p = 2$, this equation gives the usual formula for a surface $\chi(\mathcal{G}) = |\mathcal{F}| - |\mathcal{E}| + |\mathcal{V}|$. The Euler characteristic is one of the central invariants in topology. For a connected complex, the Euler characteristic may be considered as the number of linearly independent cells which are possible, but not necessarily present, in the complex (plus one). For example, we have seen that the number of linearly independent cycles in a graph are equal to $|\mathcal{E}| - |\mathcal{V}| + 1$. Therefore, if all of these cycles are included in our set of faces, then $|\mathcal{F}| = |\mathcal{E}| - |\mathcal{V}| + 1$ and the Euler characteristic equals one. If the "exterior face" is additionally included in the set of faces (see Chap. 2) then the Euler characteristic equals two, which is the classical result for a simply-connected closed surface.[2] Two examples are given in Fig. 8.3 showing a cube with Euler characteristic two and an annulus with Euler characteristic zero.

---

[2]The exterior face is a device to enable a finite graph to be defined such that it has no boundary and is therefore closed. This imparts a global topology on the graph—that of a sphere or a sphere with handles—which may be interpreted as a finite graph including a face "at infinity", in analogy to projections of the sphere into the plane (e.g., by stereographic projection) in which the coordinate at the pole of the sphere opposite the origin is mapped to the point at infinity on the flat plane.

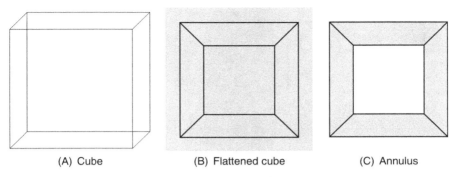

Fig. 8.3 A cube, the flattened cube (with exterior face), and an annulus. The Euler characteristic is $\chi = 6 - 12 + 8 = 2$ for the cube (flattened and not flattened), and $\chi = 4 - 12 + 8 = 0$ for the annulus. Faces included in the face set are shaded *gray* (including the exterior face for the flattened cube)

Closely related to the Euler number of a closed, orientable 2-complex is its **genus**, which may be defined in terms of the Euler characteristic via the relationship

$$\chi(\mathcal{G}) = 2 - 2\,\text{Genus}(\mathcal{G}) \qquad (8.26)$$

for a closed surface. The genus is often thought of as the number of handles on the surface. Therefore, the sphere has genus equal to zero, whereas the torus (a sphere with a handle) has a genus equal to one. The genus of a 1-complex (graph) has been defined as the minimum genus of the surface on which the graph may be embedded such that is has no edge crossings [188]. Therefore, a planar graph is considered to have genus zero.

The property of surface **orientability** describes whether or not it is possible to describe the surface as the boundary of some object. A classical example of a non-orientable structure is the Möbius strip. One test for surface orientability on a finite complex is to examine the orders of the torsion subgroups of the homology group of the complex, which are known as the **torsion coefficients** of the complex [254, 272]. These coefficients may be determined computationally by calculating the invariant factors of the incidence matrix (i.e., the diagonal of the incidence matrix after placing it in Smith Normal Form). Specifically, the kth torsion coefficients are defined as the set of invariant factors of the kth incidence matrix greater than unity [254]. Therefore, if the complex is orientable, then the complex is torsion-free and there are no invariant factors of the kth incidence matrix greater than unity [272]. For example, Fig. 8.4 shows a triangulated Klein bottle with its corresponding face–edge incidence matrix and the incidence matrix in Smith Normal Form.

8.4 Geometric Measures

In addition to the properties considered above, we may also compute classical *geometric* quantities to describe a complex. Here we follow the computer graphics

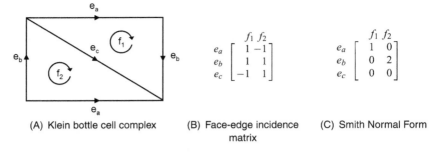

Fig. 8.4 Torsion coefficients may be extracted from the face–edge incidence matrix by placing the matrix in Smith Normal Form. In this example, we give the face–edge incidence matrix of a cell decomposition of the (flattened) Klein bottle [196] and show that the Klein bottle has a torsion coefficient equal to two since the only invariant factor greater than unity lies at face f_2

literature which defines measures of curvature for a 2-complex which is embedded in $\mathbb{R}^3$.

Geometric measures defined on simplicial complexes (i.e., triangular meshes representing surface for 2-complexes) depend on the metrics ascribed to the complexes. In the case of discrete surfaces the metric is typically derived from the embedding of the two-dimensional surface in a three-dimensional metric space, although the notion of distance may be produced from any process. Therefore, the distances along the surface are the natural Euclidean distances inherited from the ambient space of the embedding.

Here we will consider two forms of curvature defined on surfaces: Gaussian and mean curvature. **Gaussian curvature** is an example of an *intrinsic* geometric property of the surface in that Gaussian curvature is invariant under isometric transformations of the surface. That is, if the surface is deformed in a way that does not affect the distance between any pair of vertices on the surface, then the Gaussian curvature is also unaffected. For this reason, the Gaussian curvature is called intrinsic and depends only on the metric of the surface. **Mean curvature**, however, is an example of an *extrinsic* geometric property that can change under isometric transformations. For instance, a punctured sphere and a flat disk in the plane are topologically equivalent, and therefore a smooth homeomorphism exists between them, but a sphere has constant positive Gaussian curvature and a disk has zero Gaussian curvature—it is clear that the deformation between the two configurations cannot take place without geometric distortion. However, the cylindrical tube is an example of a surface with zero Gaussian curvature and non-zero mean curvature, so if the tube is cut straight down its side (imagine a piece of paper rolled so that the two short ends meet) it can be unrolled and flattened into the plane without stretching or intrinsic geometric distortion.

8.4.1 Discrete Gaussian Curvature

Gaussian curvature is typically defined in terms of the principle curvatures [110]. However, an alternate definition of Gaussian curvature is provided by a special case of the Gauss–Bonnet theorem that applies equally well to the continuous or discrete cases. This alternate definition of Gaussian curvature is based on the idea that, in the plane, at a given vertex the sum of angles between adjacent edges connecting the vertex to its neighbors always sum to 2π. For curved surfaces, this same sum can be either greater than or less than 2π, and this *angle deficiency* is the basis of Gaussian curvature.

The Gauss–Bonnet theorem establishes a deep result in differential geometry that links the intrinsic geometry of a manifold $\mathcal{M}$ to its topology. For a given metric, the integral of the Gaussian curvature K over the manifold (also known as the **total curvature**), plus a boundary term consisting of the integral of the geodesic curvature, k_g, along the manifold boundary, is related to the Euler characteristic of the manifold as

$$\iint_{\mathcal{M}} K \, dV + \int_{\partial \mathcal{M}} k_g \, ds = 2\pi \, \chi(\mathcal{M}). \tag{8.27}$$

Therefore, for any topological sphere, the total curvature is always 4π, whereas the total curvature for a 1-torus is always 0.

This general theorem can be applied locally to provide an integral definition for Gaussian curvature on two-dimensional surfaces that holds for discrete spaces. If we consider the total curvature of the dual cell surrounding each node of a graph, then for a particular node v_i at its corresponding dual cell $\mathcal{C}$ the Gauss–Bonnet theorem reduces to

$$\iint_{\mathcal{C}} K \, dV - \int_{\partial \mathcal{C}} k_g \, ds = 2\pi \tag{8.28}$$

since, by definition, $\chi(\mathcal{C}) = 1$ for the cell. For the case in which the dual cell is conveniently given by the Voronoï cell at each vertex, the integral of the geodesic curvature is zero at the piecewise linear edges of the cell boundary plus the angles at the corners of its boundary. Furthermore, these angles are equivalent to the interior angles θ at node v_i formed between each pair of edges incident to v_i in this special case of a Voronoï cell [280, 366]. As a result, the Gaussian curvature $K(v_i)$ at node v_i may be neatly defined as

$$K(v_i) = \frac{2\pi - \sum_{j \, \forall e_{ij}} \theta_j}{A_{v_i}} \tag{8.29}$$

where θ_j is given by $\theta_j \equiv \angle\{v_j, v_i, v_{(j+1) \bmod B}\}$ if B represents the number of neighboring nodes, and A_{v_i} represents the area of the Voronoï cell.

8.4.2 Discrete Mean Curvature

The definition of Gaussian curvature at a vertex in (8.29), which phrases the integral of the curvature over a small patch as the deviation from 2π of the internal

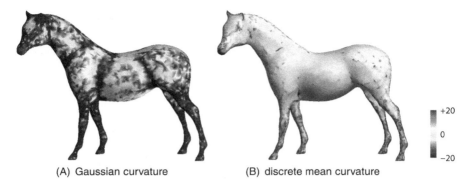

(A) Gaussian curvature (B) discrete mean curvature

Fig. 8.5 Example of discrete curvature measures applied to horse triangular mesh. (**A**) Gaussian curvature calculated from the Gauss–Bonnet theorem. (**B**) Discrete mean curvature computed from the method of Meyer et al. [280]. Both measures calculate curvature as a node quantity measured for each vertex in the polygonal mesh based on the embedding of the neighboring vertices and the incident faces. Both curvatures are visualized with a common color scale provided on the *lower right*

angle sum, holds for finer and finer mesh spacing and is equivalent to the continuous definition in the limit. In contrast, mean curvature does not possess an analogous integral definition and therefore it is less natural to express mean curvature in the discrete setting. One approach to defining mean curvature, adopted by Meyer et al. [280], is to establish a similar integral relationship in the discrete setting that approaches the continuous definition in the limit of finer mesh sampling in order to provide a robust and compatible curvature measure. The definition begins by noting that if one establishes a specific coordinate system for the surface that provides a isothermal parameterization, the mean curvature H is related to the Laplacian operator ∇^2 applied to this specific choice of coordinate functions [110], and therefore the integral of the mean curvature around a vertex v_i can be expressed as the integral of the Laplacian operator evaluated at the midpoints of all edges incident to v_i. The resulting expression for mean curvature vector is then given by

$$\tilde{H}(v_i) = \frac{1}{2A_{v_i}} \sum_{j \forall e_{ij}} (\cot \alpha_{ij} + \cot \beta_{ij})(\tilde{c}_i - \tilde{c}_j) \qquad (8.30)$$

where α_{ij} and β_{ij} represent the opposing angles in the two triangles containing edge e_{ij} at the two vertices that are neither v_i nor v_j. The requirement of the two coordinate vectors $\tilde{c}_i$ and $\tilde{c}_j$ demonstrates why the mean curvature formulation is dependent on the embedding of the graph and is therefore an extrinsic quantity.

Although the mean curvature vector is a vectorial quantity in the ambient embedding space, it is typically expressed as a scalar-valued curvature measure given by the length of the vector, $H = \|\tilde{H}\|$. Intuitively, this definition phrases the mean curvature vector at a vertex as the average edge vector of all edges incident to the vertex weighted by the angle sum around the vertex. Examples of these measures for Gaussian and mean curvature are provided in Fig. 8.5 (contained in the color plate section at the end of the book).

Although these two curvature measures are not phrased in terms of the discrete calculus operators that comprise the central theme of this book, they do provide examples of quantities that are typically considered only in the continuous setting. The definition of Gaussian curvature holds equally well in the discrete setting, whereas the translation of mean curvature to the discrete setting is not as straightforward. From a practical standpoint, these measures are also generally useful for practitioners of discrete methods.

8.5 Applications

Measures of network structure may be useful in several ways. One way to use this information would be to learn more about the network and focus attention on the most relevant nodes. This type of application would employ individual node measures more than the global measures of the network. In this section, we will explore an application in social networks as an illustration of this usage of the network measures.

A different way of using the network measures is to predict other properties of the structure being represented by the network. This type of application effectively summarizes the network structure via a collection of numbers which may then be correlated to other quantities of interest. As a representative of this usage of the network measures, we take an example from chemical graph theory.

8.5.1 Social Networks

Identification of important persons from the network structure is a typical application in a social networks (e.g., [45]). For example, important individuals in a terrorist network could be identified as the most effective leaders to capture. In Chap. 6 we studied Zachary's Karate Club network [415] in which two individuals in the network split the club into two groups. Given just the network structure, we show that the network measures allow us to predict the leaders of the two factions.

Figure 8.6 shows Zachary's Karate Club network with the two actual leaders of the factions identified. Without knowing who the actual leaders of the two factions were, we could examine various node measures. For example, we would expect that important nodes would have a large number of direct social connections to other individuals (nodes) in the network, measured as node degree. Beyond direct connections, we could reasonably expect that the leaders would be well-connected indirectly to all of the individuals (nodes) in the network, measured as a low total distance (closeness). Additionally, we would expect that the leaders would act as a conduit for connecting other individuals in their faction, measured as node betweenness. From Fig. 8.6 we see that the actual leaders of the two factions score much better than the other nodes in terms of the node degree, total distance and betweenness, and are therefore distinct and identifiable from the rest of the nodes.

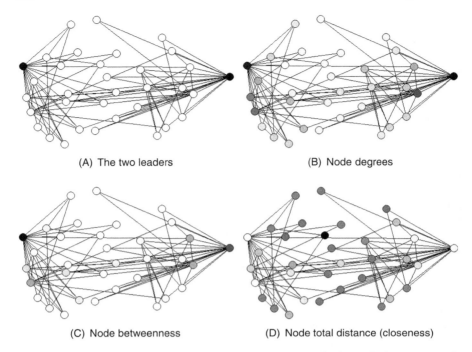

Fig. 8.6 Prediction of the leaders of the two factions in Zachary's Karate Club network [415] using node measures. Nodes are shaded darker to indicate greater values of each measure. Note that the measures in each figure are normalized to use the full white–black range. (**A**) The two actual leaders of the factions marked in *black*. (**B**) Number of neighbors for each node (degree). (**C**) Node betweenness. (**D**) Node total distance (closeness). These measures tell us that the two leaders knew the most other people (highest degrees), were hubs for other people to know each other (highest betweenness) and had the fewest average number of links between them and the rest of the network

Therefore, these three measures of node degree would provide an accurate prediction of the two leaders in Zachary's Karate Club network.

8.5.2 Chemical Graph Theory

Graph theory has a long history in chemistry, dating back to Sylvester's work in 1878 [367]. In fact, the term "graph theory" was coined by Sylvester in the context of the "graphical notation" used to describe the chemical structure of a molecule [37].

In chemistry, a graph represents a molecule by associating each atom with a node and each bond between atoms with an edge. Hydrogen atoms (with just a single bond) are conventionally removed from the graph structure. Network measures are widely used in chemical graph theory to predict **quantitative structure–property relationships** (QSPR) and **quantitative structure–activity relationships**

8.5 Applications

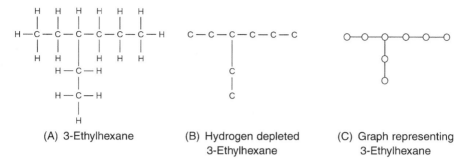

Fig. 8.7 Predicting the boiling point of 3-Ethylhexane from its structure. (**A**) 3-Ethylhexane. (**B**) The hydrogen-depleted molecule commonly used for calculation (i.e., the hydrogen nodes are removed). (**C**) The graph corresponding to the hydrogen-depleted molecule. $W(\mathcal{G}) = 72$, $P = 7$, $\alpha = 1.53$, $\beta = 5.5$, $\gamma = -30.35$. Using Wiener's formula in (8.31) we can predict the boiling point for 3-Ethylhexane as 118.4°C. The actual boiling point for 3-Ethylhexane is 118.6°C

(QSAR). Examples of such molecular properties include the boiling point, melting point, molar volume, refractive index, critical pressure, surface tension, viscosity, rate of electro-reduction and heats of isomerization, vaporization, formation and atomization [109]. Several review articles [187, 282] and books [42, 44, 170] detail the use of these measures in chemical graph theory to predict molecular properties. In addition to molecular properties, network measures have been widely used to predict pharmacological properties for purposes of drug discovery [265, 266, 325].

In this section, we do not intend to provide a comprehensive review of this vast literature, but rather to focus on the first and most important measure to appear in chemical graph theory: the Wiener index (8.6). Note that Wiener's initial definition was not phrased in graph theory language, but later formulated in this setting by Hosoya [207]. In his early papers [404, 405], Wiener used his index to predict the boiling point (b.p.) of alkanes from the formula

$$\text{b.p.} = \alpha W(\mathcal{G}) + \beta P + \gamma, \tag{8.31}$$

where α, β, γ are empirical constants and P, the "polarity number", was defined as the number of node pairs with distance equal to three. By fitting parameters to thirty-seven alkanes, Wiener determined the parameters to be $\alpha = 98/n^2$, $\beta = 5.5$ and $\gamma = -30.35$, where n represents the number of carbon atoms (number of nodes). It is interesting to observe that the only difference between the definitions of the Wiener index (8.6) and the Average Path Length (8.7) was the normalization of the Average Path Length measure by the number of node pairs, $n^2 - n$. However, in Wiener's original work, he used the parameter setting to normalize his measure by n^2, which brings these two measures very close together.

Figure 8.7 shows one of Wiener's original examples in which he predicts the boiling point of 3-Ethylhexane from the molecular structure alone (i.e., by measuring network properties). Specifically, for 3-Ethylhexane we may calculate $W(\mathcal{G}) = 72$, $P = 7$. Using Wiener's formula (8.31) and the parameter values listed above allows us to calculate the boiling point for 3-Ethylhexane as 118.4°C. The actual boiling point for 3-Ethylhexane is 118.6°C. In Wiener's original paper, he applied his formula to ninety-four compounds to predict the actual boiling points with an average deviation of 0.97°C. It remains a remarkable fact that a measure of the structure of the network representing a molecule can provide such accurate predictions of the molecule's chemical properties.

Many additional models have been developed that predict molecular properties from the Wiener index. Another example is the chromatographic retention time (CRT) of monoalkyl- and o-dialkylbenzenes which are well modeled in [51] by

$$\text{CRT} = \alpha W(\mathcal{G})^\beta + \gamma, \tag{8.32}$$

where α, β, γ are empirical constants (which are different from those appearing above in (8.31)). The variety of models in which the Wiener index appears has been explained by arguing that the Wiener index measures the van der Waals surface area of a molecule [181].

By representing a molecule as a list of numerical descriptors, these descriptors may be used in conjunction with machine learning techniques to predict many different QSPR and QSAR properties. In the present day, vast libraries of graph representations of chemical compounds have been compiled which make it possible to search for a compound with a particular set of properties or to numerically screen compounds without having to manufacture and test them [171, 213, 370]. Many structural descriptors have been devised in the chemical graph theory literature that were not reviewed in this chapter. In this section, we have simply intended to provide the reader with a glimpse into this rich literature and to point the interested reader to more comprehensive sources in this area.

8.6 Conclusion

In this chapter we considered methods of measuring different kinds of quantities that describe the structure of the graph. The distance and clustering measures approached the related intuitive concepts of connectedness and separability. In contrast, the topological and geometric measures probe other aspects of the complex.

The measures may be applied in several ways. One method for applying these measures has been to reduce a network to a series of numbers that may be used to predict the behavior of certain processes on the graph or of the object represented by the network. A central example of this usage is the success of the chemical graph theory literature in relating the distance-based measures to the chemical properties of the molecules represented by the graph. A smaller example of this usefulness is the observation [175] that graph diameter is a good predictor of the convergence of conjugate gradient applied to solving a linear system with the Laplacian matrix. As network models are increasingly explored to describe computer networks, neural connections, traffic flow, gene regulation and sociology, we believe that these measurements will provide useful predictions about the behavior of these networks.

8.6 Conclusion

Finally we note that the distance and separability descriptors that have been employed in the literature to describe networks are exclusively dependent on node connectivity and separability. We suggested some possibilities for extending these measures to higher-order connectivity, but we believe that measures defined on edge and cycle connectivity and separability present an untapped source of additional descriptors for a network or complex.

Appendix A
Representation and Storage of a Graph and Complex

In the main text we treated graphs and cell complexes without discussing the numerical representation and storage of these structures. Here we intend to fill this gap by discussing how graphs and complexes may be represented and stored, since a numerical representation is a prerequisite for performing any computation.

A.1 General Representations for Complexes

In general, chains and cochains will form both the inputs and outputs of the computations and algorithms presented. Storage of chains/cochains may be accomplished efficiently with one-dimensional *arrays* (shown in the text as column vectors) of the appropriate size (e.g., n for 0-cochains, m for 1-cochains). There are two equivalent options for representing a complex, the **cells list representation** and the **operator representation**.

A.1.1 Cells List Representation

A *cells list representation* stores a p-complex using p two-dimensional arrays. Each two-dimensional array is the length of the number of k-cells for all $0 < k \leq p$, and each array contains the node list (0-cells) of the corresponding k-cell. This list encodes the orientation of the cell via the ordering of the nodes in the list. A cells list representation stores the weights of a weighted complex as a separate array where every entry in the cells list is matched by an entry in the weights list containing the cell weight. The cells list representation is especially convenient for a graph (i.e., a 1-complex) or a triangular mesh (i.e., a 2-complex) since each edge or triangle in the complex is composed of exactly the same number of nodes. For example, the cells list representation is used in the common Geomview *Object File Format* as an ".OFF file" for storing meshes.

A general cell complex may contain k-cells that are composed of many different numbers of nodes (e.g., an irregular mesh). In these situations, a list representation requires that we additionally store the node count for each individual k-cell.

The cells list representation is simple and convenient, but may be redundant if the complex possesses structure or topological regularity. For example, if the complex represents a lattice, it is unnecessary to store the neighborhood of each cell and it is therefore more efficient to employ the operator representation, which we now describe.

A.1.2 Operator Representation

The *operator representation* stores the complex structure via two-dimensional arrays or matrices such as the incidence matrices, Laplacian matrices, or adjacency matrices discussed extensively in the main text. The primary advantage of the operator representation is that a *structured complex* will generally allow for efficient storage. For example, the Laplacian matrix for a lattice will be banded and symmetric, resulting in the need for storing only the upper triangular bands. Therefore, in the case of a 4-connected two-dimensional lattice, only $2n$ values require storage—the graph weights. In the special case where the weights are all uniform, the dimensions of the lattice completely specify the complex and no explicit storage is needed. In other words, for a lattice with uniform weights, when the corresponding Laplacian matrix is required for a computation (such as the matrix–vector product employed in previous chapters) the computation can be expressed without any explicit storage of the matrix, and the computation can be simply expressed as a function of the inputs.

Of the three matrices commonly used to represent a complex (i.e., the Laplacian, adjacency, and incidence matrices), only the incidence matrix stores the orientation assigned to each of the cells. However, since the orientation may be arbitrary we may assign the orientation of each cell as needed, so long as we maintain a consistent orientation for each cell. Figure A.1 gives an example of the list and operator representation for a small 1-complex (graph).

A secondary advantage of the operator representation is that *one data structure simultaneously represents a set of topological relations of the complex as well as one of its differential operators*. This practical advantage of the operator representation reinforces the topological nature of the differential operators, as highlighted in Chap. 2. Although the operator representation requires a generic data structure akin to a two-dimensional array to encode the incidence relations between elements of the complex, in contemporary programming environments such as MATLAB such two-dimensional arrays are also interpreted as matrix objects and are endowed with the ability to perform matrix operations such as matrix multiplication. Therefore, when using the operator representation the data structure representing the graph is also literally the differential operator for the graph.

Most graphs and complexes that arise in practice are *sparse*. Consequently, a sparse matrix format will be the most efficient storage of an operator representation [19]. The **Compressed Column (or Row) format** is the most common approach for representing a sparse matrix, due to its generality. However, a common motivation for using an operator representation is due to the structure of the matrix operator. Consequently, a **Compressed Diagonal Storage format** is often appropriate to store a complex using the operator representation, e.g., for the Laplacian matrix or adjacency matrix of a lattice.

A.2 Representation of 1-Complexes

In addition to the cells list representation and operator representation used above to represent a general complex, there are numerical representations often used in the special case where the complex is of dimension one i.e., a graph. Here we discuss one data structure for a graph which can be more efficient for some algorithms.

A.2.1 Neighbor List Representation

A common representation for a graph may be called the **neighbor list representation** since it is motivated by the common numerical operation of traversing the node neighbors of a graph. For example, shortest-path computation with Dijkstra's algorithm requires that when a node is added to the list of known nodes, all of the neighbors of the added node are also pushed onto a heap [87]. Therefore, it is necessary to know the neighbors of each node in order to apply this algorithm efficiently. Unfortunately, the cells list representation does not allow for an efficient search for the node neighbors of each node, since the cells list representation is organized by the edges. Consequently, finding the neighbors of a particular node in an edge list requires traversal of the entire edge list to search for edges that connect the node to other neighbors.

To facilitate easy neighbor finding, we can employ the neighbor list representation which stores the graph using a two-dimensional array of length $n \times d_{\max}$. Each entry in the array contains the neighboring nodes of the indexed node. A neighbor list representation stores the weights of a weighted complex using a second two-dimensional array of equal size where every entry in the second set of arrays stores the weight of the edge connecting the neighboring nodes. In general, the neighbor list representation is not as efficient as storage of the cells lists unless the graph is regular. If the graph is irregular, then some entries of the two-dimensional array will not be used, but memory for these entries must still be allocated (or the degree of each node must be stored). Figure A.1 also displays the neighbor list representation of the nodes in the example graph.

Fig. A.1 An example graph (**A**) and its numerical storage with the representations discussed in the text: (**B**) The neighbor list representation, (**C**) cells list representation for edges, (**D**) cells list representation for faces, (**E**) operator representation for the edge–node incidence matrix, (**F**) operator representation of the face–edge incidence matrix

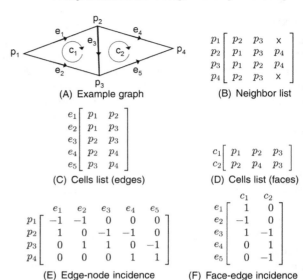

Appendix B
Optimization

The main text has provided the mathematical tools for formulating continuous concepts in a discrete setting and provided applications in which this machinery may be applied to solve real-world problems. The primary mechanism for applying these discrete calculus operators was to formulate algorithms as *variational* or energy minimization problems. We believe that an important motivation for the use of discrete calculus is the availability of powerful optimization tools to allow efficient energy minimization. Although some details and algorithmics were discussed in the text for solving these optimization problems, it was generally assumed that the reader was familiar with optimization techniques. In this appendix we provide more information about the optimization techniques which appear in the text in an effort to enable the reader to implement the techniques. This appendix is by no means intended as a comprehensive treatment of the vast topic of optimization, but rather as a tour of pertinent techniques and pointers to literature in which more information may be found.

Optimization of both real-valued and integer-valued functions appears commonly in practical applications of discrete calculus. In the framework presented in the text, the underlying spaces are discrete but the fields defined on these spaces are free to be continuous-valued. Consequently, there is an asymmetry between the treatment of chains and cochains. In general, cochains are defined with real-valued variables (e.g., image intensity, vertex coordinates) and optimization for a cochain is intended to find the optimum real-valued function to associate with the domain. Examples of this kind of optimization are mesh filtering or dimensionality reduction. Conversely, chains are expressed as lists of cell indices and thus are generally defined with integer (or binary) variables, therefore optimization of chain quantities is usually intended to identify a set of nodes, edges or higher-dimensional structures. For example, segmentation and clustering are examples of chain optimization in which the goal is to identify each node with its corresponding label.

In this appendix, we will treat the topic of optimization with both real-valued variables and integer-valued variables. Additionally, we will devote particular attention to linear and quadratic objective functions since these are very common in mathematical physics and because they frequently appear in the optimization of an energy defined on a graph.

B.1 Real-Valued Optimization

Optimization of real-valued variables is generally much easier than the optimization of integer- or binary-valued variables because it is possible to take a *derivative* of a real-valued function, even if the function is defined as a cochain on a discrete domain. Our focus in this section will primarily be on *convex optimization*. Before addressing convex optimization, we begin with a few facts about derivatives of finite real-valued functions that will be helpful in subsequent sections.

Assume that we have a real-valued vector of N variables, $\mathbf{x} \in \mathbb{R}^N$. Given an energy functional $\mathcal{Q}[\mathbf{x}]$ evaluating to a scalar, $\partial \mathcal{Q}/\partial \mathbf{x}$ is defined as the vector quantity

$$\frac{\partial \mathcal{Q}}{\partial \mathbf{x}} = \begin{bmatrix} \frac{\partial \mathcal{Q}}{\partial x_1} \\ \frac{\partial \mathcal{Q}}{\partial x_2} \\ \vdots \\ \frac{\partial \mathcal{Q}}{\partial x_N} \end{bmatrix}, \tag{B.1}$$

which is the derivative of the energy functional $\mathcal{Q}[\mathbf{x}]$ with respect to each of the N components of the vector $\mathbf{x}$. Similarly, the derivative of a vector $\mathbf{y}$ with respect to another vector $\mathbf{x}$ gives the matrix of all possible combinations of derivatives

$$\frac{\partial \mathbf{y}}{\partial \mathbf{x}} = \begin{bmatrix} \frac{\partial y_1}{\partial x_1} & \frac{\partial y_1}{\partial x_2} & \cdots & \frac{\partial y_1}{\partial x_N} \\ \frac{\partial y_2}{\partial x_1} & \frac{\partial y_2}{\partial x_2} & \cdots & \frac{\partial y_2}{\partial x_N} \\ \vdots & \vdots & \ddots & \vdots \\ \frac{\partial y_N}{\partial x_1} & \frac{\partial y_N}{\partial x_2} & \cdots & \frac{\partial y_N}{\partial x_N} \end{bmatrix}. \tag{B.2}$$

Due to their prevalence in physical systems, it is beneficial to pay particular attention to quadratic energy functionals. Applying the above rules to the quadratic energy $\mathcal{Q}[\mathbf{x}] = \mathbf{x}^T \mathbf{H} \mathbf{x}$ evaluates to

$$\frac{\partial \mathcal{Q}}{\partial \mathbf{x}} = (\mathbf{H} + \mathbf{H}^T)\mathbf{x}. \tag{B.3}$$

If $\mathbf{H}$ is symmetric, then $\partial \mathcal{Q}/\partial \mathbf{x} = 2\mathbf{H}\mathbf{x}$. Taking a second derivative of $\mathcal{Q}$ with respect to $\mathbf{x}$ gives us the **Hessian matrix** for $\mathcal{Q}$ with respect to $\mathbf{x}$, which for a quadratic energy $\mathcal{Q}[\mathbf{x}] = \mathbf{x}^T \mathbf{H} \mathbf{x}$ is

$$\frac{\partial^2 \mathcal{Q}}{\partial \mathbf{x}^2} = \mathbf{H} + \mathbf{H}^T. \tag{B.4}$$

If $\mathcal{Q}[\mathbf{x}]$ is a linear functional defined by $\mathcal{Q}[\mathbf{x}] = \mathbf{x}^T \mathbf{y}$, the derivative with respect to $\mathbf{x}$ is simply

$$\frac{\partial \mathcal{Q}}{\partial \mathbf{x}} = \mathbf{y}. \tag{B.5}$$

With these preliminaries in place, we may now provide a tour of convex optimization. The **convexity** of a functional (and the solution constraints) has long been

B.1 Real-Valued Optimization

recognized as the most important property for determining the difficulty of finding an optimal solution [319]. The principal advantage of an unconstrained convex optimization problem is that *any local minimum is also a global minimum* [52]. A function, $f(\mathbf{x})$, defined on an interval (or convex subset of a vector space) is convex if

$$f(\theta \mathbf{x} + (1-\theta)\mathbf{y}) \leq \theta f(\mathbf{x}) + (1-\theta) f(\mathbf{y}), \tag{B.6}$$

where θ is any value in the range [0, 1] and $\mathbf{x}$ and $\mathbf{y}$ represent any points in the domain of f. If the inequality in the convexity definition (B.6) is strict for any valid θ, then the function is said to be **strictly convex**. Strictly convex functions are important for optimization because a strictly convex function has a unique global minimum. In contrast, a convex function which is not strictly convex will have multiple global minima and therefore an optimization of the functional will require additional regularization to produce a unique solution. It follows from the definition of convexity that the sum of convex functions is a convex function and, importantly, that any sum of functions will be strictly convex if *any* of the functions in the sum are strictly convex.

A quadratic functional, $\mathcal{Q}[\mathbf{x}] = \mathbf{x}^T \mathbf{H} \mathbf{x}$, is convex if $\mathbf{H}$ is a symmetric positive semidefinite matrix. If $\mathbf{H}$ is symmetric positive definite, then $\mathcal{Q}[\mathbf{x}]$ is strictly convex. The linear functional $\mathcal{Q}[\mathbf{x}] = \mathbf{w}^T \mathbf{x}$ is always convex for any $\mathbf{w}$.

B.1.1 Unconstrained Direct Solutions

In the absence of solution constraints, the simplest way of finding an optimum solution to a convex functional is to find a critical point. Therefore, if a solution can be found to the equation

$$\frac{\partial \mathcal{Q}[\mathbf{x}]}{\partial \mathbf{x}} = 0, \tag{B.7}$$

then the $\mathbf{x}$ satisfying this equation is a global optimum.

Consider the minimization problem given by

$$\min_{\mathbf{x}} \mathcal{Q}[\mathbf{x}] = \min_{\mathbf{x}} \tfrac{1}{2}\mathbf{x}^T \mathbf{H} \mathbf{x} - \mathbf{x}^T \mathbf{f}, \tag{B.8}$$

for a symmetric positive definite (SPD) matrix $\mathbf{H}$ and some vector $\mathbf{f}$. The functional $\mathcal{Q}[\mathbf{x}]$ is convex and therefore the critical point (minimum) for $\mathcal{Q}[\mathbf{x}]$ is given by

$$\mathbf{H}\mathbf{x} = \mathbf{f}. \tag{B.9}$$

This equation may be efficiently solved for $\mathbf{x}$ using any technique for solving a system of linear equations from standard linear algebra [154].

If $\mathbf{H}$ is a symmetric positive semidefinite matrix, then an optimum for $\mathcal{Q}[\mathbf{x}]$ will still satisfy (B.9), but this set of equations is no longer straightforward to solve because $\mathbf{H}$ is singular. In this circumstance, a regularization may be imposed to provide a unique answer. The most common regularization to employ in this circumstance is the **Tikhonov regularization**, which replaces the $\mathcal{Q}[\mathbf{x}]$ in (B.8) with

$$\tilde{\mathcal{Q}}[\mathbf{x}] = \tfrac{1}{2}\mathbf{x}^T\mathbf{H}\mathbf{x} - \mathbf{x}^T\mathbf{f} + \lambda \mathbf{x}^T\mathbf{x}, \tag{B.10}$$

where λ is a free parameter controlling the strength of the regularization. The purpose of the new term $\mathbf{x}^T\mathbf{x}$ is to penalize solutions with a large norm. Since the regularization term is strictly convex, the new $\tilde{\mathcal{Q}}[\mathbf{x}]$ will be strictly convex, producing a unique optimum. The regularized optimum for $\tilde{\mathcal{Q}}[\mathbf{x}]$ is now given by the solution to

$$(\mathbf{H} + \lambda \mathbf{I})\mathbf{x} = \mathbf{f}, \tag{B.11}$$

where $\mathbf{I}$ represents the identity matrix.

A quadratic functional with an SPD matrix always has a global minimum which is achieved for some $\mathbf{x}$, even if the $\mathbf{x}$ is not unique. In contrast, an unconstrained optimization of a linear functional, $\mathcal{Q}[\mathbf{x}] = \mathbf{w}^T\mathbf{x}$, is unbounded since there is no critical point for this functional.

B.1.2 Constrained Direct Solutions

Most practical optimization problems impose constraints on the solution which must be satisfied. Here we consider three commonly-encountered types of constraints: boundary conditions, equality constraints and inequality constraints.

B.1.2.1 Optimization with Boundary Conditions

Given that many of the problems considered in the text have adopted a variational framework and are solved using tools from the theory of partial differential equations, the first natural category of constraints to consider are **boundary conditions**. We will treat only the most common boundary conditions on scalar fields, which are **Dirichlet boundary conditions** and **Neumann boundary conditions**.

In the context of discrete calculus, we define a Dirichlet boundary condition on a set of nodes, $\mathcal{S} \subset \mathcal{V}$, by specifying the values of the function, b_i, at the location of those nodes, so that $x_i = b_i$, $\forall v_i \in \mathcal{S}$. Although we employ the traditional term of a "boundary" condition, it is important to note that the set $\mathcal{S}$ may be arbitrary and need not lie on the boundary of the complex. Allowing $\mathcal{S}$ to be arbitrary is analogous in the continuum analysis to permitting "internal" boundary conditions.

Finding an optimum solution with respect to the Dirichlet boundary conditions is equivalent to optimization over the nodes which are not fixed. Optimizing over the nodes which are not fixed to a Dirichlet boundary condition may be accomplished by separating $\mathbf{x}$ into two distinct components. One component corresponds to the known boundary conditions, $\mathbf{x}_\mathcal{S}$, and the other component is defined wherever no boundary conditions have been asserted, $\mathbf{x}_{\bar{S}}$. Minimization of a quadratic functional specified by an SPD matrix $\mathbf{H}$, such as

$$\min_{\mathbf{x}} \mathcal{Q}[\mathbf{x}] = \min_{\mathbf{x}} \tfrac{1}{2}\mathbf{x}^T\mathbf{H}\mathbf{x} - \mathbf{x}^T\mathbf{f}, \tag{B.12}$$

with respect to $\mathbf{x}_{\bar{S}}$, may be accomplished by rewriting $\mathcal{Q}$ in block form to distinguish the two distinct components,

$$\mathcal{Q}[\mathbf{x}] = \tfrac{1}{2}\begin{bmatrix} \mathbf{x}_S^T & \mathbf{x}_{\bar{S}}^T \end{bmatrix}\begin{bmatrix} \mathbf{H}_S & \mathbf{R}^T \\ \mathbf{R} & \mathbf{H}_{\bar{S}} \end{bmatrix}\begin{bmatrix} \mathbf{x}_S \\ \mathbf{x}_{\bar{S}} \end{bmatrix} - \begin{bmatrix} \mathbf{x}_S^T & \mathbf{x}_{\bar{S}}^T \end{bmatrix}\begin{bmatrix} \mathbf{f}_S \\ \mathbf{f}_{\bar{S}} \end{bmatrix}, \tag{B.13}$$

where $\mathbf{R}$ represents the off-diagonal block of $\mathbf{H}$. Differentiating with respect to $\mathbf{x}_{\bar{S}}$ gives

$$\frac{\partial \mathcal{Q}[\mathbf{x}]}{\partial \mathbf{x}_{\bar{S}}} = \mathbf{H}_{\bar{S}}\mathbf{x}_{\bar{S}} + \mathbf{R}\mathbf{x}_S - \mathbf{f}_{\bar{S}}. \tag{B.14}$$

Therefore, a critical point for the functional is achieved for a $\mathbf{x}_{\bar{S}}$ satisfying

$$\mathbf{H}_{\bar{S}}\mathbf{x}_{\bar{S}} = \mathbf{f}_{\bar{S}} - \mathbf{Rb}, \tag{B.15}$$

where the boundary function $\mathbf{b}$ has replaced $\mathbf{x}_S$.

Sometimes our Dirichlet boundary conditions appear in initial-value problems with asymmetric operators (e.g., as in the advection equation treated in Chap. 2). These asymmetric operators are not conveniently placed in quadratic form. For example, consider the equation

$$\frac{\partial \mathbf{x}}{\partial t} = \mathbf{U}\mathbf{x} - \mathbf{f}, \tag{B.16}$$

in which $\mathbf{U}$ is asymmetric. At steady-state, the solution $\mathbf{x}$ must satisfy

$$0 = \mathbf{U}\mathbf{x} - \mathbf{f} = \begin{bmatrix} \mathbf{U}_S & \mathbf{P} \\ \mathbf{R} & \mathbf{U}_{\bar{S}} \end{bmatrix}\begin{bmatrix} \mathbf{x}_S \\ \mathbf{x}_{\bar{S}} \end{bmatrix} - \begin{bmatrix} \mathbf{f}_S \\ \mathbf{f}_{\bar{S}} \end{bmatrix}. \tag{B.17}$$

Therefore, at steady-state our unknowns $\mathbf{x}_{\bar{S}}$ must satisfy

$$\mathbf{U}_{\bar{S}}\mathbf{x}_{\bar{S}} = \mathbf{f}_{\bar{S}} - \mathbf{Rb}, \tag{B.18}$$

which gives us the same form as the boundary-value problem stated in (B.15).

A similar procedure allows us to optimize with respect to Neumann boundary conditions. Recall the standard definition of a Neumann boundary condition

$$\vec{n} \cdot \nabla x = b. \tag{B.19}$$

In other words, the gradient of a function on the normal vector to some surface (represented by $\vec{n}$) is prescribed by b. Given a set $\mathcal{T} \subset \mathcal{E}$ of edges over which the gradient is prescribed, we may rewrite a discrete formulation of (B.19) as

$$\mathbf{G}_{\mathcal{T}}^{-1}\mathbf{A}_{\mathcal{T}}\mathbf{x} = \mathbf{b}, \tag{B.20}$$

where $\mathbf{A}_{\mathcal{T}}$ indicates the rows of the edge–node incidence matrix corresponding to the set $\mathcal{T}$. For example, if we minimize the Dirichlet energy $\mathcal{Q}[\mathbf{x}] = \frac{1}{2}\mathbf{x}^\mathsf{T}\mathbf{A}^\mathsf{T}\mathbf{G}^{-1}\mathbf{A}\mathbf{x}$ subject to $\mathbf{G}_{\mathcal{T}}^{-1}\mathbf{A}_{\mathcal{T}}\mathbf{x} = \mathbf{b}$, we obtain the constrained cost function

$$\mathcal{Q}[\mathbf{x}] = \tfrac{1}{2}\mathbf{x}^\mathsf{T}\mathbf{A}^\mathsf{T}\mathbf{G}^{-1}\mathbf{A}\mathbf{x} = \tfrac{1}{2}\mathbf{x}^\mathsf{T}\begin{bmatrix}\mathbf{A}_{\mathcal{T}}^\mathsf{T} & \mathbf{A}_{\bar{\mathcal{T}}}^\mathsf{T}\end{bmatrix}\begin{bmatrix}\mathbf{G}_{\mathcal{T}}^{-1} & 0 \\ 0 & \mathbf{G}_{\bar{\mathcal{T}}}^{-1}\end{bmatrix}\begin{bmatrix}\mathbf{A}_{\mathcal{T}} \\ \mathbf{A}_{\bar{\mathcal{T}}}\end{bmatrix}\mathbf{x}$$
$$= \tfrac{1}{2}\left(\mathbf{x}^\mathsf{T}\mathbf{A}_{\bar{\mathcal{T}}}^\mathsf{T}\mathbf{G}_{\bar{\mathcal{T}}}^{-1}\mathbf{A}_{\bar{\mathcal{T}}}\mathbf{x} + \mathbf{x}^\mathsf{T}\mathbf{A}_{\mathcal{T}}^\mathsf{T}\mathbf{b}\right). \tag{B.21}$$

Therefore, taking a variation of $\mathcal{Q}$ with respect to $\mathbf{x}$ yields

$$\frac{\partial \mathcal{Q}}{\partial \mathbf{x}} = \mathbf{A}_{\bar{\mathcal{T}}}^\mathsf{T}\mathbf{G}_{\bar{\mathcal{T}}}^{-1}\mathbf{A}_{\bar{\mathcal{T}}}\mathbf{x} + \tfrac{1}{2}\mathbf{A}_{\mathcal{T}}^\mathsf{T}\mathbf{b}, \tag{B.22}$$

which is minimized when

$$\mathbf{L}_{\bar{\mathcal{T}}}\mathbf{x} = -\tfrac{1}{2}\mathbf{A}_{\mathcal{T}}^\mathsf{T}\mathbf{b}, \tag{B.23}$$

where we have replaced $\mathbf{A}_{\bar{\mathcal{T}}}^\mathsf{T}\mathbf{G}_{\bar{\mathcal{T}}}^{-1}\mathbf{A}_{\bar{\mathcal{T}}}$ with the Laplacian matrix $\mathbf{L}_{\bar{\mathcal{T}}}$ defined on the subset $\bar{\mathcal{T}}$. Note that a set of Neumann boundary conditions may not permit a solution to (B.23). For example, if all of the edges incident on a single node are fixed by Neumann boundary conditions, then $\mathbf{L}_{\bar{\mathcal{T}}}$ is singular.

For a physical understanding of these boundary conditions in a discrete setting, we may revisit circuit theory (see Chap. 3). In the context of circuit theory, fixing a Dirichlet boundary condition at a node is equivalent to fixing the electrical potential of the node by establishing a voltage source between the node and ground with a voltage equal to the desired Dirichlet boundary condition. If we consider the circuit example in Fig. 3.2, then node v_4 has been fixed with a Dirichlet boundary condition, since its potential has been fixed at 2V (with respect to ground). The circuit interpretation of Neumann boundary conditions may also be found in Chap. 3, which showed that the gradient between two nodes corresponds to the current flowing between the nodes. Therefore, fixing a Neumann boundary condition between two nodes may be interpreted as fixing the current between the nodes. If we again reference the example circuit in Fig. 3.2, then a Neumann boundary condition exists between nodes v_2 and v_5 (ground) as a result of the current source.

In our treatment of Neumann boundary conditions, we considered the important special case of optimizing the Dirichlet energy with respect to the boundary conditions. However, we now show how to optimize a more general quadratic functional with respect to Neumann boundary conditions. Consider the solution to the more general minimization problem

$$\min_{\mathbf{x}} \mathcal{Q}[\mathbf{x}] = \min_{\mathbf{x}} \tfrac{1}{2}\mathbf{x}^\mathsf{T}\mathbf{H}\mathbf{x}$$
$$\text{s.t.} \quad \mathbf{G}^{-1}\mathbf{A}\mathbf{x} = \mathbf{b} \tag{B.24}$$

with SPD matrix $\mathbf{H}$. In order to solve this problem, we may incorporate $\mathbf{G}^{-1}\mathbf{A}\mathbf{x} = \mathbf{b}$ into $\mathcal{Q}$ via a Lagrange multiplier to yield

$$\mathcal{Q}[\mathbf{x}] = \tfrac{1}{2}\mathbf{x}^T\mathbf{H}\mathbf{x} + \lambda(\mathbf{G}^{-1}\mathbf{A}\mathbf{x} - \mathbf{b}). \tag{B.25}$$

Taking a variation of $\mathcal{Q}$ with respect to $\mathbf{x}$ and setting the result to zero yields

$$\mathbf{H}\mathbf{x} + \mathbf{G}^{-1}\mathbf{A}^T\lambda = \mathbf{0}. \tag{B.26}$$

Therefore, the optimum solution to our problem may be obtained by solving the linear system

$$\begin{bmatrix} \mathbf{H} & \mathbf{G}^{-1}\mathbf{A}^T \\ \mathbf{G}^{-1}\mathbf{A} & 0 \end{bmatrix} \begin{bmatrix} \mathbf{x} \\ \lambda \end{bmatrix} = \begin{bmatrix} 0 \\ \mathbf{b} \end{bmatrix}. \tag{B.27}$$

This solution to the quadratic optimization problem in the presence of Neumann boundary conditions may now be generalized to the solution to the quadratic optimization problem in the presence of any linear equality constraints.

B.1.2.2 Optimization with Linear Equality Constraints

Optimization with respect to Dirichlet and Neumann boundary conditions are special cases of the more general constrained optimization problem in the presence of linear equality constraints. Specifically, the above procedure for quadratic optimization in the presence of Neumann boundary conditions may be applied to any set of linear equality constraints $\mathbf{R}\mathbf{x} = \mathbf{b}$. Therefore, the quadratic optimization problem

$$\min_{\mathbf{x}} \mathcal{Q}[\mathbf{x}] = \min_{\mathbf{x}} \tfrac{1}{2}\mathbf{x}^T\mathbf{H}\mathbf{x} \tag{B.28}$$
$$\text{s.t.} \quad \mathbf{R}\mathbf{x} = \mathbf{b}$$

may be solved by finding $\mathbf{x}$ and λ satisfying

$$\begin{bmatrix} \mathbf{H} & \mathbf{R}^T \\ \mathbf{R} & 0 \end{bmatrix} \begin{bmatrix} \mathbf{x} \\ \lambda \end{bmatrix} = \begin{bmatrix} 0 \\ \mathbf{b} \end{bmatrix}. \tag{B.29}$$

Although a direct solution of (B.29) is possible, there are two unpleasantries with a direct solution: (i) the matrix is not positive definite and therefore iterative methods are not usable; (ii) one is generally concerned with only the variable $\mathbf{x}$ and not the values of the Lagrange multipliers λ. There are three approaches to overcoming these limitations: a Schur factorization, a nullspace method, and a penalty method. We now review each of these approaches to avoid a direct solution to (B.29).

Schur Factorization

Schur factorization is essentially Gaussian elimination of the block matrix. Applying a Schur factorization to (B.29) produces

$$\begin{bmatrix} H & R^T \\ 0 & -RH^{-1}R^T \end{bmatrix} \begin{bmatrix} x \\ \lambda \end{bmatrix} = \begin{bmatrix} 0 \\ b \end{bmatrix}. \quad (B.30)$$

Therefore, the optimization (B.29) may be solved in two steps

$$-RH^{-1}R^T\lambda = b, \quad (B.31a)$$

$$Hx = -R^T\lambda. \quad (B.31b)$$

The Schur factorization approach to solving (B.29) is the preferred approach when H is an easily invertible matrix, such as a diagonal matrix.

Nullspace Approach

The **nullspace approach** to solving (B.29) proceeds by observing that if we knew a basis, P, for the nullspace of R so that $RP = 0$, then

$$R(x + P\bar{x}) = b. \quad (B.32)$$

Therefore, if we had some initial guess x_0 satisfying $Rx_0 = b$, then we can change variables to $\bar{x}$ by allowing $x = x_0 + P\bar{x}$. This change of variables seeks a solution for the $\bar{x}$ that allows us to minimize $x^T H x$ knowing that the constraint $Rx = b$ will always be satisfied. Specifically, substituting $x = x_0 + P\bar{x}$ yields the unconstrained optimization problem

$$\mathcal{Q}[x] = x_0^T H x_0 + 2\bar{x}^T P^T H x_0 + \bar{x}^T P^T H P \bar{x}, \quad (B.33)$$

taking a minimum at

$$P^T H P \bar{x} = -P^T H x_0. \quad (B.34)$$

The nullspace method is quite useful when the nullspace of R takes a convenient, known form and it is possible to easily generate an x_0. The node–edge and edge–face incidence matrices are a good example of when this approach is useful because of the convenient form of the nullspace operator.

Penalty Method

The final approach that we discuss to solving a constrained quadratic algorithm is a **penalty method** in which the formal constraints are replaced by a modified objective function that penalizes solutions which do not satisfy the constraints. The

B.1 Real-Valued Optimization

replacement of the constraints then allows for an unconstrained optimization of the modified objective function. Specifically, we may transform the original problem

$$\min_{\mathbf{x}} \mathbf{x}^T \mathbf{H} \mathbf{x},$$
$$\text{s.t.} \quad \mathbf{R}\mathbf{x} = \mathbf{b}, \tag{B.35}$$

into the unconstrained minimization problem

$$\min_{\mathbf{x}} \mathbf{x}^T \mathbf{H}\mathbf{x} + \alpha \|\mathbf{R}\mathbf{x} - \mathbf{b}\| = \min_{\mathbf{x}} \mathbf{x}^T \mathbf{H}\mathbf{x} + \alpha (\mathbf{x}^T \mathbf{R}^T \mathbf{R}\mathbf{x} + 2\mathbf{x}^T \mathbf{R}^T \mathbf{b}), \tag{B.36}$$

taking a minimum at

$$(\mathbf{H} + \alpha \mathbf{R}^T \mathbf{R})\mathbf{x} = \alpha \mathbf{R}^T \mathbf{b}. \tag{B.37}$$

Although this form is generally easy to solve, there are two competing determinations in choosing α: (1) the constraints are not enforced exactly unless $\alpha \to \infty$, (2) a reasonable, finite, α value must be chosen to cause the problem to be nonsingular (since $\mathbf{R}^T \mathbf{R}$ is generally singular). In practice, it is generally possible to find an α that produces a usable solution for (B.37), but the constraints will not be enforced exactly. A summary of each of the different approaches to solving the equality-constrained quadratic problem may be found in Table B.1.

When the objective function is linear rather than quadratic, then the equality-constrained optimization problem is more challenging, and in the absence of any further constraints, the solution to this problem is often unbounded. We begin with the more difficult case of linear *inequality* constraints for both linear and quadratic objective functions.

B.1.2.3 Optimization with Linear Inequality Constraints

Linear Energy Functionals

Optimization with respect to linear inequality constraints is generally much more difficult than optimization with respect to linear equality constraints. Incorporation of linear inequality constraints into the optimization of a linear functional leads us to the generic **linear programming** problem in the form

$$\min_{\mathbf{x}} \mathcal{Q}[\mathbf{x}] = \min_{\mathbf{x}} \mathbf{w}^T \mathbf{x},$$
$$\text{s.t.} \quad \mathbf{R}\mathbf{x} \geq \mathbf{b}, \tag{B.38a}$$
$$\mathbf{x} \geq 0.$$

We may write the inequality constraint using Lagrange multipliers to give us the *Lagrangian* equation for the linear programming problem,

$$\min_{\mathbf{x}} \mathcal{Q}[\mathbf{x}] = \min_{\mathbf{x}} \mathbf{w}^T \mathbf{x} - \lambda^T (\mathbf{R}\mathbf{x} - \mathbf{b}) = \min_{\mathbf{x}} \mathbf{w}^T \mathbf{x} - \lambda^T \mathbf{R}\mathbf{x} + \lambda^T \mathbf{b}.$$

Table B.1 Summary of methods for solving a quadratic optimization problem with linear equality constraints

Solution method	When to use	Pros	Cons
Direct	Problem is small and none of the other methods apply	General, exact solution	Extra variables due to Lagrange multipliers, matrix not SPD so that iterative methods are unavailable
Schur factorization	Use when **H** is easily invertible	Fast, exact solution with an SPD matrix, no extra variables from Lagrange multipliers	Rare that **H** is invertible
Nullspace method	Use when nullspace of **R** is known and convenient or when rank of nullspace of **R** is small (resulting in few nullspace variables)	Fast, exact solution with an SPD matrix, no extra variables from Lagrange multipliers	Must understand nullspace of **R**
Penalty method	Use when Schur and nullspace do not apply and exact satisfaction of constraint not important	Fast solution with an SPD matrix, applies to general problem	Requires choice of α, constraints not exactly satisfied

Written in the Lagrangian form, we see that **x** may alternatively be viewed as the *Lagrange multiplier* enforcing a constraint on the "variable" λ, i.e.,

$$\min_{\mathbf{x}} \mathcal{Q}[\mathbf{x}] = \min_{\mathbf{x}} \mathbf{x}^T(\mathbf{w} - \mathbf{R}^T\lambda) + \mathbf{b}^T\lambda.$$

Therefore we may rewrite the entire optimization problem in terms of the variable λ as

$$\max_{\lambda} \mathcal{G}[\lambda] = \max_{\lambda} \mathbf{b}^T\lambda,$$

$$\text{s.t.} \quad \mathbf{R}^T\lambda \leq \mathbf{w}, \tag{B.38b}$$

$$\lambda \geq 0.$$

The form given in (B.38b) is known as the *dual formulation* and in this context the Lagrange multipliers, λ, are known as the *dual variables*.

The general linear programming problem expressed in primal form (B.38a), or dual form (B.38b), is generally solvable with a simplex algorithm or interior point algorithm (see [52, 304, 312] for an introduction to the variety of techniques to solve the general linear programming problem). For any particular problem it is often possible to design a very efficient algorithm that alternately optimizes the primal and dual problems, using the current solution of one problem to perform a constrained optimization of the other problem. This methodology leads to a **primal–dual algorithm**. However, the development of a primal–dual algorithm specific to a particular problem requires insight into the problem construction (see [304] for more details on the development of primal-dual algorithms).

It is not our purpose here to further cover the excellent, extensive literature on linear programming. However, we wish to show that, in the case of mixed equality and inequality constraints, a version of the nullspace method also applies in this context.

Consider the problem

$$\min_{\mathbf{x}} \mathbf{w}^T\mathbf{x},$$
$$\text{s.t.} \quad \mathbf{Ax} = \mathbf{b}, \quad (\text{B.39})$$
$$\mathbf{Hx} \geq \mathbf{h},$$

where a known operator $\mathbf{B}$ spans the nullspace of operator $\mathbf{A}$, i.e., $\mathbf{AB} = \mathbf{0}$. In this situation, we can eliminate the equality constraints by again changing variables to $\mathbf{x} = \mathbf{x}_0 + \mathbf{By}$, under the new variable the constraint is satisfied for any $\mathbf{y}$ provided we pick $\mathbf{x}_0$ such that $\mathbf{Ax}_0 = \mathbf{b}$. The linear functional problem then becomes

$$\min_{\mathbf{y}} \mathbf{w}^T \mathbf{By},$$
$$\text{s.t.} \quad \mathbf{HBy} \geq \mathbf{h} - \mathbf{Hx}_0, \quad (\text{B.40})$$

which removes the equality constraints. We now proceed back to the optimization of quadratic energy functionals in the presence of inequality constraints.

Quadratic Energy Functionals

Optimizing a quadratic energy functional with respect to inequality constraints is harder than optimization with respect to equality constraints. The optimization of a quadratic energy functional with respect to inequality constraints is known as the **quadratic programming** problem, which may be written as

$$\min_{\mathbf{x}} \mathbf{x}^T\mathbf{Hx} + \mathbf{x}^T\mathbf{f},$$
$$\text{s.t.} \quad \mathbf{Rx} \geq \mathbf{b}. \quad (\text{B.41})$$

When $\mathbf{H}$ is SPD, the quadratic programming may be solved in polynomial time. We refer the readers to [295] for more information on the optimization of a quadratic programming problem.

B.1.2.4 Ratio Optimization

Before concluding this section on the direct optimization of a convex functional, we briefly cover the optimization of a ratio of convex functionals. Specifically, consider the ratio

$$Q[\mathbf{x}] = \frac{\mathcal{P}[\mathbf{x}]}{\mathcal{R}[\mathbf{x}]}, \tag{B.42}$$

where $\mathcal{P}$ and $\mathcal{R}$ are convex functionals. Unfortunately, there is no guarantee that Q is convex simply because $\mathcal{P}$ and $\mathcal{R}$ are convex. However, we may treat two important cases in which Q is convex. In both cases, $\mathcal{P}[\mathbf{x}] = \mathbf{x}^T \mathbf{H} \mathbf{x}$ for some SPD $\mathbf{H}$, but in the first case, $\mathcal{R}[\mathbf{x}] = \mathbf{x}^T \mathbf{x}$ and in the second case $\mathcal{R}[\mathbf{x}] = \mathbf{x}^T \mathbf{f}$.

When $\mathcal{R}[\mathbf{x}] = \mathbf{x}^T \mathbf{x}$, then Q becomes

$$Q[\mathbf{x}] = \frac{\mathbf{x}^T \mathbf{H} \mathbf{x}}{\mathbf{x}^T \mathbf{x}}, \tag{B.43}$$

which is known as the **Rayleigh quotient** of $\mathbf{H}$. Before we can consider finding an $\mathbf{x}$ to optimize $Q[\mathbf{x}]$ we need to avoid the problem at $\mathbf{x} = 0$. This problem may be avoided by requiring that $\mathbf{x}^T \mathbf{x} = k$ for some $k > 0$. If we conveniently choose $k = 1$, then we may find an $\mathbf{x}$ optimizing $Q[\mathbf{x}]$ by solving the problem

$$\begin{aligned} \min_{\mathbf{x}} \; & \mathbf{x}^T \mathbf{H} \mathbf{x}, \\ \text{s.t.} \; & \mathbf{x}^T \mathbf{x} = 1. \end{aligned} \tag{B.44}$$

Imposing the constraint via a Lagrange multiplier gives the problem

$$\min_{\mathbf{x}} \; \mathbf{x}^T \mathbf{H} \mathbf{x} - \lambda(\mathbf{x}^T \mathbf{x} - 1), \tag{B.45}$$

which takes a minimum at

$$\mathbf{H}\mathbf{x} = \lambda \mathbf{x}. \tag{B.46}$$

Consequently, the optimal $\mathbf{x}$ is an eigenvector of $\mathbf{H}$ and λ is the corresponding eigenvalue. Furthermore, since

$$\lambda = \frac{\mathbf{x}^T \mathbf{H} \mathbf{x}}{\mathbf{x}^T \mathbf{x}}, \tag{B.47}$$

we see that the minimum $\mathbf{x}$ is given by the eigenvector corresponding to the smallest eigenvalue. Therefore, the optimization of the Rayleigh quotient in (B.43) is achieved by the eigenvector of $\mathbf{H}$ corresponding to the smallest eigenvalue. Optimization of the Rayleigh quotient can be used to generate eigenvectors corresponding to larger eigenvalues if the additional constraint is imposed that the optimization is performed over the space of vectors orthogonal to those eigenvectors corresponding to the smaller eigenvalues. Additionally, optimization of the **generalized Rayleigh quotient**

$$Q[\mathbf{x}] = \frac{\mathbf{x}^T \mathbf{H} \mathbf{x}}{\mathbf{x}^T \mathbf{C} \mathbf{x}}, \tag{B.48}$$

produces the generalized eigenvector of $\mathbf{H}$ and $\mathbf{C}$,

$$\mathbf{H}\mathbf{x} = \lambda \mathbf{C} \mathbf{x}, \tag{B.49}$$

B.1 Real-Valued Optimization

corresponding to the smallest generalized eigenvalue.

The **Lanczos algorithm** [154] is the standard method for computing eigenvector/eigenvalue pairs for a sparse SPD matrix. However, since the **power method** is used in the pagerank and HITS algorithms of Chap. 7, we briefly review this method. The power method is a procedure for finding the eigenvector corresponding to the largest eigenvalue, which consists of iterating

$$\mathbf{x}^{[k+1]} = \frac{\mathbf{H}\mathbf{x}^{[k]}}{\|\mathbf{H}\mathbf{x}^{[k]}\|}, \tag{B.50}$$

until convergence. In order to converge to the largest eigenvector, the initial solution $\mathbf{x}^{[0]}$ must not be orthogonal to the largest eigenvector. The division by $\|\mathbf{H}\mathbf{x}^{[k]}\|$ in the power method effectively normalizes the largest eigenvalue of $\mathbf{H}$ to unity and then each iteration drives to zero every component of $\mathbf{x}$ which projects onto the other eigenvalues. The power method is easy to implement and each iteration is fast when $\mathbf{H}$ is sparse, however the convergence rate depends on the spread between the largest eigenvalue and the second largest eigenvalue. In practice, the power method often converges quickly for the pagerank and HITS algorithms presented in Chap. 7.

The second convex ratio problem that we consider here is the optimization of

$$\mathcal{Q}[\mathbf{x}] = \frac{\mathbf{x}^T \mathbf{H} \mathbf{x}}{\mathbf{x}^T \mathbf{f}}, \tag{B.51}$$

for some arbitrary $\mathbf{f}$. This problem appears in some clustering applications (see Chap. 6). Unfortunately, the minimum of $\mathcal{Q}$ is generally unbounded. However, we may again modify this optimization problem by constraining the denominator to require that $\mathbf{x}^T \mathbf{f} = 1$. Given this constraint, the optimization problem becomes

$$\begin{aligned} \min_{\mathbf{x}} \ & \mathbf{x}^T \mathbf{H} \mathbf{x}, \\ \text{s.t.} \ & \mathbf{x}^T \mathbf{f} = 1, \end{aligned} \tag{B.52}$$

which may be formulated as an unconstrained optimization problem using a Lagrange multiplier

$$\min_{\mathbf{x}} \ \mathbf{x}^T \mathbf{H} \mathbf{x} - \lambda (\mathbf{x}^T \mathbf{f} - 1). \tag{B.53}$$

This optimization problem takes a minimum at

$$\mathbf{H}\mathbf{x} = \tfrac{1}{2} \lambda \mathbf{f}. \tag{B.54}$$

Solving for the value of the Lagrange multiplier produces

$$\lambda = \frac{2}{\mathbf{f}^T \mathbf{H}^{-1} \mathbf{f}}, \tag{B.55}$$

giving the optimum $\mathbf{x}$ as

$$\mathbf{x} = \frac{\mathbf{H}^{-1} \mathbf{f}}{\mathbf{f}^T \mathbf{H}^{-1} \mathbf{f}}. \tag{B.56}$$

B.1.3 Descent Methods

In the previous section we reviewed the direct methods for solving a convex optimization problem, with a particular focus on quadratic and linear functionals. A significant problem with direct methods is the requirement that we can identify a critical point of the solution, which may not always be possible. In contrast, descent methods do not require that we can identify a critical point of the functional. In this section we describe descent methods for solving a convex optimization problem.

B.1.3.1 Gradient Descent

The most common descent method for optimizing a differentiable convex functional, $\mathcal{Q}[\mathbf{x}]$, is **gradient descent**. The method of gradient descent starts with an initial solution, $\mathbf{x}^{[0]}$, and updates the solution with the iteration

$$\mathbf{x}^{[k+1]} = \mathbf{x}^{[k]} - \alpha \nabla \mathcal{Q}[\mathbf{x}^{[k]}], \tag{B.57}$$

where α is a parameter whose value is chosen at each iteration to preserve $\mathcal{Q}[\mathbf{x}^{[k+1]}] < \mathcal{Q}[\mathbf{x}^{[k]}]$. The iteration is typically stopped when the value of α necessary to continue decreasing the energy falls below some tolerance.

Sometimes it is possible to compute the value of α that will produce optimal improvement of the solution. For example, if we apply the method of gradient descent to solve $\mathcal{Q}[\mathbf{x}] = \frac{1}{2}\mathbf{x}^T\mathbf{H}\mathbf{x} - \mathbf{x}^T\mathbf{f}$ for some SPD $\mathbf{H}$, then

$$\frac{\partial \mathcal{Q}[\mathbf{x}]}{\partial \mathbf{x}} = \mathbf{H}\mathbf{x} - \mathbf{f} = \mathbf{r}, \tag{B.58}$$

where $\mathbf{r}$ is defined as the *residual*. In this case, the optimal α can be calculated by taking a derivative of $\mathcal{Q}[\mathbf{x}^{[k+1]}]$ with respect to α,

$$\frac{\partial \mathcal{Q}[\mathbf{x}^{[k+1]}]}{\partial \alpha} = \alpha \mathbf{r}^T\mathbf{H}\mathbf{r} - \mathbf{r}^T\mathbf{H}\mathbf{x}^{[k]} + \mathbf{r}^T\mathbf{b}. \tag{B.59}$$

Setting this expression to zero shows that the optimal step value is

$$\alpha = \frac{\mathbf{r}^T\mathbf{r}}{\mathbf{r}^T\mathbf{H}\mathbf{r}}. \tag{B.60}$$

B.1.3.2 Newton's Method

Gradient descent often has a poor convergence rate. However, we can improve the convergence rate of the descent algorithm if the functional has a positive definite Hessian by using **Newton's method**. Newton's method may be derived from the

B.1 Real-Valued Optimization

Taylor series, which describes the approximation for the change in the energy functional $\mathcal{Q}[\mathbf{x}]$ produced by a perturbation of the current solution $\mathbf{x}$

$$\mathcal{Q}[\mathbf{x} + \Delta\mathbf{x}] \approx \mathcal{Q}[\mathbf{x}] + \Delta\mathbf{x}^T \mathcal{Q}'[\mathbf{x}] + \tfrac{1}{2}\Delta\mathbf{x}^T \mathcal{Q}''[\mathbf{x}]\Delta\mathbf{x}, \tag{B.61}$$

(note the use of "prime notation" in this section to denote derivatives of $\mathcal{Q}$ with respect to $\mathbf{x}$). Consequently, a minimum of $\mathcal{Q}[\mathbf{x} + \Delta\mathbf{x}]$ is obtained when $\Delta\mathbf{x}$ satisfies

$$\mathcal{Q}''[\mathbf{x}]\Delta\mathbf{x} = -\mathcal{Q}'[\mathbf{x}]. \tag{B.62}$$

With this optimal update value for $\Delta\mathbf{x}$, we may write the iterative optimization method as

$$\mathbf{x}^{[k+1]} = \mathbf{x}^{[k]} + \Delta\mathbf{x} = \mathbf{x}^{[k]} - (\mathcal{Q}''[\mathbf{x}])^{-1}\mathcal{Q}'[\mathbf{x}]. \tag{B.63a}$$

This iterative optimization method is known as Newton's method. In practice, the iteration is often modified by introducing a free parameter α which controls the step size,

$$\mathbf{x}^{[k+1]} = \mathbf{x}^{[k]} - \alpha(\mathcal{Q}''[\mathbf{x}])^{-1}\mathcal{Q}'[\mathbf{x}]. \tag{B.63b}$$

When the Hessian matrix is well-conditioned, this method generally converges much faster than gradient descent.

Newton's method is the easiest procedure for solving several of the optimization problems that appear in this book. Before showing how Newton's method may be applied to these optimization problems, we begin with a simpler example of using Newton's method. Consider the problem of fitting a set of unknown variables $\mathbf{x}$ to a series of conflicting measurements $\mathbf{f}$, where the number of measurements exceeds the number of variables. A common approach to fitting the variables to the data is to use a p-norm penalty to find the optimal $\mathbf{x}$ variables which fit the $\mathbf{f}$ measurements,

$$\min_{\mathbf{x}} \mathcal{Q}[\mathbf{x}] = \min_{\mathbf{x}} \sum |\mathbf{R}\mathbf{x} - \mathbf{f}|^p, \tag{B.64}$$

where $\mathbf{R}$ represents a matrix that matches x_i variables with f_j measurements. To produce a least-squares estimate of $\mathbf{f}$, we would employ $p = 2$ in the energy formulation above. A choice of $p = 2$ would allow us to easily find the optimum $\mathbf{x}$ in (B.64) by solving

$$\mathbf{R}^T\mathbf{R}\mathbf{x} = \mathbf{R}^T\mathbf{f}. \tag{B.65}$$

However, a least-squares estimate is often too sensitive to outliers in the measurements $\mathbf{f}$. In order to avoid the undue influence of outliers, we may choose a more robust value of p in the range $1 \leq p < 2$. Unfortunately, choosing a p in this range does not admit the straightforward solution that we saw for $p = 2$. However, Newton's method provides a good alternative for this optimization. In order to apply Newton's method, we must find the gradient $\mathcal{Q}'[\mathbf{x}]$ and the Hessian $\mathcal{Q}''[\mathbf{x}]$

$$\mathcal{Q}'[\mathbf{x}] = p\mathbf{R}^T |\mathbf{R}\mathbf{x} - \mathbf{f}|^{p-1}, \tag{B.66}$$

$$\mathcal{Q}''[\mathbf{x}] = p(p-1)\mathbf{R}^T \operatorname{diag}\bigl(|\mathbf{R}\mathbf{x} - \mathbf{f}|^{p-2}\bigr)\mathbf{R}. \tag{B.67}$$

Therefore, the update rule for the kth iteration $\mathbf{x}^{[k]}$ to $\mathbf{x}^{[k+1]} = \mathbf{x}^{[k]} + \Delta\mathbf{x}$ is generated by solving for $\Delta\mathbf{x}$ given in (B.62) via

$$(p-1)\mathbf{R}^\mathsf{T} \operatorname{diag}\left(\left|\mathbf{R}\mathbf{x}^{[k]} - \mathbf{f}\right|^{p-2}\right)\mathbf{R}\Delta\mathbf{x} = -\mathbf{R}^\mathsf{T}\left|\mathbf{R}\mathbf{x}^{[k]} - \mathbf{f}\right|^{p-1}. \tag{B.68}$$

Therefore, each iteration of Newton's method requires solving a system of linear equations. However, by comparing the solution for $p = 2$ in (B.65) with the update given for an arbitrary p in (B.68), we can see that the only difference in the left hand side is a diagonal *weighting* matrix which weights each equation by $|\mathbf{R}\mathbf{x}^{[k]} - \mathbf{f}|^{p-2}$. We may pursue this idea further by rewriting the gradient of $\mathcal{Q}$ as

$$\mathcal{Q}'[\mathbf{x}] = p\,\mathbf{R}^\mathsf{T} \operatorname{diag}\left(|\mathbf{R}\mathbf{x} - \mathbf{f}|^{p-2}\right)(\mathbf{R}\mathbf{x} - \mathbf{f}). \tag{B.69}$$

Formulating the gradient in this way allows us to rewrite the update as

$$\mathbf{R}^\mathsf{T} \operatorname{diag}\left(\left|\mathbf{R}\mathbf{x}^{[k]} - \mathbf{f}\right|^{p-2}\right)\mathbf{R}(\mathbf{x}^{[k]} + (p-1)\Delta\mathbf{x}) = \mathbf{R}^\mathsf{T} \operatorname{diag}\left(\left|\mathbf{R}\mathbf{x}^{[k]} - \mathbf{f}\right|^{p-2}\right)\mathbf{f}. \tag{B.70}$$

Letting $\alpha = (p-1)$ in (B.63b) allows us to view the equation above as solving directly for $\mathbf{x}^{[k+1]} = \mathbf{x}^{[k]} + \alpha\Delta\mathbf{x}$. This direct solution for $\mathbf{x}^{[k+1]}$ given by

$$\mathbf{R}^\mathsf{T} \operatorname{diag}\left(\left|\mathbf{R}\mathbf{x}^{[k]} - \mathbf{f}\right|^{p-2}\right)\mathbf{R}\mathbf{x}^{[k+1]} = \mathbf{R}^\mathsf{T} \operatorname{diag}\left(\left|\mathbf{R}\mathbf{x}^{[k]} - \mathbf{f}\right|^{p-2}\right)\mathbf{f}, \tag{B.71}$$

is known as **iteratively reweighted least squares** (IRLS) because (B.71) can be thought of as a weighted version of the least-squares solution in (B.65) in which the weights are generated from the solution at the previous iteration $\mathbf{x}^{[k]}$. Consequently, the IRLS method is just Newton's method with a particular step size parameter. The IRLS method is the most common method for optimizing the fitting problem introduced in (B.64) due to its simplicity. However, care must be taken with IRLS due to the fact that it will converge only for $1 < p < 3$ (see [299]). Additionally, when $p < 2$, the "weights" given by $|\mathbf{R}\mathbf{x}^{[k]} - \mathbf{f}|^{p-2}$ must be regularized to avoid division by zero.

In the filtering Chap. 5 we described the *Basic Energy Model*. Optimization of the Basic Energy Model allowed us to produce filtered data from noisy data associated with an arbitrary graph. The filtered data was given as a minimum of $\mathcal{Q}[\mathbf{x}]$ where

$$\mathcal{Q}[\mathbf{x}] = \sum_{e_{ij}} w_{ij}^p |x_i - x_j|^p + \lambda \sum_i w_i^p |x_i - f_i|^p$$

$$= \mathbf{1}^\mathsf{T}(\mathbf{G}_1^{-1})^p |\mathbf{A}\mathbf{x}|^p + \lambda \mathbf{1}^\mathsf{T}(\mathbf{G}_0)^p |\mathbf{x} - \mathbf{f}|^p, \tag{B.72}$$

where $\mathbf{A}$ represents the graph edge–node incidence matrix and $\mathbf{f}$ indicates the vector of noisy data with one element for each node. The above procedure may be used to optimize $\mathcal{Q}[\mathbf{x}]$ for $1 < p < 3$ via IRLS by solving for the update $\mathbf{x}^{[k+1]}$ with

$$\left(\mathbf{L}(\mathbf{x}^{[k]}) + \lambda\mathbf{H}(\mathbf{x}^{[k]})\right)\mathbf{x}^{[k+1]} = \lambda\mathbf{H}(\mathbf{x}^{[k]})\mathbf{f}, \tag{B.73}$$

B.1 Real-Valued Optimization

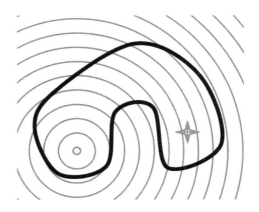

Fig. B.1 Convex constraints are necessary for a descent algorithm to achieve the global minimum. The iso-energy lines are shown in *gray* and the feasible region (allowable solutions) is indicated by the *thick black line*. A solution starting at the cross cannot descend the energy profile and achieve the global minimum while remaining inside the feasible region

where

$$\mathsf{H}(\mathbf{x}^{[k]}) = \mathrm{diag}\Big((\mathbf{G}_1^{-1})^{p-2}\big|\mathbf{x}^{[k]} - \mathbf{f}^{p-2}\big|\Big), \quad (B.74)$$

and similarly, $\mathsf{L}(\mathbf{x}^{[k]})$ indicates the Laplacian matrix with weights equal to $w_{ij}^* = w_{ij}^{p-2}|x_i^{[k]} - x_j^{[k]}|^{p-2}$. Consequently, the iteration in (B.73) allows for a straightforward solution of the Basic Energy Model presented in Chap. 5. The same model appears in the clustering Chap. 6 with Dirichlet boundary conditions for some target nodes. Incorporating these Dirichlet boundary conditions is a straightforward modification of the iteration in (B.73) that is left to the reader. However, we note that the boundary conditions could also be enforced approximately using the unmodified iteration in (B.73) by setting a subset of the measurements f_i to the Dirichlet boundary condition at node v_i and setting w_i to a large number.

B.1.3.3 Descent Methods for Constrained Optimization

Finally, we consider the application of descent methods to the constrained optimization of a convex functional. Consider the problem

$$\min_{\mathbf{x}} \; \mathcal{Q}[\mathbf{x}], \\ \text{s.t.} \quad g(\mathbf{x}) \geq \mathbf{b}, \quad (B.75)$$

where $\mathcal{Q}[\mathbf{x}]$ is convex. The ability of descent methods to solve this problem is dependent on whether the *constraints* are also convex. If the constraints are convex, then any of the *interior point methods* are available to perform the optimization [312]. Figure B.1 illustrates why convex constraints are necessary. If the constraint set is not convex, then the only way to arrive at the minimum from some initial points may be to move *away* from the solution. Interior point methods are generally good optimization techniques (see [52, 312] for details on implementation of an interior point method), but the generality of interior point methods causes them to be less competitive with optimization algorithms developed for specific problems.

B.1.4 Nonconvex Energy Optimization over Real Variables

The solution of nonconvex energy optimization problems is, in general, NP-Hard [27]. The easiest (and most common) approach in real-world applications is to somehow find a good initialization and then apply a descent algorithm to produce a reasonable local minimum. However, if a good initialization is difficult to find, a gradient of the functional cannot be computed or if a near-global optimum solution is desired, there exist a few general strategies for optimization in this context. When all other optimization approaches have been exhausted, the most general class of nonconvex, global energy minimization algorithms may be employed, such as simulated annealing [67, 233, 251], tabu search [150] and genetic algorithms [153, 192]. For more information about approaches to the general nonconvex optimization problem, we refer the reader to [204, 205].

B.2 Integer-Valued Optimization

The real-valued problems studied above allow us to take derivatives of the energy with respect to the unknown variables, often permitting us to find optimal solutions. Unfortunately, the general integer programming problem in which the domain is the set of integers is considerably harder. In fact, the integer programming problem

$$\min_{\mathbf{x}} \mathbf{w}^T\mathbf{x},$$
$$\text{s.t.} \quad \mathbf{Hx} \geq \mathbf{b}, \quad (B.76)$$
$$\mathbf{x} \in \mathbb{Z}^N,$$

is generally NP-Hard [291]. A full treatment of this problem is beyond the scope of the present work (we refer the reader to the excellent work of Nemhauser and Woosley [291] for more information). However, there are some problems relevant to this book for which we can still find an optimal solution efficiently.

B.2.1 Linear Objective Functions

One approach to solving the integer optimization problem is to simply relax the constraint $\mathbf{x} \in \mathbb{Z}^N$ to $\mathbf{x} \in \mathbb{R}^N$, in which case the problem becomes a linear programming problem with a polynomial-time solution. Unfortunately, the optimum of this relaxed problem will generally not produce an integer solution for $\mathbf{x}$. This real-valued solution for each element x_i could be rounded to the nearest integer, but this integer solution may be arbitrarily far away from the true integer-valued minimum of the cost function.

In some circumstances, we may relax $\mathbf{x} \in \mathbb{Z}^N$ to $\mathbf{x} \in \mathbb{R}^N$ and still guarantee that the solution $\mathbf{x}$ is integer. The solution of the relaxed problem is integer if and only

B.2 Integer-Valued Optimization

if **H** has the property of being *totally unimodular* and the constraint $\mathbf{b} \in \mathbb{Z}^M$ is also integer-valued [304]. Consequently, if **H** is totally unimodular, then we may relax to $\mathbf{x} \in \mathbb{R}^N$ and find the optimal solution via traditional LP methods. A matrix is defined to be totally unimodular if every submatrix has a determinant equal to an element of the set $\{-1, 0, +1\}$. The edge–node incidence matrix (of any graph) is the prototypical matrix which is totally unimodular [291]. In fact, it is known that almost every totally unimodular matrix can be considered as the edge–node incidence matrix for some graph (with two strange exceptions) [291]. Although the edge–node incidence matrix is always totally unimodular, the face–edge incidence matrix is not totally unimodular in general (e.g., [296, 338]). However, the face–edge incidence matrix is totally unimodular when each edge is included in exactly two faces which traverse the edge in opposite directions (e.g., a planar graph with a minimum cycle basis). Total unimodularity of the face–edge incidence matrix in this circumstance is because the face–edge incidence matrix is the edge–node incidence matrix of the dual graph (see Chap. 2).

Total unimodularity of the constraint matrix **H** is both necessary and sufficient for finding an integer solution to the integer optimization problem in (B.76). However, if the inequality constraints in (B.76) are replaced with equality constraints

$$\min_{\mathbf{x}} \mathbf{w}^T\mathbf{x},$$

$$\text{s.t.} \quad \mathbf{H}\mathbf{x} = \mathbf{b}, \tag{B.77}$$

$$\mathbf{x} \geq 0,$$

$$\mathbf{x} \in \mathbb{Z}^N,$$

then the total unimodularity condition on **H** is simply sufficient (i.e., not necessary) to ensure an integer solution. This problem with equality constraints has received much less attention than the problem with inequality constraints described in (B.76). The original work on the equality-constrained problem was performed by Truemper [383] and was expanded in later work by Grady [162]. Specifically, Grady introduced the notion of *pre-unimodular matrices* for the case of equality constraints described in (B.77) which are sufficient to produce an integer solution for **x** if **x** is relaxed to optimize over real values.

Let us call matrix **A** pre-unimodular if

$$\min_{\mathbf{x}} \mathbf{w}^T\mathbf{x},$$

$$\text{s.t.} \quad \mathbf{A}\mathbf{x} = \mathbf{A}\mathbf{x}_0, \tag{B.78}$$

$$\mathbf{x} \geq 0,$$

has integer solution **x** for all integer $\mathbf{x}_0$.

Let **A** be of size $M \times N$, rank$(\mathbf{A}) = r$. Grady [162] proved that the following conditions are equivalent:

(a) **A** is pre-unimodular.
(b) There exists totally unimodular matrix **U** such that $\text{Ker}\,\mathbf{A} = \text{Im}\,\mathbf{U}$.

(c) **A** can be converted to a totally unimodular matrix by elementary row operations (namely: adding a (possibly fractional) multiple of one row to another, adding/removing a zero row).

The interested reader may find the proof of this theorem in [162]. Note that a totally unimodular matrix is always pre-unimodular, but not *vice versa*. This definition for pre-unimodularity implies that the face–edge incidence matrix for a complex with trivial homology will be pre-unimodular. Therefore, the discrete Plateau's problem motivating the work on pre-unimodular matrices by Grady is solvable using standard linear programming.

B.2.2 Quadratic Objective Functions

In the previous section we addressed the question of when an integer programming problem with linear objective functions could be solved using standard methods in linear programming. This section provides some guidance for addressing the more difficult problem of unconstrained optimization of quadratic or pairwise objective functions. Once again this problem is generally NP-Hard. However, there are important special cases in which the problem is solvable exactly.

B.2.2.1 Pure Quadratic

Consider a problem of optimizing a binary-valued quantity **x** of the form

$$\min_{\mathbf{x}} \mathcal{Q}[\mathbf{x}] = \min_{\mathbf{x}} \mathbf{g}^T \mathbf{x} - \mathbf{x}^T \mathbf{W} \mathbf{x}, \quad (B.79)$$

in which $\mathbf{x} \in \mathbb{B}^N$, $g_i \geq 0$, $W_{ij} \geq 0$, $W_{ij} = W_{ji}$ and **W** has a zero diagonal (i.e., $W_{ii} = 0$). It is known that functionals of this form are submodular and may be optimized with the max-flow/min-cut method [291]. Recall that a functional is **submodular** if it satisfies

$$\mathcal{Q}[\mathbf{a} \wedge \mathbf{b}] + \mathcal{Q}[\mathbf{a} \vee \mathbf{b}] \leq \mathcal{Q}[\mathbf{a}] + \mathcal{Q}[\mathbf{b}], \quad (B.80)$$

where $\wedge$ and $\vee$ represent componentwise minimum/maximum respectively. The somewhat surprising fact that (B.79) may be optimized with a max-flow/min-cut method is seen more clearly if $\mathcal{Q}[\mathbf{x}]$ is written as

$$\begin{aligned} \mathcal{Q}[\mathbf{x}] &= \mathbf{x}^T (\mathbf{T} - \mathbf{W}) \mathbf{x} = \mathbf{x}^T \mathbf{L} \mathbf{x} + \mathbf{x}^T (\mathbf{T} - \mathbf{D}) \mathbf{x} \\ &= \sum_{ij} W_{ij} (x_i - x_j)^2 + \sum_i (t_i - d_i) x_i^2, \end{aligned} \quad (B.81)$$

where $\mathbf{T} = \text{diag}(\mathbf{t})$ and d_i is the degree of node v_i (i.e., $d_i = \sum_j W_{ij}$). Written in this form, the value of $\mathcal{Q}[\mathbf{x}]$ can be seen to represent the cost of a cut in the graph represented by (the sparsity pattern) of **W**. Specifically, the value of $\mathcal{Q}[\mathbf{x}]$ represents the cost of the cut between a set $\mathcal{S} \subset \mathcal{V}$ and its complement $\bar{\mathcal{S}}$ when **x** is a binary indicator vector that defines membership in $\mathcal{S}$. Since $\mathcal{Q}[\mathbf{x}]$ represents the cost of a cut, the minimum of $\mathcal{Q}[\mathbf{x}]$ is achieved by the minimum cut in the graph. The minimum cut between two terminals in a graph with nonnegative weights is solvable by any algorithm for computing max-flow/min-cut. In this case, the "terminals" are a result of the second term in which each node is attached to a terminal with weight $a_i = |t_i - d_i|$ with the node connected to the source terminal ($x_i = 1$) if $a_i < 0$ and the node is connected to the sink terminal ($x_i = 0$) if $a_i > 0$.

A max-flow/min-cut computation is a key component of many binary and integer optimization procedures. The utility of max-flow/min-cut in optimization is derived from the ability to isomorphically map the *state space* of a binary-valued function to a cut. When the state space is identified with the space of possible cuts and the objective function is submodular, then an algorithm to find the minimum cut will also find the minimum energy state. The mapping of a cut to a binary-valued state is accomplished by associating each variable with a particular node and treating the resulting binary-valued vector **x** as an indicator vector representing a set of nodes. The indicator vector then represents a cut between nodes inside the set and nodes outside the set. The edge weights between each node are then defined so that the cost of each cut equals the value of the energy for the corresponding state. Note that scalar Dirichlet boundary conditions for some of the x_i variables (i.e., $\mathbf{x}_\mathcal{S} = \mathbf{0}$, $\mathbf{x}_\mathcal{U} = \mathbf{1}$) can easily be enforced through manipulation of **T** to strongly penalize these variables from deviating from their prescribed values.

If we are trying to solve a problem of the form (B.79) which is not submodular (i.e., $W_{ij} \geq 0$ and $t_i \geq 0$ are not satisfied everywhere), then it will not be possible to transform our state space into a minimum-cut problem on a graph with nonnegative weights. However, it may still be possible to find a fast optimization. Specifically, if the sparsity pattern of **W** represents the adjacency matrix of a *planar graph*, then it is possible to produce an exact solution to (B.79) even if (B.79) is not submodular. This surprising result is due to the fact that a minimum-cut problem with negative weights may be cast as a maximum-cut problem, which is solvable for planar graphs by equivalence with *optimal matching* [183, 346]. Fast, polynomial-time algorithms exist for the solution of this matching problem (e.g., the Blossom algorithm [120]), although better matching algorithms is an area of active research [241]. Problems on planar graphs do appear often in practice, e.g., two-dimensional image processing or surface mesh processing (when the surface has zero genus).

If (B.79) is not submodular and the sparsity pattern of **W** does not represent a planar graph, then the solution (i.e., min-cut on a nonplanar graph with negative weights) may be well-approximated in polynomial time using semidefinite programming (SDP) [152]. Instead of approximating the solution with SDP, a very different approach to finding an optimum solution was proposed by Hammer et al. [46, 186]. Hammer et al. proposed a method for optimizing functionals of the form in (B.79) (more generally, to optimize *quadratic pseudo-boolean functions*) which showed that, by solving a max-flow/min-cut problem on a modified graph, it is possible to obtain a *partial solution* to the problem. In other words, the output of the optimization by Hammer et al. is to definitively label some variables x_i as $x_i = 1$ or $x_i = 0$, but some variables may be left unlabeled. The surprising aspect of this method is that all of the variables which are output with labels are labeled optimally (i.e., these variables are labeled the same as they would be in an optimal solution). In some practical problems in image processing, it has been demonstrated [243] that a large number of variables are definitively labeled using this procedure and suggest that the remaining unlabeled variables can be decided using a heuristic procedure. Additionally, it was observed in [243] that the number of unlabeled variables after this optimization procedure depended on the difference between **t** and the node degree vector **d**. Specifically, if **t** = **d**, then all of the nodes will be unlabeled and if the difference between **t** and **d** is large, then there will be fewer unlabeled nodes (possibly none).

B.2.2.2 General Pairwise Terms

In the previous section we considered the energy for the purely quadratic optimization which consisted of the sum of a linear term and a quadratic term with a binary variable **x**. A more generalized way of viewing the quadratic term is to see it as a term which incurs a cost based on the values of pair of variables x_i and x_j. Therefore, we may write a more generalized "quadratic" optimization problem as finding a minimum of $\mathcal{Q}[\mathbf{x}]$ where

$$\mathcal{Q}[\mathbf{x}] = \sum_{e_{ij}} \mathcal{E}_1[x_i, x_j] + \sum_{v_i} \mathcal{E}_0[x_i]. \qquad (B.82)$$

Therefore, our original quadratic optimization problem is a special case of this generalized energy when $\mathcal{E}_1[x_i, x_j] = W_{ij} x_i x_j + W_{ji} x_i x_j$ and $\mathcal{E}_0[x_i] = t_i x_i$.

We now consider the circumstances under which (B.82) may be solved for a global optimum when our variable **x** is integer-valued, i.e., $x_i \in \mathbb{Z}$ and $0 \leq x_i < k$. Ishikawa [211] showed that if

$$\mathcal{E}_1[x_i, x_j] = f(x_i - x_j), \qquad (B.83)$$

for some convex function f, then (B.82) may be solved for a global optimum [211] via a single max-flow/min-cut computation on a graph. This graph has a number of nodes equal to Nk where N represents the number of variables and k is the number of integers that each variable is permitted to take. The key idea for this optimization is to map each state of **x** to one cut in the graph, which causes the minimum cut to represent the minimum state of **x**.

When $\mathcal{E}_1[x_i, x_j]$ is not a convex function, then it is generally NP-Hard to optimize (B.82) over the set $x_i \in \mathbb{Z}$, $0 \leq x_i < k$ (see [331] for a more thorough characterization of when these multilabel problems are solvable with a polynomial time algorithm). Unfortunately, the descent methods discussed previously do not apply to this case because we cannot take gradients of the energy with respect to integer-valued variables. However, several good methods do exist for finding a local minimum, given an initial state $\mathbf{x}^{[0]}$. The simplest method to generate an improved state

$\mathbf{x}^{[k+1]}$ from a previous state $\mathbf{x}^{[k]}$ is a greedy algorithm known as **Iterated Conditional Modes** [32]. One iteration of the Iterated Conditional Modes algorithm updates only a single variable, such that the variable chosen for update produces a label change with the best improvement to the energy functional in (B.82). This iteration is then applied until convergence.

Iterated Conditional Modes is straightforward to implement, but the algorithm has a tendency to quickly get stuck in an undesirable local minimum [56]. Boykov et al. [56] noted that this tendency was due to the fact that the algorithm updated just one variable at a time. Consequently, Boykov et al. [56] proposed different algorithms that produced a much better local minimum if $\mathcal{E}_1[x_i, x_j]$ satisfied certain conditions. Specifically, the first of these algorithms applied to the case when $\mathcal{E}_1[x_i, x_j]$ defines a semi-metric, i.e.

$$\mathcal{E}_1[x_i, x_j] = 0, \quad \text{if } x_i = x_j, \tag{B.84a}$$

$$\mathcal{E}_1[x_i, x_j] = \mathcal{E}_1[x_j, x_i] \geq 0. \tag{B.84b}$$

This algorithm, called the α–β-swap algorithm, iteratively proposes to change any variable with value $x_i = \alpha$ to $x_i = \beta$ and vice versa, with all other variables keeping their label. The iterative update in the α–β-swap algorithm is cast as a max-flow/min-cut problem which finds the optimal α–β-swap from one iteration to the next. The advantage of this algorithm is that a large number of variables are updated simultaneously. This simultaneous update is in contrast to Iterated Conditional Modes which updates each variable independently. Boykov et al. [56] additionally provide a stronger algorithm when $\mathcal{E}_1[x_i, x_j]$ defines a metric, meaning that $\mathcal{E}_1[x_i, x_j]$ is a semi-metric and additionally satisfies the triangle inequality $\mathcal{E}_1[x_i, x_k] \leq \mathcal{E}_1[x_i, x_j] + \mathcal{E}_1[x_j, x_k]$. This second algorithm, called α-expansion, operates similarly to the α–β-swap algorithm, except that each variable $x_i \neq \alpha$ is given the opportunity to switch to $x_i = \alpha$. The optimal update for all variables to switch to $x_i = \alpha$ from the current state is also accomplished by solving a max-flow/min-cut problem. Boykov et al. [56] showed that the Potts Model [311] is a special case of (B.82) with a metric $\mathcal{E}_1[v_i, v_j]$ defined by

$$\mathcal{E}_1[x_i, x_j] = \begin{cases} 1 & \text{if } x_i \neq x_j, \\ 0 & \text{otherwise.} \end{cases} \tag{B.85}$$

Consequently, the α-expansion algorithm may be applied to find a good local optimum of the Potts Model.

Since the general pairwise energy optimization problem in (B.82) is NP-Hard, there is ongoing work into finding algorithms that produce better local minima in practice. In particular, substantial effort has been recently devoted to fast approximation algorithms to this problem in the computer vision literature [245, 369]. Other examples of this type of optimization algorithm are belief propagation [411] and tree-reweighted message passing [390].

B.2.2.3 Higher-Order Terms

The general pairwise energy in (B.82) could be described as assigning energy contribution for individual variables (nodes) and pairs of variables (edges). Consequently, it is natural to consider an energy which assigns contributions for higher-order structures, such as triplets (faces), e.g.,

$$\mathcal{Q}[\mathbf{x}] = \sum_{f_{ijk}} \mathcal{E}_2[x_i, x_j, x_k] + \sum_{e_{ij}} \mathcal{E}_1[x_i, x_j] + \sum_{v_i} \mathcal{E}_0[x_i]. \qquad (B.86)$$

These energies frequently appear in the study of Markov Random Fields, for which the Hammersley–Clifford theorem states that the probability of any configuration $\mathbf{x}$ may be decomposed into the probabilities defined for each clique [231].

Our intention for covering the optimization of higher-order terms for integer-valued variables is intended primarily to direct the reader to the ongoing work on this topic (see [212, 245, 324] for the most recent work in the context of computer vision). Kolmogorov and Zabih [244] showed that an optimal solution for (B.86) may be obtained by solving a max-flow/min-cut problem if $\mathbf{x}$ is binary-valued and fixing any one value in every $\mathcal{E}_2[x_i, x_j, x_k]$ term is submodular in the remaining two variables. Kohli et al. [239] extended the α-expansion algorithm to higher-order energies, such as (B.86). The belief propagation method has also been extended to address the minimization of a higher-order energy functional [252, 309].

B.2.3 General Integer Programming Problems

We have now reviewed some of the known strategies for solving the integer programming problem in which an exact solution is obtainable in polynomial time. However, there are many real-world problems in which the integer programming problem is not addressed by the techniques discussed previously.

When we considered the optimization of real-valued variables, we eventually ran out of good optimization choices. If none of the reviewed techniques are applicable, then the most general set of optimization tools may be applied to the integer optimization problem. One technique to avoid an exhaustive search for an integer solution is known as *branch-and-bound* in which a search tree is defined to partition the space of possible solutions into leaves of a search tree and the energies of the solutions in various branches are bounded to avoid traversing the entire tree [62, 253]. In addition to the branch-and-bound methodology, one can always employ the most general set of optimization techniques such as simulated annealing [67, 233, 251], tabu search [150] and genetic algorithms [153, 192] for the optimization of a general energy functional. For more information about approaches to the general integer optimization problem, we refer the reader to Nemhauser and Wolsey [291].

Appendix C
The Hodge Theorem: A Generalization of the Helmholtz Decomposition

In this appendix, the basic Helmholtz decomposition theorem is reviewed, and the generalization of the decomposition known as the Hodge decomposition is presented.

Consider a closed, compact, n-dimensional Riemannian manifold $\mathcal{M}^n$. The central result of Hodge theory is that for any p-form $\tilde{\alpha}$ there is a uniquely determined cohomologous p-form $\tilde{\beta}$, $\tilde{\beta} \sim \tilde{\alpha}$, $\tilde{\beta} \in \bigwedge^p$, such that $\star\tilde{\beta}$ is closed, and if $\tilde{\alpha}$ is closed then $\tilde{\beta}$ is harmonic [201, 401]. Simply put,

$$\operatorname{Ker}\{\Delta_p\} \cong H^p(\mathcal{M}^n; \mathbb{G}) \tag{C.1}$$

for some Abelian group $\mathbb{G}$, where the isomorphism results from the uniqueness of a harmonic representative for each de Rham class with coefficients in $\mathbb{G}$.

In addition to this isomorphism, Hodge theory presents a unique decomposition of a p-form into the sum of an exact form, a coexact form, and a harmonic part. Viewed in this way, the Hodge decomposition is the natural generalization of the Helmholtz decomposition, which is defined only for 1-forms on noncompact three-dimensional manifolds with the trivial first cohomology group.

C.1 The Helmholtz Theorem

In its original form, the Helmholtz theorem guaranteed that the infinitesimal deformation of a deformable object or perfect fluid could be decomposed into a translation, rotation, uniform dilation, and two orthogonal components of shear [195]. In vector notation, the Helmholtz theorem can be restated as follows:

Theorem 1 (Helmholtz theorem) *Any vector field $\vec{F}$ on $\mathbb{R}^3$ approaching a limit at ∞*

(a) *is determined by its curl and divergence, and*
(b) *can be written as the sum of a curl and a gradient.*

In vector calculus, the Helmholtz theorem can be reconstructed from the definition of the Laplacian operator on three-dimensional vector fields. Recall that for a vector field $\vec{F} \in \mathbb{R}^3$, the vector Laplacian is given by

$$\Delta \vec{F} = \nabla(\nabla \cdot \vec{F}) - \nabla \times (\nabla \times \vec{F}) \tag{C.2}$$

provided that $\vec{F}$ is twice differentiable. Suppose $\phi = \nabla \cdot \vec{F}$ and $\vec{A} = \nabla \times \vec{F}$. Thus, the Laplacian operator on $\vec{F}$ may be rewritten as

$$\Delta \vec{F} = \nabla \phi - \nabla \times \vec{A}, \tag{C.3}$$

thus our goal is to solve the Poisson equation in (C.3). In Cartesian coordinates, one can label the components c_i of the Laplacian of $\vec{F}$ as

$$c_i = (\nabla \phi - \nabla \times \vec{A})_i \quad \text{for } i = 1, 2, 3, \tag{C.4}$$

in which case the vectorized Poisson equation becomes $\Delta \vec{F} = \vec{c}$. If we specify that F_i goes as $\frac{1}{|\vec{r}|}$ as $|\vec{r}| \to \infty$ as a boundary condition, the solution to the scalar Poisson equation is computable via the *Green's function*. In three dimensions, the Green's function $G(\vec{r}, \vec{r}')$ corresponding to the scalar Laplacian operator is defined by the equation $\Delta G(\vec{r}, \vec{r}') = \delta^3(\vec{r} - \vec{r}')$, thus the Green's function for the Poisson equation is given by

$$G(\vec{r}, \vec{r}') = -\frac{1}{4\pi |\vec{r} - \vec{r}'|}, \tag{C.5}$$

representing the kernel of an integral transform. We may define a integral operator based on the Green's function kernel that we shall call the *Green's operator* as

$$\mathcal{G}[u(\vec{r})] = \int_{\mathbb{R}^3} G(\vec{r}, \vec{r}') u(\vec{r}') \mathrm{d}V' \tag{C.6}$$

for an arbitrary function $u(\vec{r})$ defined over $\mathbb{R}^3$.

For the scalar Poisson equation, the solution to (C.4) is given by

$$F_i(\vec{r}) = \mathcal{G}[c_i(\vec{r})] = \int_{\mathbb{R}^3} G(\vec{r}, \vec{r}') c_i(\vec{r}') \mathrm{d}V'$$

$$= -\frac{1}{4\pi} \int_{\mathbb{R}^3} \frac{c_i(\vec{r}')}{|\vec{r} - \vec{r}'|} \mathrm{d}V', \quad i = 1, 2, 3 \tag{C.7}$$

using the Green's function given in (C.5). From the vector field components, the full vectorial representation of the field,

$$\vec{F}(\vec{r}) = \int_{\mathbb{R}^3} G(\vec{r}, \vec{r}') \vec{c}(\vec{r}') \mathrm{d}V' = -\frac{1}{4\pi} \int_{\mathbb{R}^3} \frac{\vec{c}(\vec{r}')}{|\vec{r} - \vec{r}'|} \mathrm{d}V', \tag{C.8}$$

C.1 The Helmholtz Theorem

the Green's function solution to the Poisson equation in (C.4), follows immediately. From here, substitution of the gradient and curl that comprise $\vec{c}$ finally yields

$$\vec{F}(\vec{r}) = -\frac{1}{4\pi} \int_{\mathbb{R}^3} \frac{\nabla \phi(\vec{r}') - \nabla \times \vec{A}(\vec{r}')}{|\vec{r} - \vec{r}'|} dV' \tag{C.9}$$

demonstrating how $\vec{F}$ is determined by its curl and divergence, thus proving the validity of the first half of the Helmholtz theorem.

To prove uniqueness, consider two solutions to (C.3), $\vec{F}_1$ and $\vec{F}_2$, such that both $\Delta \vec{F}_1 = \nabla \phi - \nabla \times \vec{A}$ and $\Delta \vec{F}_2 = \nabla \phi - \nabla \times \vec{A}$. By the linearity of the Laplacian operator, their difference $\tilde{\vec{F}} = \vec{F}_1 - \vec{F}_2$ would satisfy $\Delta \tilde{\vec{F}} = \Delta \vec{F}_1 - \Delta \vec{F}_2 = 0$. We may extract conditions on the curl and divergence of $\tilde{\vec{F}}$ from its harmonic character by noting that

$$0 = \int_{\mathbb{R}^3} \tilde{\vec{F}} \cdot \Delta \tilde{\vec{F}} dV = \int_{\mathbb{R}^3} \left(\|\nabla \times \tilde{\vec{F}}\|^2 + \|\nabla \cdot \tilde{\vec{F}}\|^2 \right) dV + \text{boundary terms}, \tag{C.10}$$

where the boundary terms vanish since $\tilde{\vec{F}}$ is bounded at infinity. Therefore, both $\nabla \times \tilde{\vec{F}} = 0$ and $\nabla \cdot \tilde{\vec{F}} = 0$ in $\mathbb{R}^3$. Since the curl of $\tilde{\vec{F}}$ vanishes in $\mathbb{R}^3$, the Poincaré lemma guarantees that $\tilde{\vec{F}} = \nabla \psi$ for some ψ, and since the divergence of $\tilde{\vec{F}}$ also vanishes,

$$0 = \nabla \cdot \tilde{\vec{F}} = \nabla \cdot \nabla \psi = \nabla^2 \psi, \tag{C.11}$$

showing that ψ must also be harmonic. It follows from the vanishing of $\tilde{\vec{F}}$ at the boundary that ψ is bounded at $|\vec{r}| \to \infty$, and by the Maximum Modulus theorem ψ must be a constant. Then $\tilde{\vec{F}} = 0$ and $\vec{F}_1 = \vec{F}_2$, thus the solution is unique.

Now that it has been proven that the vector field $\vec{F}$ is uniquely determined by its curl and divergence on noncompact, but topologically trivial, $\mathbb{R}^3$ provided it is well-behaved at the boundary, it remains to be shown that $\vec{F}$ can be written as the sum of a divergence-free component and a curl-free component, i.e., the sum of a curl and a gradient. Consider again the expression of the vector field $\vec{F}(\vec{r})$ given in (C.8). By assumption, the vector Laplacian of $\vec{F}(\vec{r})$ may be expressed in terms of a gradient and a curl as in (C.3), and thus serves to define $\vec{F}(\vec{r})$ in terms of a vector Poisson equation such that $\vec{F}(\vec{r})$ may in turn be represented as

$$\vec{F}(\vec{r}) = \int_{\mathbb{R}^3} G(\vec{r}, \vec{r}')(\nabla \phi(\vec{r}') - \nabla \times \vec{A}(\vec{r}')) dV' \tag{C.12a}$$

$$= \int_{\mathbb{R}^3} [G(\vec{r}, \vec{r}')\nabla \phi(\vec{r}') - G(\vec{r}, \vec{r}')(\nabla \times \vec{A}(\vec{r}'))] dV' \tag{C.12b}$$

where again $G(\vec{r}, \vec{r}')$ is the Green's function of the Poisson equation. It is our goal to manipulate the above expression for $\vec{F}(\vec{r})$ as the sum of a gradient with a curl in given only that $\vec{F}$ approaches a constant at the boundary of the space in which it is defined, $\mathbb{R}^3$, and therefore both $\phi(\vec{r}) = 0$ and $\vec{A}(\vec{r}) = 0$ on $\partial(\mathbb{R}^3)$.

First we shall consider the term involving $\phi(\vec{r}')$. Integrating by parts—or invoking the product rule—shows that

$$\nabla(G(\vec{r},\vec{r}')\phi(\vec{r}')) = (\nabla G(\vec{r},\vec{r}'))\phi(\vec{r}') + G(\vec{r},\vec{r}')(\nabla\phi(\vec{r}')) \tag{C.13}$$

where the gradient operator is with respect to the primed coordinates, i.e., with respect to $\vec{r}'$. To extract a boundary term out of this relation, we require an integrand with which we may invoke the Divergence theorem to translate a volume integral over $\mathbb{R}^3$ to a vanishing surface integral on its boundary. One method is to let $\vec{\alpha}(\vec{r}') = \vec{\alpha} \in \mathbb{R}^3$ be a constant vector, then we may integrate by parts the scalar product of $\vec{\alpha}$ with $G(\vec{r},\vec{r}')\phi(\vec{r}')$ to yield

$$\nabla \cdot (\vec{\alpha} G(\vec{r},\vec{r}')\phi(\vec{r}')) = (\nabla \cdot \vec{\alpha}) G(\vec{r},\vec{r}')\phi(\vec{r}') + \vec{\alpha} \cdot \nabla(G(\vec{r},\vec{r}')\phi(\vec{r}')) \tag{C.14a}$$

but since $\vec{\alpha}$ is constant, $\nabla \cdot \vec{\alpha} = 0$,

$$\nabla \cdot (\vec{\alpha} G(\vec{r},\vec{r}')\phi(\vec{r}')) = \vec{\alpha} \cdot \nabla(G(\vec{r},\vec{r}')\phi(\vec{r}')) \tag{C.14b}$$

then substituting in (C.13) into the left hand side term, we see that

$$\nabla \cdot (\vec{\alpha} G(\vec{r},\vec{r}')\phi(\vec{r}')) = \vec{\alpha} \cdot \left[(\nabla G(\vec{r},\vec{r}'))\phi(\vec{r}') + G(\vec{r},\vec{r}')(\nabla\phi(\vec{r}')) \right]. \tag{C.14c}$$

The volume integral of the term $\nabla \cdot (\vec{\alpha} G(\vec{r},\vec{r}')\phi(\vec{r}'))$ may be re-expressed as a boundary integral via the Divergence theorem, and since by construction $\phi(\vec{r}') = 0$ on $\partial(\mathbb{R}^3)$,

$$0 = \int_{\partial(\mathbb{R}^3)} (\vec{\alpha} \cdot \hat{n}) G(\vec{r},\vec{r}')\phi(\vec{r}') dS' \tag{C.15a}$$

$$= \int_{\mathbb{R}^3} \nabla \cdot (\vec{\alpha} G(\vec{r},\vec{r}')\phi(\vec{r}')) dV' \tag{C.15b}$$

$$= \vec{\alpha} \cdot \left[\int_{\mathbb{R}^3} (\nabla G(\vec{r},\vec{r}'))\phi(\vec{r}') dV' + \int_{\mathbb{R}^3} G(\vec{r},\vec{r}')(\nabla\phi(\vec{r}')) dV' \right] \tag{C.15c}$$

therefore

$$\int_{\mathbb{R}^3} (\nabla G(\vec{r},\vec{r}'))\phi(\vec{r}') dV' = -\int_{\mathbb{R}^3} G(\vec{r},\vec{r}')(\nabla\phi(\vec{r}')) dV', \tag{C.16}$$

where the right-hand side of (C.16) is equivalent to the $\phi(\vec{r}')$ term of the expression for $\vec{F}(\vec{r})$ in (C.12). This construction has made it possible to transfer the gradient operator acting on $\phi(\vec{r}')$ to a gradient operator acting on the Green's function $G(\vec{r},\vec{r}')$. We will now attempt to make a similar transfer of the curl operator on $\vec{A}(\vec{r}')$.

C.1 The Helmholtz Theorem

Again, for the $\vec{A}(\vec{r}')$ term we wish to take advantage of the vanishing of $\vec{A}(\vec{r}')$ on the boundary of $\mathbb{R}^3$ via an application of the Divergence theorem. By the product rule we see that the $G(\vec{r},\vec{r}')(\nabla \times \vec{A}(\vec{r}'))$ term may be viewed as a term of

$$\nabla \times (G(\vec{r},\vec{r}')\vec{A}(\vec{r}')) = (\nabla G(\vec{r},\vec{r}')) \times \vec{A}(\vec{r}') + G(\vec{r},\vec{r}')(\nabla \times \vec{A}(\vec{r}')). \quad \text{(C.17)}$$

Following the procedure above, consider a constant vector $\vec{\alpha}(\vec{r}') = \vec{\alpha} \in \mathbb{R}^3$, such that integrating by parts the vector product of $\vec{\alpha}$ with $G(\vec{r},\vec{r}')\vec{A}(\vec{r}')$ yields

$$\nabla \cdot (\vec{\alpha} \times G(\vec{r},\vec{r}')\vec{A}(\vec{r}')) = (\nabla \times \vec{\alpha}) \cdot G(\vec{r},\vec{r}')\vec{A}(\vec{r}')$$
$$- \vec{\alpha} \cdot [\nabla \times (G(\vec{r},\vec{r}')\vec{A}(\vec{r}'))] \quad \text{(C.18a)}$$

but since $\vec{\alpha}$ is constant, $\nabla \cdot \vec{\alpha} = 0$,

$$\nabla \cdot (\vec{\alpha} \times G(\vec{r},\vec{r}')\vec{A}(\vec{r}')) = -\vec{\alpha} \cdot [\nabla \times (G(\vec{r},\vec{r}')\vec{A}(\vec{r}'))] \quad \text{(C.18b)}$$

then substituting in (C.17) into the left hand side term, we see that

$$\nabla \cdot (\vec{\alpha} \times G(\vec{r},\vec{r}')\vec{A}(\vec{r}')) = -\vec{\alpha} \cdot \Big[(\nabla G(\vec{r},\vec{r}')) \times \vec{A}(\vec{r}')$$
$$+ G(\vec{r},\vec{r}')(\nabla \times \vec{A}(\vec{r}'))\Big]. \quad \text{(C.18c)}$$

The volume integral of the term $\nabla \cdot (\vec{\alpha} \times G(\vec{r},\vec{r}')\vec{A}(\vec{r}'))$ may also be re-expressed as a boundary integral via the Divergence theorem, and since by construction $\vec{A}(\vec{r}') = 0$ on $\partial(\mathbb{R}^3)$,

$$0 = \int_{\partial(\mathbb{R}^3)} (\vec{\alpha} \times G(\vec{r},\vec{r}')\vec{A}(\vec{r}'))dS' \quad \text{(C.19a)}$$

$$= \int_{\mathbb{R}^3} \nabla \cdot (\vec{\alpha} \times G(\vec{r},\vec{r}')\vec{A}(\vec{r}'))dV' \quad \text{(C.19b)}$$

$$= -\vec{\alpha} \cdot \Big[\int_{\mathbb{R}^3} ((\nabla G(\vec{r},\vec{r}')) \times \vec{A}(\vec{r}'))dV'$$
$$+ \int_{\mathbb{R}^3} G(\vec{r},\vec{r}')(\nabla \times \vec{A}(\vec{r}'))dV'\Big] \quad \text{(C.19c)}$$

therefore,

$$\int_{\mathbb{R}^3} [(\nabla G(\vec{r},\vec{r}')) \times \vec{A}(\vec{r}')]dV' = \int_{\mathbb{R}^3} [G(\vec{r},\vec{r}')(\nabla \times \vec{A}(\vec{r}'))]dV', \quad \text{(C.20)}$$

where the right-hand side of (C.20) is equivalent to the $\phi(\vec{r}')$ term of the expression for $\vec{F}(\vec{r})$ in (C.12). This construction has made it possible to transfer the curl operator acting on $\vec{A}(\vec{r}')$ to a gradient operator acting on the Green's function $G(\vec{r},\vec{r}')$.

The vector field $\vec{F}(\vec{r})$ can now be seen as the sum of two integrals,

$$\vec{F} = -\int_{\mathbb{R}^3} (\nabla_{\vec{r}'} G(\vec{r}, \vec{r}')) \phi(\vec{r}') dV' - \int_{\mathbb{R}^3} [(\nabla_{\vec{r}'} G(\vec{r}, \vec{r}')) \times \vec{A}(\vec{r}')] dV', \quad (C.21)$$

each with an integrand in terms of the gradient of the Green's function. We have introduced the notation $\nabla_{\vec{r}'}$ to specify that the gradient operator in this case is with respect to the primed coordinates given by the vector $\vec{r}'$. We may now exploit the natural *symmetry* of the Green's operator relative to the two coordinate frames, one given by $\vec{r}$ and the other by $\vec{r}'$. Explicitly evaluating the gradient of the Green's function of the Poisson equation with respect to the primed coordinates shows that the result of the gradient operation with respect to *unprimed* coordinates is *identical* to the gradient operation with respect to *primed* coordinates, i.e.,

$$\nabla_{\vec{r}'} G(\vec{r}, \vec{r}') = \nabla_{\vec{r}'} \frac{1}{4\pi |\vec{r} - \vec{r}'|} = \frac{(\vec{r} - \vec{r}')}{4\pi |\vec{r} - \vec{r}'|^3} = \nabla_{\vec{r}} \frac{-1}{4\pi |\vec{r} - \vec{r}'|}$$
$$= -\nabla_{\vec{r}} G(\vec{r}, \vec{r}'), \quad (C.22)$$

where the notation $\nabla_{\vec{r}}$ specifies the gradient with respect to the unprimed coordinates.

In the integral expression for $\vec{F}(\vec{r})$, by the symmetry of the gradient operator we may freely substitute the gradient operators with respect to the primed coordinates with the gradient operator with respect to unprimed coordinates, that is,

$$\vec{F} = \int_{\mathbb{R}^3} (\nabla_{\vec{r}} G(\vec{r}, \vec{r}')) \phi(\vec{r}') dV' + \int_{\mathbb{R}^3} [(\nabla_{\vec{r}} G(\vec{r}, \vec{r}')) \times \vec{A}(\vec{r}')] dV', \quad (C.23)$$

without loss of generality. Since the fields $\phi(\vec{r}')$ and $\vec{A}(\vec{r}')$ of the integrands are both functions defined over the primed coordinates, they are constant with respect to the unprimed coordinates and therefore their derivatives with respect to $\vec{r}$ vanish identically. Therefore, in the first integral term containing $\phi(\vec{r}')$,

$$(\nabla_{\vec{r}} G(\vec{r}, \vec{r}')) \phi(\vec{r}') = \nabla_{\vec{r}} (G(\vec{r}, \vec{r}') \phi(\vec{r}')) - G(\vec{r}, \vec{r}')(\nabla_{\vec{r}} \phi(\vec{r}'))$$
$$= \nabla_{\vec{r}} (G(\vec{r}, \vec{r}') \phi(\vec{r}')) \quad (C.24)$$

so the gradient operator may be applied to the product of $G(\vec{r}, \vec{r}') \phi(\vec{r}')$. Further, in the second integral term containing $\vec{A}(\vec{r}')$,

$$(\nabla_{\vec{r}} G(\vec{r}, \vec{r}')) \times \vec{A}(\vec{r}') = \nabla_{\vec{r}} \times (G(\vec{r}, \vec{r}') \vec{A}(\vec{r}')) - G(\vec{r}, \vec{r}')(\nabla_{\vec{r}} \times \vec{A}(\vec{r}'))$$
$$= \nabla_{\vec{r}} \times (G(\vec{r}, \vec{r}') \vec{A}(\vec{r}')) \quad (C.25)$$

so here the gradient operator may be extended to a curl operation on the product of $G(\vec{r}, \vec{r}') \vec{A}(\vec{r}')$. Given that the two gradient operators may be expressed as a gradient and a curl operating on the full integrand, and the operators operate on the unprimed

C.1 The Helmholtz Theorem

coordinates whereas the integration takes place over the primed coordinates, each operator may be pulled out from the integrand.[1] Therefore, substituting the results from (C.24) and (C.25) into the integral expression for $\vec{F}(\vec{r})$, we see that

$$\vec{F}(\vec{r}) = \int_{\mathbb{R}^3} [\nabla_{\vec{r}} (G(\vec{r},\vec{r}')\phi(\vec{r}'))] dV' + \int_{\mathbb{R}^3} [\nabla_{\vec{r}} \times (G(\vec{r},\vec{r}')\vec{A}(\vec{r}'))] dV' \quad \text{(C.27a)}$$

$$= \nabla_{\vec{r}} \left[\int_{\mathbb{R}^3} G(\vec{r},\vec{r}')\phi(\vec{r}') dV' \right] + \nabla_{\vec{r}} \times \left[\int_{\mathbb{R}^3} (G(\vec{r},\vec{r}')\vec{A}(\vec{r}')) dV' \right] \quad \text{(C.27b)}$$

and thus $\vec{F}(\vec{r})$ is finally expressed as the sum of a gradient and a curl. Using the Green's operator $\mathcal{G}$ in place of the integrals with the Green's function as the kernel, we may express the sum more suggestively as

$$\boxed{\vec{F}(\vec{r}) = \nabla \mathcal{G}[\phi(\vec{r})] + \nabla \times \mathcal{G}[\vec{A}(\vec{r})] = \nabla \mathcal{G}[\nabla \cdot \vec{F}(\vec{r})] + \nabla \times \mathcal{G}[\nabla \times \vec{F}(\vec{r})]} \quad \text{(C.28)}$$

where the curl, gradient, and divergence operators are defined with respect to the unprimed coordinates, i.e., $\vec{r}$. Commonly, the arguments of the gradient and curl operators are given explicitly as

$$\varphi(\vec{r}) \triangleq \mathcal{G}[\nabla \cdot \vec{F}(\vec{r})] = -\int_{\mathbb{R}^3} \frac{\nabla \cdot \vec{F}(\vec{r}')}{4\pi |\vec{r} - \vec{r}'|} dV', \quad \text{(C.29a)}$$

$$\vec{A}(\vec{r}) \triangleq \mathcal{G}[\nabla \times \vec{F}(\vec{r})] = \int_{\mathbb{R}^3} \frac{\nabla \times \vec{F}(\vec{r}')}{4\pi |\vec{r} - \vec{r}'|} dV' \quad \text{(C.29b)}$$

---

[1] As an aside for those familiar with electromagnetics, the exchange of the order of integration and differentiation that takes place in (C.27) can be seen as example of the connection between Coulomb's law and Gauss's law in the case of the scalar integrand, and as an example of the connection between the Biot–Savart law and Ampère's law in the case of the vector integrand [214].

Assuming a constant permittivity, the integral form of the electric displacement field $\vec{D}(\vec{r})$ is given by Coulomb's law

$$\vec{D}(\vec{r}) = \nabla \mathcal{G}[\rho(\vec{r})] = \int_{\mathbb{R}^3} \frac{\rho(\vec{r}')(\vec{r} - \vec{r}')}{4\pi |\vec{r} - \vec{r}'|^3} dV', \quad \text{(C.26a)}$$

where $\rho(\vec{r})$ represents the charge density. Gauss's law states that $\nabla \cdot \vec{D} = \rho$.

For a constant permeability, the integral form of the magnetic field $\vec{H}(\vec{r})$ is given by the Biot–Savart law

$$\vec{H}(\vec{r}) = \nabla \times \mathcal{G}[\vec{J}(\vec{r})] = \int_{\mathbb{R}^3} \frac{\vec{J}(\vec{r}') \times (\vec{r} - \vec{r}')}{4\pi |\vec{r} - \vec{r}'|^3} dV', \quad \text{(C.26b)}$$

where $\vec{J}(\vec{r})$ is the current density. Ampère's law states that $\nabla \times \vec{H} = \vec{J}$.

such that (C.28) may be re-written as

$$\vec{F}(\vec{r}) = -\nabla \varphi(\vec{r}) + \nabla \times \vec{A}(\vec{r}) \tag{C.29c}$$

thus both parts (a) and (b) of the Helmholtz theorem have been proven.

The limitations of the Helmholtz decomposition are apparent. Uniqueness of the decomposition only holds on non-compact $\mathbb{R}^3$ when the vector field $\vec{F}$ approaches a limit on the boundary. Immediately, one might question how the theorem breaks down on a *closed* 3-manifold $\mathcal{M}^3$.

In addition to the requirement that the vector field behave "nicely" at the boundary, the Helmholtz decomposition, due to its vector calculus representation, is constrained to apply only in three-dimensional space. The Hodge decomposition generalizes the Helmholtz decomposition for n-dimensional manifolds.

It is important to note that the Helmholtz theory outlined above provides an *orthogonal* decomposition of the vector field $\vec{F}(\vec{r})$. The set of irrotational vectors are orthogonal in the Hilbert space sense to the set of solenoidal vectors. The orthogonality of the decomposition also translates into the differential forms setting in which we shall consider the Hodge decomposition.

C.2 The Hodge Decomposition

To motivate the discovery of the Hodge decomposition—the generalization of the Helmholtz decomposition for arbitrary p-forms in n dimensions—as the beginning of continuous Hodge theory and harmonic integral theory, we shall attempt to follow the logical path of the previous section.

Thus, we begin with a statement of the generalized Helmholtz theorem that will grow into the central result of Hodge theory.

Theorem 2 (Generalized Helmholtz theorem) *Consider an orientable, compact, n-dimensional Riemannian manifold $\mathcal{M}^n$. Consider the Laplace–de Rham operator $\Delta_p = \mathrm{dd}^* + \mathrm{d}^*\mathrm{d}$ on p-forms. Then, given a p-form $\tilde{\omega}$,*

(a) *$\tilde{\omega}$ is determined by $\tilde{\eta}_1 = \mathrm{d}\tilde{\omega}$ and $\tilde{\eta}_2 = \mathrm{d}^*\tilde{\omega}$ if a finite number of* compatibility conditions *are met, and*

(b) *$\tilde{\omega} = \mathrm{d}\tilde{\lambda}_1 + \mathrm{d}^*\tilde{\lambda}_2 + \tilde{\lambda}_H$, where $\mathrm{d}^*\tilde{\lambda}_1 = 0$, $\mathrm{d}\tilde{\lambda}_2 = 0$, and $\mathrm{d}\tilde{\lambda}_H = \mathrm{d}^*\tilde{\lambda}_H = 0$, and the space of all $\tilde{\lambda}_H$ is finite-dimensional.*

On closed manifolds, the p-forms $\tilde{\lambda}_H$ satisfying $\mathrm{d}\tilde{\lambda}_H = \mathrm{d}^*\tilde{\lambda}_H = \Delta_p \tilde{\lambda}_H = 0$ are called *harmonic p-forms*. The space of harmonic forms on $\mathcal{M}^n$ will be denoted as $\mathcal{H}^p(\mathcal{M}^n)$ or equivalently as $\mathrm{Ker}\{\Delta_p\}$. Before proceeding, the following definitions are required.

Definition 1 A **separable Hilbert space** is an infinite-dimensional inner product space where any vector v can be written as $v = \sum_{i=1}^{\infty} \alpha_i v_i$ where $\{v_i\}_{i=1}^{\infty}$ is a set of basis vectors.

C.2 The Hodge Decomposition

Definition 2 A **Fredholm operator** is a linear operator whose spectrum is discrete and all eigenvalues have finite multiplicity.

The Laplace–de Rham operator defined on p-forms in $T^*\mathcal{M}^n$, Δ_p, is an elliptic operator and if $\mathcal{M}^n$ is compact then Δ_p is Fredholm.

The central result required to establish the generalized Helmholtz theorem is the **Fredholm Alternative**, which provides a definition of the *rank* of an infinite-dimensional linear operator and motivates a decomposition in terms of the range and nullspace of the operator.

Theorem 3 (Fredholm Alternative) *If a bounded operator L is Fredholm and if u and f are elements of separable Hilbert spaces, then the equation*

$$Lu = f \tag{C.30}$$

has a solution if and only if f is orthogonal to all solutions v of $L^*v = 0$, i.e., $\langle f, v \rangle = 0$, where L^* is the adjoint operator of L.

The Fredholm Alternative states that solutions $\hat{\tilde{\omega}}$ to the Poisson equation $\Delta\tilde{\omega} = \tilde{\eta}$ must live in a space orthogonal to the space of solutions $\Delta^*\tilde{\xi} = 0$.

However, the Laplace–de Rham operator is *formally self-adjoint* [356].

Theorem 4 (Formal self-adjointness of Δ) *Given that d^* is the adjoint of d, $\mathrm{d}^* \equiv \mathrm{d}^*$, with respect to the Hodge inner product defined on the n-dimensional (Riemannian) manifold $\mathcal{M}^n$, and either (1) $\mathcal{M}^n$ is closed or (2) the Laplace–de Rham operator is constrained to operate on the set of p-forms that are bounded on $\partial\mathcal{M}^n$, then $\Delta = \Delta^*$, i.e.,*

$$\langle \Delta\tilde{\mu}, \tilde{v} \rangle = \langle \mathrm{d}\mathrm{d}^*\tilde{\mu} + \mathrm{d}^*\mathrm{d}\tilde{\mu}, \tilde{v} \rangle \tag{C.31a}$$

$$= \langle \mathrm{d}\mathrm{d}^*\tilde{\mu}, \tilde{v} \rangle + \langle \mathrm{d}^*\mathrm{d}\tilde{\mu}, \tilde{v} \rangle \tag{C.31b}$$

$$= \langle \mathrm{d}^*\tilde{\mu}, \mathrm{d}^*\tilde{v} \rangle + \langle \mathrm{d}\tilde{\mu}, \mathrm{d}\tilde{v} \rangle \tag{C.31c}$$

$$= \langle \tilde{\mu}, \mathrm{d}\mathrm{d}^*\tilde{v} \rangle + \langle \tilde{\mu}, \mathrm{d}^*\mathrm{d}\tilde{v} \rangle \tag{C.31d}$$

$$= \langle \tilde{\mu}, \mathrm{d}\mathrm{d}^*\tilde{v} + \mathrm{d}^*\mathrm{d}\tilde{v} \rangle \tag{C.31e}$$

$$= \langle \tilde{\mu}, \Delta\tilde{v} \rangle. \tag{C.31f}$$

Combining $\Delta = \Delta^*$ with the Fredholm alternative, it follows that $\Delta\tilde{\omega} = \tilde{\eta}$ has a solution if and only if $\tilde{\eta} \perp \tilde{\lambda}_H$ for all harmonic $\tilde{\lambda}_H$ satisfying $\Delta\tilde{\lambda}_H = 0$. In other words, $\Delta\tilde{\omega} = \tilde{\eta}$ has a solution if and only if $\tilde{\eta}$ is orthogonal to the space of harmonic forms. This also implies that if $\Delta\tilde{\omega} = \tilde{\eta}$ has a solution it is unique up to a harmonic form, i.e., if $\hat{\tilde{\omega}}$ is a solution, then $\hat{\tilde{\omega}} + \tilde{\lambda}_H$ is also a valid solution. In this case, $\hat{\tilde{\omega}}$ may be considered the inhomogeneous solution and $\tilde{\lambda}_H$ the homogeneous solution to the Poisson equation.

Consider the Poisson equation $\Delta\tilde{\omega} = \tilde{v}$ where here $\tilde{v}$ is a p-form chosen as orthogonal to the space of harmonic forms ($\tilde{v} \perp \mathrm{Ker}\{\Delta\}$).

Fact 1 *There exists a* Green's *operator* $\mathcal{G} : \bigwedge^p(\mathcal{M}^n) \to (\mathcal{H}^p)^\perp$ *such that* $\mathcal{G}\tilde{v} = \tilde{\omega}$ [394].

(Note that $\mathcal{G}\tilde{v}$ is equivalent to the right-hand side of (C.8) in the vector calculus setting.)

To show that the Green's operator commutes with the differential and codifferential operators, we must first show that the differential and codifferential operators commute with the Laplace–de Rham operator. The commutativity of the two derivative operators is clear from the following manipulations:

$$d\Delta = d(dd^* + d^*d) = d^2d^* + dd^*d = dd^*d + d^*d^2$$
$$= (dd^* + d^*d)d = \Delta d, \tag{C.32a}$$
$$d^*\Delta = d^*(dd^* + d^*d) = d^*dd^* + d^{*2}d = dd^{*2} + d^*dd^*$$
$$= (dd^* + d^*d)d^* = \Delta d^* \tag{C.32b}$$

as both $d^2 = 0$ and $d^{*2} = 0$. Therefore, if in the orthogonal complement of the space of harmonic forms $(\mathcal{H}^p(\mathcal{M}^n))^\perp$ the composition of the Laplace–de Rham operator with the Green's operator is the identity operator, i.e., $\mathcal{G}\Delta = \mathsf{Id}$ on $(\mathcal{H}^p(\mathcal{M}^n))^\perp$, then on $(\mathcal{H}^p(\mathcal{M}^n))^\perp$, by the commutativity of d and Δ,

$$d = \mathcal{G}\Delta d = \mathcal{G}d\Delta, \tag{C.33a}$$
$$d = d\mathcal{G}\Delta \tag{C.33b}$$

so, subtracting (C.33a) and (C.33b),

$$(d\mathcal{G} - \mathcal{G}d)\Delta = 0 \quad \text{on } (\mathcal{H}^p(\mathcal{M}^n))^\perp. \tag{C.34}$$

Since we are operating on the orthogonal complement of the kernel of the Laplace–de Rham operator, the left hand side term $(d\mathcal{G} - \mathcal{G}d)$ must vanish. It is clear that the commutativity relations

$$d\mathcal{G} = \mathcal{G}d \quad \text{and, similarly,} \quad d^*\mathcal{G} = \mathcal{G}d^* \quad \text{on } (\mathcal{H}^p(\mathcal{M}^n))^\perp \tag{C.35}$$

hold. So, if the inhomogeneous solution $\hat{\tilde{\omega}} \in (\mathcal{H}^p(\mathcal{M}^n))^\perp$ of $\Delta\tilde{\omega} = \tilde{\eta}$, where by necessity $\tilde{\eta} \in (\mathcal{H}^p(\mathcal{M}^n))^\perp$ for the solution to exist, can be expressed as $\Delta\hat{\tilde{\omega}} = \tilde{\eta} = d\tilde{\eta}_1 + d^*\tilde{\eta}_2$ for some p-forms $\tilde{\eta}_1$ and $\tilde{\eta}_2$ such that $d\tilde{\eta}_1, d^*\tilde{\eta}_2 \in (\mathcal{H}^p(\mathcal{M}^n))^\perp$ and thus $\tilde{\eta}_1, \tilde{\eta}_2 \in (\mathcal{H}^p(\mathcal{M}^n))^\perp$, then one can decompose the p-form $\tilde{\omega}$ into a component in $(\mathcal{H}^p(\mathcal{M}^n))^\perp$ and a component in $\mathcal{H}^p(\mathcal{M}^n)$ as in

$$\tilde{\omega} = \hat{\tilde{\omega}} + \tilde{\lambda}_H = \mathcal{G}\Delta\hat{\tilde{\omega}} + \tilde{\lambda}_H = \mathcal{G}d\tilde{\eta}_1 + \mathcal{G}d^*\tilde{\eta}_2 + \tilde{\lambda}_H$$
$$= d(\mathcal{G}\tilde{\eta}_1) + d^*(\mathcal{G}\tilde{\eta}_2) + \tilde{\lambda}_H \tag{C.36}$$

by the commutativity of d and d* with $\mathcal{G}$ on $(\mathcal{H}^p(\mathcal{M}^n))^\perp$. This decomposition of the p-form $\tilde{\omega}$ into a *closed, coclosed* and *harmonic* form is known as the **Hodge**

C.2 The Hodge Decomposition

decomposition on p-forms, the generalization of the Helmholtz decomposition to n dimensions.

The expression (C.36) provides an orthogonal decomposition of the elements of $\bigwedge^p$ defined over $\mathcal{M}^n$ equivalent to

$$\bigwedge^p = \left(\Delta \bigwedge^p\right) \oplus \mathcal{H}^p = \left(dd^* \bigwedge^p\right) \oplus \left(d^*d \bigwedge^p\right) \oplus \mathcal{H}^p$$

$$= \left(d \bigwedge^{p-1}\right) \oplus \left(d^* \bigwedge^{p+1}\right) \oplus \mathcal{H}^p \tag{C.37}$$

where $\oplus$ denotes the direct sum between spaces.

The existence proof of the Hodge decomposition is difficult (see Refs. [139, 394]), but to demonstrate uniqueness of this result, suppose we have a Hodge decomposition of the p-form $\tilde{\beta}$ given by

$$\tilde{\beta} = d\tilde{\alpha} + d^*\tilde{\gamma} + \tilde{h} = 0,$$

where $\tilde{\alpha} \in \bigwedge^{p-1}$, $\tilde{\gamma} \in \bigwedge^{p+1}$, and $\tilde{h} \in \mathcal{H}^p$. It follows that $d(d\tilde{\alpha}) = 0$ and that $d\tilde{h} = 0$. Therefore,

$$dd^*\tilde{\gamma} = 0, \tag{C.38a}$$

$$\langle dd^*\tilde{\gamma}, \tilde{\gamma} \rangle = 0, \tag{C.38b}$$

$$\langle d^*\tilde{\gamma}, d^*\tilde{\gamma} \rangle = 0, \tag{C.38c}$$

$$d^*\tilde{\gamma} = 0, \quad \text{thus} \tag{C.38d}$$

$$d\tilde{\alpha} + \tilde{h} = 0. \tag{C.38e}$$

Similarly, one can show that both $d\tilde{\alpha} = 0$ and $\tilde{h} = 0$ [139, p. 139]. Note that the uniqueness is guaranteed for the summands of the Hodge decomposition only, e.g., $d\tilde{\alpha}$ is unique, but $\tilde{\alpha}$ is not, as we may add to $\tilde{\alpha}$ any closed $(p-1)$ form. We are now in a position to state Hodge's theorem in its entirety.

Theorem 5 (Hodge's theorem [142]) *Let $\mathcal{M}^n$ be a closed Riemannian manifold. The space of harmonic p-forms, defined by*

$$\mathcal{H}^p(\mathcal{M}^n) \triangleq \left\{ \tilde{h} \mid \tilde{h} \in \bigwedge^p, \ d\tilde{h} = d^*\tilde{h} = 0 \right\}, \tag{C.39}$$

is finite-dimensional, and Poisson's equation

$$\Delta \tilde{\omega}^p = \tilde{\eta}^p \tag{C.40}$$

has a solution if and only if $\tilde{\eta}$ is orthogonal to $\mathcal{H}^p$, i.e.,

$$\langle \tilde{\eta}^p, \tilde{h}^p \rangle = 0 \quad \text{for all } \tilde{h}^p \in \mathcal{H}^p. \tag{C.41}$$

The requirement of orthogonality between $\tilde{\eta}$ and $\tilde{h}$ can be seen as a necessity for the Poisson equation to have a solution since if $\Delta\tilde{\omega} = \tilde{\eta}$ and $\tilde{h} \in \mathcal{H}^p$,

$$\langle \tilde{\eta}, \tilde{h} \rangle = \langle \Delta\tilde{\omega}, \tilde{h} \rangle = \langle \tilde{\omega}, \Delta\tilde{h} \rangle = 0$$

as $\tilde{h} \in \text{Ker}\{\Delta\}$.

The finite dimensionality of $\mathcal{H}^p$ is a deep result that follows from the properties of elliptic operators. It is unclear why $[d(\bigwedge^{p-1}) \oplus (d^* \bigwedge^{p+1})]^\perp = \mathcal{H}^p$ given that $\bigwedge^p$ is infinite-dimensional, but the finite dimensionality of the space of harmonic forms can help to prove that the de Rham cohomology groups defined on $\mathcal{M}^n$ are finite-dimensional despite the fact that the closed and exact forms also form infinite-dimensional spaces.

Consider the Hodge decomposition $\tilde{\beta} = d\tilde{\alpha} + d^*\tilde{\gamma} + \tilde{h}$ as presented above. If $\tilde{\beta}^p$ is closed, $d\tilde{\beta}^p = 0$, thus $dd^*\tilde{\gamma}^{p+1} = 0$, and $\tilde{\beta}$ is expressible in terms of the remaining components as

$$\tilde{\beta}^p = d\tilde{\alpha}^{p-1} + \tilde{h}^p \tag{C.42}$$

where $\tilde{h}^p$ is harmonic and is thus closed. Therefore, $\tilde{\beta}$ and $\tilde{\beta} - d\tilde{\alpha} = \tilde{h}$ are in the same de Rham class, i.e., $\tilde{\beta} \sim \tilde{h}$. From this observation, it is apparent that in each de Rham class $[\tilde{\beta}]$ there is a unique harmonic representative $\text{harm}(\tilde{\beta})$. Further, since

$$\|\tilde{\beta}^p\|^2 = \|d\tilde{\alpha}^{p-1}\|^2 + \|\tilde{h}^p\|^2 \tag{C.43}$$

the harmonic representative $\tilde{h}$ has the smallest norm within any de Rham class $[\tilde{\beta}]$.

The pth Betti number $b_p = \dim\{H^p(\mathcal{M}^n)\}$ is the dimension of the space of compatibility conditions and the space of all $\tilde{\lambda}_H$. Each vanishing period is a "compatibility condition", all of which must be satisfied to guarantee a trivial pth cohomology group. Thus there exists a unique harmonic p-form with b_p prescribed periods on a homology basis for the real p-cycles on $\mathcal{M}^n$. Thus, if the compatibility conditions are met, then for noncompact manifolds, such as $\mathbb{R}^n$, the harmonic p-forms are closed and coclosed, thus any p-form $\tilde{\omega}$ is determined from $\tilde{\eta}_1 = d\tilde{\omega}$ and $\tilde{\eta}_2 = d^*\tilde{\omega}$.

As a corollary, it is apparent that the de Rham cohomology groups for a compact, orientable, differentiable manifold are all finite-dimensional [394]. This result stands out against intuition as there is no *a priori* reason to think that the quotient space of two infinite-dimensional spaces should be finite. It appears that this result is equivalent to the result on the finite dimensionality of the space of harmonic p-forms at the center of Hodge theory.

Summary of Notation

We have attempted throughout this book to provide a consistent set of terminology and notation. The list below proves a summary of this notation. Please note that in a handful of circumstances we have deviated from this notation but have clearly highlighted these exceptions in the text.

$\mathbf{s}$	A vector
s_i	The ith element of the vector $\mathbf{s}$
$\mathbf{x}$	A 0-cochain (function defined on nodes)
$\mathbf{y}$	A 1-cochain (function defined on edges)
$\mathbf{z}$	A p-cochain (function defined on p-cells), for $p > 1$
z_i	The ith element of the cochain $\mathbf{z}$
$\mathbf{0}$	The vector for which each entry contains the value '0'
$\mathbf{1}$	The vector for which each entry contains the value '1'
$\|\mathbf{x}\|_q^p$	$\|\mathbf{x}\|_q^p = (\sum_i |x_i|^q)^{\frac{p}{q}}$; if p or q are missing, it is assumed that $p=1$ and $q=2$
$|\mathbf{x}|^p$	The element-wise exponentiation of a vector, i.e., $|\mathbf{x}|_i^p = |\mathbf{x}_i|^p$; if p is missing, it is assumed that $p=1$
$\text{sign}\{\mathbf{x}\}$	The sign operator applied to vector $\mathbf{x}$, that returns a vector that is the same length as $\mathbf{x}$ where each entry is either $+1$, -1, or 0 depending on whether the corresponding entry in $\mathbf{x}$ is strictly positive, strictly negative, or zero-valued
$\text{diag}(\mathbf{d})$	The diagonal matrix operator applied to vector $\mathbf{d}$ that forms a diagonal matrix with the entries of $\mathbf{d}$ along the diagonal
$\mathbf{M}$	A matrix
$M_{i,j}$	The element in the ith row and jth column of matrix $\mathbf{M}$
$\mathbf{M}(i,j)$	The element in the ith row and jth column of matrix $\mathbf{M}$
$|\mathbf{M}|$	Determinant of matrix $\mathbf{M}$
$\mathbf{M}^*$	The adjoint of $\mathbf{M}$
$\mathbf{I}$	The identity matrix
$\tilde{s}$	A data tuple
$\mathbb{R}$	Space of real numbers
$\mathbb{R}^N$	N-dimensional Euclidean space

L.J. Grady, J.R. Polimeni, *Discrete Calculus*,
DOI 10.1007/978-1-84996-290-2, © Springer-Verlag London Limited 2010

$\mathbb{Z}$	Space of integer numbers
$\mathbb{B}$	Space of binary numbers, i.e., 0 and 1
$\mathcal{C}$	A set
$\bar{\mathcal{C}}$	The complement of set $\mathcal{C}$
$|\mathcal{C}|$	Cardinality of set $\mathcal{C}$
$\mathcal{V}$	Set of vertices or nodes
$\mathcal{E}$	Set of edges
$\mathcal{G}$	A graph, consisting of a node and edge set, i.e., $\mathcal{G} = (\mathcal{V}, \mathcal{E})$
$\mathcal{F}$	Set of faces or cycles
e_{ij}	The oriented edge connecting v_i to v_j
w_{ij}	Weight of edge e_{ij}—may be a distance or affinity weight, depending on context
d_i	Degree of node v_i, $d_i = \sum_{e_{ij} \in \mathcal{E}} w_{ij}$
A	Edge–node incidence matrix
B	Face–edge incidence matrix
D	Diagonal matrix of node degree
L	Node Laplacian matrix
$\mathbf{L}^\dagger$	Pseudoinverse of the Laplacian matrix
$\mathbf{L}_p$	The Laplacian matrix for p-cells (note that the subscript is omitted for nodes or 0-cells)
$\mathbf{L}_0$	The reduced node Laplacian matrix, in which one row/column has been removed
$\mathbf{N}_p$	The incidence matrix mapping p-cells to $(p-1)$-cells
$\mathbf{x}_{\text{nbhd}(i)}$	The set of values in 0-cochain **x** that are in the neighborhood of node v_i
G	The primal metric tensor matrix, representing distance weights
$\mathbf{G}_p$	the primal metric tensor matrix, representing distance weights for the set of p-cells (when the subscript is neglected, assume that $p = 1$)
g	Vectorized distance weights, $\mathbf{g} = \mathbf{G1}$
$\mathbf{G}^{-1}$	The dual metric tensor, affinity weights
$\mathbf{G}_p^{-1}$	The dual metric tensor, affinity weights for the set of p-cells (when the subscript is neglected, assume that $p = 1$)
$\mathbf{g}^{-1}$	Vectorized affinity weights, $\mathbf{g}^{-1} = \mathbf{G}^{-1}\mathbf{1}$
$\mathcal{M}^n$	An n-dimensional manifold
$T\mathcal{M}_q^n$	The tangent space of the n-dimensional manifold $\mathcal{M}^n$ at point q
$T^*\mathcal{M}_q^n$	The cotangent space of the n-dimensional manifold $\mathcal{M}^n$ at point q
$\vec{F}$	A vector in Euclidean space
$\bar{x}$	A vector in general vector space
$\tilde{\omega}$	A form or differential form
∂	The boundary operator
d	The exterior derivative operator
d*	The codifferential operator
$\nabla \cdot, \nabla, \nabla \times$	The divergence, gradient, and curl operators in Cartesian coordinates and a Euclidean metric
∇^2	The scalar Laplacian operator

Summary of Notation

∇^2	The vector Laplacian operator
Δ	The Laplace–de Rham operator
$\star$	The Hodge star operator
$\langle \mathbf{u}, \mathbf{v} \rangle$	The inner product of chains $\mathbf{u}$ and $\mathbf{v}$
$\langle\langle \mathbf{u}, \mathbf{v} \rangle\rangle$	The inner product of chains on the dual complex $\mathbf{u}$ and $\mathbf{v}$
$[\![\tilde{\omega}, \mathcal{R}]\!]$	The integral of form $\tilde{\omega}$ over chain $\mathcal{R}$, expressed as a bilinear pairing
$(\tilde{a}, \tilde{b})$	The scalar product of p-forms $\tilde{a}$ and $\tilde{b}$
vol^n	Canonical volume element
$\mathbf{vol}^n$	Volume cochain
$\text{volume}(\tau)$	Volume of the chain τ
$\mathcal{D}(a, b)$	The distance operator, measuring distance between points a and b
$\text{dist}_{\mathbf{G}_p}(\mathbf{a}, \mathbf{b})$	The distance operator on chains, measuring the distance between chains $\mathbf{a}$ and $\mathbf{b}$ with respect to the metrix tensor matrix $\mathbf{G}_p$
$x^{[k]}$	Variable x at iteration k
$\mathcal{Q}[f(x)]$	A functional of the function $f(x)$
$\delta(\cdot)$	The Kronecker delta function

References

1. Agarwal, S., Branson, K., Belongie, S.: Higher order learning with graphs. In: Proc. of the 23rd Int. Conf. on Mach. Learn., vol. 148, pp. 17–24 (2006)
2. Agrawal, A., Raskar, R., Chellappa, R.: What is the range of surface reconstructions from a gradient field? In: Proc. of ECCV. Lecture Notes in Computer Science, vol. 3954, pp. 578–591. Springer, Berlin (2006)
3. Albert, R., Jeong, H., Barabasi, A.: Diameter of the world wide web. Nature **401**(6749), 130–131 (1999)
4. Allène, C., Audibert, J.Y., Couprie, M., Cousty, J., Keriven, R.: Some links between min cuts, optimal spanning forests and watersheds. In: Proc. of ISMM'07, vol. 2, pp. 253–264 (2007)
5. Aloise, D., Deshpande, A., Hansen, P., Popat, P.: NP-hardness of Euclidean sum-of-squares clustering. Machine Learning **75**(2), 245–248 (2009)
6. Alon, N.: Eigenvalues and expanders. Combinatorica **6**, 83–96 (1986)
7. Alon, N., Milman, V.: λ_1, isoperimetric inequalities for graphs and superconcentrators. Journal of Combinatorial Theory. Series B **38**, 73–88 (1985)
8. Alvino, C.V., Unal, G.B., Slabaugh, G., Peny, B., Fang, T.: Efficient segmentation based on Eikonal and diffusion equations. International Journal of Computer Mathematics **84**(9), 1309–1324 (2007)
9. Anandkumar, A., Tong, L., Swami, A.: Detection of Gauss–Markov random field on nearest-neighbor graph. In: IEEE International Conference on Acoustics, Speech and Signal Processing, ICASSP 2007, vol. 3 (2007)
10. Apostol, T.M.: Calculus, vol. 1, 2nd edn. Wiley, New York (1967)
11. Appleton, B., Talbot, H.: Globally optimal surfaces by continuous maximal flows. IEEE Transactions on Pattern Analysis and Machine Intelligence **28**(1), 106–118 (2006)
12. Arenas, A., Díaz-Guilera, A., Kurths, J., Moreno, Y., Zhou, C.: Synchronization in complex networks. Physics Reports **469**(3), 93–153 (2008)
13. Aronsson, G., Crandall, M.G., Juutinen, P.: A tour of the theory of absolutely minimizing functions. Bulletin of the American Mathematical Society **41**(4), 439–505 (2004)
14. Arora, S., Rao, S., Vazirani, U.: Geometry, flows, and graph-partitioning algorithms. Communications of the ACM **51**(10), 96–105 (2008)
15. Awate, S.P., Whitaker, R.T.: Unsupervised, information-theoretic, adaptive image filtering for image restoration. IEEE Transactions on Pattern Analysis and Machine Intelligence **28**(3), 364–376 (2006)
16. Bader, D., Kintali, S., Madduri, K., Mihail, M.: Approximating betweenness centrality. In: Proc. of WAW, Lecture Notes in Computer Science, vol. 4863, pp. 124–137. Springer, Berlin (2007)

17. Bae, E., Tai, X.C.: Graph cut optimization for the piecewise constant level set method applied to multiphase image segmentation. In: Proc. of the International Conference of Scale Space and Variational Methods in Computer Vision, pp. 1–13 (2009)
18. Bai, X., Sapiro, G.: A geodesic framework for fast interactive image and video segmentation and matting. In: ICCV (2007)
19. Bai, Z., Demmel, J., Dongarra, J., Ruhe, A., van der Vorst, H.: Templates for the Solution of Algebraic Eigenvalue Problems: A Practical Guide (Software, Environments and Tools). SIAM, Philadelphia (1987). Chap. 10.1, Sparse matrix storage formats
20. Barabasi, A., Albert, R.: Emergence of scaling in random networks. Science **286**(5439), 509–512 (1999)
21. Barabasi, A.L.: Linked: How Everything Is Connected to Everything Else and What It Means for Business, Science, and Everyday Life. Plume, Cambridge (2003)
22. Barber, C.B., Dobkin, D.P., Huhdanpaa, H.: The quickhull algorithm for convex hulls. ACM Transactions on Mathematical Software **22**(4), 469–483 (1996)
23. Belkin, M., Niyogi, P.: Laplacian eigenmaps and spectral techniques for embedding and clustering. In: Advances in Neural Information Processing Systems, vol. 14, pp. 585–591. MIT Press, Cambridge (2001)
24. Bell, N., Hirani, A.N.: PyDEC: A Python library for Discrete Exterior Calculus. http://code.google.com/p/pydec/ (2008)
25. Bell, N., Hirani, A.N.: PyDEC: Algorithms and software for Discretization of Exterior Calculus (2010, in preparation)
26. Belykh, I., Hasler, M., Lauret, M., Nijmeijer, H.: Synchronization and graph topology. International Journal of Bifurcation and Chaos in Applied Sciences and Engineering **15**(11), 3423–3433 (2005)
27. Ben-Tal, A., Nemirovskiĭ, A.: Lectures on Modern Convex Optimization: Analysis, Algorithms, and Engineering Applications. SIAM, Philadelphia (2001)
28. Bengio, Y., Delalleau, O., Roux, N.L., Paiement, J.F., Vincent, P., Ouimet, M.: Learning eigenfunctions links spectral embedding and kernel PCA. Neural Computation **16**(10), 2197–2219 (2004)
29. Bengio, Y., Paiement, J., Vincent, P., Delalleau, O., Le Roux, N., Ouimet, M.: Out-of-sample extensions for LLE, Isomap, MDS, eigenmaps, and spectral clustering. In: Proc. of NIPS, pp. 177–184 (2004)
30. Berger, F., Gritzmann, P., de Vries, S.: Minimum cycle bases for network graphs. Algorithmica **40**(1), 51–62 (2004)
31. Besag, J.: Spatial interaction and the statistical analysis of lattice systems. Journal of the Royal Statistical Society. Series B **36**(2), 192–236 (1974)
32. Besag, J.: On the statistical analysis of dirty pictures. Journal of the Royal Statistical Society. Series B **48**(3), 259–302 (1986)
33. Bezem, G.J., van Leeuwen, J.: Enumeration in graphs. Technical Report RUU-CS-87-7, Rijksuniversiteit Utrecht (1987)
34. Biggs, N.: Algebraic Graph Theory. Cambridge Tracts in Mathematics, vol. 67. Cambridge University Press, Cambridge (1974)
35. Biggs, N.: Algebraic Graph Theory, 2nd edn. Cambridge University Press, Cambridge (1994)
36. Biggs, N.: Algebraic potential theory on graphs. Bulletin of the London Mathematical Society **29**, 641–682 (1997)
37. Biggs, N., Lloyd, E., Wilson, R.: Graph Theory, 1736–1936. Clarendon, Oxford (1986)
38. Black, M.J., Sapiro, G., Marimont, D.H., Heeger, D.: Robust anisotropic diffusion. IEEE Transactions on Image Processing **7**(3), 421–432 (1998)
39. Blake, A., Zisserman, A.: Visual Reconstruction. MIT Press, Cambridge (1987)
40. Bochev, P.B., Hyman, J.M.: Principles of mimetic discretizations of differential operators. In: Arnold, D.N., Bochev, P.B., Lehoucq, R.B., Nicolaides, R.A., Shashkov, M. (eds.) Compatible Spatial Discretizations. The IMA Volumes in Mathematics and Its Applications, vol. 142, pp. 89–119. Springer, New York (2006)

41. Bohland, J., Bokil, H., Pathak, S., Lee, C., Ng, L., Lau, C., Kuan, C., Hawrylycz, M., Mitra, P.: Clustering of spatial gene expression patterns in the mouse brain and comparison with classical neuroanatomy. Methods **50**(2), 105–112 (2010)
42. Bonchev, D.: Information Theoretic Indices for Characterization of Chemical Structure. Research Studies Press, Chichester (1983)
43. Bonchev, D., Balaban, A., Liu, X., Klein, D.: Molecular cyclicity and centricity of polycyclic graphs. I. Cyclicity based on resistance distances or reciprocal distances. International Journal of Quantum Chemistry **50**(1), 1–20 (1994)
44. Bonchev, D., Rouvray, D.H. (eds.): Chemical Graph Theory—Introduction and Fundamentals. Gordon & Breach, New York (1991)
45. Borgatti, S.: Identifying sets of key players in a social network. Computational and Mathematical Organization Theory **12**(1), 21–34 (2006)
46. Boros, E., Hammer, P.L., Sun, X.: Network flows and minimization of quadratic pseudo-boolean functions. Technical Report RRR 17-1991, Rutgers RUTCOR Research Report (1991)
47. Bossavit, A.: Computational Electromagnetism. Academic Press, San Diego (1998)
48. Bossavit, A.: Applied differential geometry—a compendium. http://butler.cc.tut.fi/~bossavit/BackupICM/Compendium.html (2005)
49. Bougleux, S., Elmoataz, A., Melkemi, M.: Discrete regularization on weighted graphs for image and mesh filtering. In: Proc. of SSVM. Lecture Notes in Computer Science, vol. 4485, pp. 128–139. Springer, Berlin (2007)
50. Bouman, C., Sauer, K.: A generalized Gaussian image model for edge-preserving MAP estimation. IEEE Transactions on Image Processing **2**(3), 296–310 (1993)
51. Bošnjak, N., Mihalić, Z., Trinajstić, N.: Application of topographic indices to chromatographic data: Calculation of the retention indices of alkanes. Journal of Chromatography **540**(1–2), 430–440 (1991)
52. Boyd, S., Vandenberghe, L.: Convex Optimization. Cambridge University Press, Cambridge (2004)
53. Boykov, Y., Jolly, M.P.: Interactive graph cuts for optimal boundary and region segmentation of objects in N-D images. In: Proc. of ICCV 2001, pp. 105–112 (2001)
54. Boykov, Y., Kolmogorov, V.: Computing geodesics and minimal surfaces via graph cuts. In: Proceedings of International Conference on Computer Vision, vol. 1 (2003)
55. Boykov, Y., Kolmogorov, V.: An experimental comparison of min-cut/max-flow algorithms for energy minimization in vision. IEEE Transactions on Pattern Analysis and Machine Intelligence **26**(9), 1124–1137 (2004)
56. Boykov, Y., Veksler, O., Zabih, R.: Fast approximate energy minimization via graph cuts. IEEE Transactions on Pattern Analysis and Machine Intelligence **23**(11), 1222–1239 (2001)
57. Brandes, U.: A faster algorithm for betweenness centrality. Journal of Mathematical Sociology **25**(2), 163–177 (2001)
58. Branin, F.H. Jr.: The inverse of the incidence matrix of a tree and the formulation of the algebraic-first-order differential equations of an RLC network. IEEE Transactions on Circuit Theory **10**(4), 543–544 (1963)
59. Branin, F.H. Jr.: The algebraic-topological basis for network analogies and the vector calculus. In: Proc. of Conf. on Generalized Networks, pp. 453–491, Brooklyn, NY (1966)
60. Brin, S., Page, L.: The anatomy of a large-scale hypertextual web search engine. Computer Networks and ISDN Systems **30**(1–7), 107–117 (1998)
61. Bruckstein, A.M., Netravali, A.N., Richardson, T.J.: Epi-convergence of discrete elastica. Applicable Analysis **79**(1–2), 137–171 (2001)
62. Brusco, M., Stahl, S.: Branch-and-Bound Applications in Combinatorial Data Analysis. Springer, Berlin (2005)
63. Buades, A., Coll, B., Morel, J.M.: A review of image denoising algorithms, with a new one. Multiscale Modeling and Simulation **4**(2), 490–530 (2005)
64. Buchanan, M.: Nexus: Small Worlds and the Groundbreaking Theory of Networks. Norton, New York (2003)

65. Burges, J.C.: Geometric methods for feature extraction and dimensional reduction. In: Data Mining and Knowledge Discovery Handbook, pp. 59–91. Springer, Berlin (2005). Chap. 1
66. Cassell, A.C., Henderson, J.C., Ramachandran, K.: Cycle bases of minimal measure for the structural analysis of skeletal structures by the flexibility method. Proceedings of the Royal Society of London. Series A, Mathematical and Physical Sciences **350**, 61–70 (1976)
67. Černý, V.: A thermodynamical approach to the travelling salesman problem: An efficient simulation algorithm. Journal of Optimization Theory and Applications **45**, 41–51 (1985)
68. Chambolle, A.: An algorithm for total variation minimization and applications. Journal of Mathematical Imaging and Vision **20**(1–2), 89–97 (2004)
69. Chambolle, A., Lions, P.L.: Image recovery via total variation minimization and related problems. Numerische Mathematik **76**(2), 167–188 (1997)
70. Chan, T., Vese, L.: Active contours without edges. IEEE Transactions on Image Processing **10**(2), 266–277 (2001)
71. Chan, T.F., Gilbert, J.R., Teng, S.H.: Geometric spectral partitioning. PARC Technical Report CSL-94-15, Xerox (1995)
72. Chandra, A., Raghavan, P., Ruzzo, W., Smolensky, R., Tiwari, P.: The electrical resistance of a graph captures its commute and cover times. Computational Complexity **6**(4), 312–340 (1996)
73. Chapelle, O., Schölkopf, B., Zien, A. (eds.): Semi-Supervised Learning. MIT Press, Cambridge (2006)
74. Cheeger, J.: A lower bound for the smallest eigenvalue of the Laplacian. In: Gunning, R. (ed.) Problems in Analysis, pp. 195–199. Princeton University Press, Princeton (1970)
75. Chen, W.: Applied Graph Theory. North-Holland, Amsterdam (1971)
76. Chen, Y., Dong, M., Rege, M.: Gene expression clustering: A novel graph partitioning approach. In: Proceedings of International Joint Conference on Neural Networks (2007)
77. Christakis, N.A., Fowler, J.H.: Connected: The Surprising Power of Our Social Networks and How They Shape Our Lives. Little, Brown and Company, London (2009)
78. Chua, L.O., Chen, L.: On optimally sparse cycle and coboundary basis for a linear graph. IEEE Transactions on Circuit Theory **CT-20**, 495–503 (1973)
79. Chung, F.: Laplacians and the Cheeger inequality for directed graphs. Annals of Combinatorics **9**(1), 1–19 (2005)
80. Chung, F.R.K.: The Laplacian of a hypergraph. In: Proc. of a DIMACS Workshop, Discrete Math. Theoret. Comput. Sci., vol. 10, pp. 21–36. Am. Math. Soc., Providence (1993)
81. Chung, F.R.K.: Spectral Graph Theory. Regional Conference Series in Mathematics, vol. 92. Am. Math. Soc., Providence (1997)
82. Cliff, A.D., Ord, J.K.: Spatial Processes: Models and Applications. Pion, London (1981)
83. Cohen, L.D.: On active contour models and balloons. CVGIP: Image Understanding **53**(2), 211–218 (1991)
84. Cohen, L., Cohen, I.: Finite-element methods for active contour models and balloons for 2-D and 3-D images. IEEE Transactions on Pattern Analysis and Machine Intelligence **15**(11), 1131–1147 (1993)
85. Coifman, R., Lafon, S., Lee, A., Maggioni, M., Nadler, B., Warner, F., Zucker, S.: Geometric diffusions as a tool for harmonic analysis and structure definition of data: Diffusion maps. Proceedings of the National Academy of Sciences of the United States of America **102**(21), 7426–7431 (2005)
86. Constantine, P.G., Gleich, D.F.: Using polynomial chaos to compute the influence of multiple random surfers in the PageRank model. In: Bonato, A., Chung, F.R.K. (eds.) Proc. of WAW. Lecture Notes in Computer Science, vol. 4863, pp. 82–95. Springer, Berlin (2007)
87. Cormen, T., Leiserson, C., Rivest, R., Stein, C.: Introduction to Algorithms. MIT Press, Cambridge (2001)
88. Cosgriff, R.L.: Identification of shape. Technical Report 820-11 ASTIA AD 254 792, Ohio State Univ. Res. Foundation (1960)
89. Costa, L., Rodrigues, F., Travieso, G., Boas, P.: Characterization of complex networks: A survey of measurements. Advances in Physics **56**(1), 167–242 (2007)

90. Couprie, C., Grady, L., Najman, L., Talbot, H.: Power watersheds: A new image segmentation framework extending graph cuts, random walker and optimal spanning forest. In: Proc. of ICCV, pp. 731–738 (2009)
91. Courant, R.L.: Variational methods for the solution of problems of equilibrium and vibration. Bulletin of the American Mathematical Society **43**, 1–23 (1943)
92. Courant, R.L., Friedrichs, K., Lewy, H.: Über die partiellen Differenzengleichungen der mathematischen Physik. Mathematische Annalen **100**(1), 32–74 (1928)
93. Cremers, D., Grady, L.: Statistical priors for efficient combinatorial optimization via graph cuts. In: Leonardis, A., Bischof, H., Pinz, A. (eds.) Computer Vision—ECCV 2006. Lecture Notes in Computer Science, vol. 3, pp. 263–274. Springer, Berlin (2006)
94. Criminisi, A., Sharp, T., Blake, A.: GeoS: Geodesic image segmentation. In: Proc. of ECCV, pp. 99–112 (2008)
95. Crowe, M.J.: A History of Vector Analysis: The Evolution of the Idea of a Vectorial System. Dover, New York (1994)
96. Dale, A.M., Fischl, B., Sereno, M.I.: Cortical surface-based analysis. I. Segmentation and surface reconstruction. NeuroImage **9**(2), 179–194 (1999)
97. Darbon, J.: A note on the discrete binary Mumford–Shah model. In: Proc. of the 3rd Int. Conf. on Computer Vision/Computer Graphics Collaboration Techniques. Lecture Notes in Computer Science, pp. 283–294. Springer, Berlin (2007)
98. Darbon, J., Sigelle, M.: Image restoration with discrete constrained total variation part I: Fast and exact optimization. Journal of Mathematical Imaging and Vision **26**(3), 261–276 (2006)
99. Darling, R.: Differential Forms and Connections. Cambridge University Press, Cambridge (1994)
100. de Pina, J.: Applications of shortest path methods. Ph.D. thesis, University of Amsterdam, Netherlands (1995)
101. Demir, C., Gultekin, S., Yener, B.: Learning the topological properties of brain tumors. In: Proc. of IEEE/ACM Transactions on Computational Biology and Bioinformatics (TCBB), vol. 2, pp. 262–270 (2005)
102. Desbrun, M., Hirani, A.N., Leok, M., Marsden, J.E.: Discrete exterior calculus. arXiv:math.DG/0508341 (2005)
103. Desbrun, M., Kanso, E., Tong, Y.: Discrete differential forms for computational modeling. In: Proc. of SIGGRAPH (2005)
104. Desbrun, M., Meyer, M., Schröder, P., Barr, A.H.: Implicit fairing of irregular meshes using diffusion and curvature flow. In: Proceedings of the 26th Annual Conference on Computer Graphics and Interactive Techniques, pp. 317–324. ACM Press/Addison-Wesley, New York (1999)
105. Desbrun, M., Polthier, K.: Discrete differential geometry. Computer Aided Geometric Design **24**(8–9), 427 (2007)
106. Dheeraj Singaraju, L.G., Vidal, R.: P-brush: Continuous valued MRFs with normed pairwise distributions for image segmentation. In: Proc. of CVPR 2009. IEEE Comput. Soc., Los Alamitos (2009)
107. Dieudonné, J.: A History of Algebraic and Differential Topology, 1900–1960. Springer, Berlin (1989)
108. Dijkstra, E.W.: A note on two problems in connexion with graphs. Numerische Mathematik **1**(1), 269–271 (1959)
109. Diudea, M.V., Gutman, I.: Wiener-type topological indices. Croatica Chemica Acta **71**(1), 21–51 (1998)
110. do Carmo, M.P.: Differential Geometry of Curves and Surfaces. Prentice-Hall, Englewood Cliffs (1976)
111. Dodziuk, J.: Difference equations, isoperimetric inequality and the transience of certain random walks. Transactions of the American Mathematical Society **284**, 787–794 (1984)
112. Dodziuk, J., Kendall, W.S.: Combinatorial Laplacians and isoperimetric inequality. In: Ellworthy, K.D. (ed.) From Local Times to Global Geometry, Control and Physics. Pitman Research Notes in Mathematics Series, vol. 150, pp. 68–74. Longman, Harlow (1986)

113. Doi, E., Lewicki, M.S.: Relations between the statistical regularities of natural images and the response properties of the early visual system. In: Japanese Cognitive Science Society, SIG P&P, pp. 1–8 (2005)
114. Donath, W., Hoffman, A.: Algorithms for partitioning of graphs and computer logic based on eigenvectors of connection matrices. IBM Technical Disclosure Bulletin **15**, 938–944 (1972)
115. Doyle, P., Snell, L.: Random Walks and Electric Networks. Carus Mathematical Monographs, vol. 22. Math. Assoc. Am., Washington (1984)
116. DuBois, E.: The sampling and reconstruction of time varying imagery with application in video systems. Proceedings of the IEEE **73**(4), 502–522 (1985)
117. Duda, R.O., Hart, P.E., Stork, D.G.: Pattern Classification, 2nd edn. Wiley-Interscience, New York (2000)
118. Eckmann, B.: Harmonische Funktionen und Randwertaufgaben in einem Komplex. Commentarii Mathematici Helvetici **17**, 240–245 (1945)
119. Edelsbrunner, H.: Geometry and Topology of Mesh Generation. Cambridge University Press, Cambridge (2001)
120. Edmonds, J., Johnson, E., Lockhart, S.: Blossom I: A computer code for the matching problem. Unpublished report, IBM TJ Watson Research Center, Yorktown Heights, New York (1969)
121. El-Zehiry, N., Xu, S., Sahoo, P., Elmaghraby, A.: Graph cut optimization for the Mumford–Shah model. In: Proc. of VIIP (2007)
122. El-Zehiry, N.Y., Elmaghraby, A.: Brain MRI tissue classification using graph cut optimization of the Mumford–Shah functional. In: Proceedings of Image and Vision Computing, New Zealand, pp. 321–326 (2007)
123. Elmoataz, A., Lézoray, O., Bougleux, S.: Nonlocal discrete regularization on weighted graphs: A framework for image and manifold processing. IEEE Transactions on Image Processing **17**(7), 1047–1060 (2008)
124. Erdős, P., Rényi, A.: On random graphs. I. Publicationes Mathematicae Debrecen **6**, 290–297 (1959)
125. Erdős, P., Rényi, A.: On the evolution of random graphs. Közlemények Publications **5**, 17–61 (1960)
126. Euler, L.: Solutio problematis ad geometriam situs pertinentis. Commentarii Academiae Scientiarum Petropolitanae **8**, 128–140 (1736)
127. Evans, L.C.: Partial Differential Equations. Graduate Studies in Mathematics, vol. 19. Am. Math. Soc., Providence (2000)
128. Falcão, A.X., Lotufo, R.A., Araujo, G.: The image foresting transformation. IEEE Transactions on Pattern Analysis and Machine Intelligence **26**(1), 19–29 (2004)
129. Falcão, A.X., Udupa, J.K., Samarasekera, S., Sharma, S., Elliot, B.H., de Lotufo, A.R.: User-steered image segmentation paradigms: Live wire and live lane. Graphical Models and Image Processing **60**(4), 233–260 (1998)
130. Farahat, A., LoFaro, T., Miller, J., Rae, G., Ward, L.: Authority rankings from HITS, Pagerank, and SALSA: Existence, uniqueness, and effect of initialization. SIAM Journal on Scientific Computing **27**(4), 1181–1201 (2006)
131. Fattal, R., Lischinski, D., Werman, M.: Gradient domain high dynamic range compression. In: Proc. of SIGGRAPH (2002)
132. Feynman, R.P., Leighton, R., Sands, M.: Lectures on Physics. Addison-Wesley, Reading (1964)
133. Fiedler, M.: Algebraic connectivity of graphs. Czechoslovak Mathematical Journal **23**(98), 298–305 (1973)
134. Fiedler, M.: Eigenvectors of acyclic matrices. Czechoslovak Mathematical Journal **25**(100), 607–618 (1975)
135. Fiedler, M.: A property of eigenvectors of nonnegative symmetric matrices and its applications to graph theory. Czechoslovak Mathematical Journal **25**(100), 619–633 (1975)
136. Fischl, B., Sereno, M.I., Dale, A.M.: Cortical surface-based analysis. II: Inflation, flattening, and a surface-based coordinate system. NeuroImage **9**(2), 195–207 (1999)

137. Fischl, B., Sereno, M.I., Tootell, R.B., Dale, A.M.: High-resolution intersubject averaging and a coordinate system for the cortical surface. Human Brain Mapping **8**(4), 272–284 (1999)
138. Fisher, R.A.: The use of multiple measurements in taxonomic problems. Annals of Eugenics **7**, 179–188 (1936)
139. Flanders, H.: Differential Forms. Academic Press, New York (1963)
140. Fowler, J., Jeon, S.: The authority of Supreme Court precedent. Social Networks **30**(1), 16–30 (2008)
141. Fowler, J., Johnson, T., Spriggs, J., Jeon, S., Wahlbeck, P.: Network analysis and the law: Measuring the legal importance of precedents at the US Supreme Court. Political Analysis **15**(3), 324–346 (2007)
142. Frankel, T.: The Geometry of Physics: An Introduction, 2nd edn. Cambridge University Press, Cambridge (2004)
143. Friedman, J.: Computing Betti numbers via combinatorial Laplacians. Algorithmica **21**(4), 331–346 (1998)
144. Geman, D., Reynolds, G.: Constrained restoration and the discovery of discontinuities. IEEE Transactions on Pattern Analysis and Machine Intelligence **14**(3), 367–383 (1992)
145. Geman, S., Geman, D.: Stochastic relaxation, Gibbs distributions and the Bayesian restoration of images. IEEE Transactions on Pattern Analysis and Machine Intelligence **6**(6), 721–741 (1984)
146. Geman, S., McClure, D.: Statistical methods for tomographic image reconstruction. In: Proc. 46th Sess. Int. Stat. Inst. Bulletin ISI, vol. 52, pp. 4–21 (1987)
147. Gibbons, A.: Algorithmic Graph Theory. Cambridge University Press, Cambridge (1989)
148. Girvan, M., Newman, M.: Community structure in social and biological networks. Proceedings of the National Academy of Sciences of the United States of America **99**(12), 7821–7826 (2002)
149. Gleich, D.F.: Models and algorithms for pagerank sensitivity. Ph.D. thesis, Stanford University (2009)
150. Glover, F., Laguna, M.: Tabu Search. Springer, Berlin (1997)
151. Godsil, C., McKay, B.: Constructing cospectral graphs. Aequationes Mathematicae **25**(1), 257–268 (1982)
152. Goemans, M., Williamson, D.: Improved approximation algorithms for maximum cut and satisfiability problems using semidefinite programming. Journal of the ACM **42**(6), 1115–1145 (1995)
153. Goldberg, D.: Genetic Algorithms in Search, Optimization and Machine Learning. Addison-Wesley, Boston (1989)
154. Golub, G., Van Loan, C.: Matrix Computations, 3rd edn. Johns Hopkins University Press, Baltimore (1996)
155. Golynski, A., Horton, J.: A polynomial time algorithm to find the minimum cycle basis of a regular matroid. In: Lecture Notes in Computer Science, pp. 200–209. Springer, Berlin (2002)
156. Gordon, C., Webb, D., Wolpert, S.: You cannot hear the shape of a drum. Bulletin of the American Mathematical Society **27**, 134–138 (1992)
157. Gotsman, C., Kaligosi, K., Mehlhorn, K., Michail, D., Pyrga, E.: Cycle bases of graphs and sampled manifolds. Computer Aided Geometric Design **24**(8–9), 464–480 (2007)
158. Grady, L.: Space-variant computer vision: A graph-theoretic approach. Ph.D. thesis, Boston University, Boston, MA (2004)
159. Grady, L.: Multilabel random walker image segmentation using prior models. In: Proc. of 2005 IEEE Computer Society Conference on Computer Vision and Pattern Recognition. CVPR, vol. 1, pp. 763–770. IEEE Press, San Diego (2005)
160. Grady, L.: Fast, quality, segmentation of large volumes — Isoperimetric distance trees. In: Leonardis, A., Bischof, H., Pinz, A. (eds.) Computer Vision—ECCV 2006. Lecture Notes in Computer Science, vol. 3, pp. 449–462. Springer, Berlin (2006)
161. Grady, L.: Random walks for image segmentation. IEEE Transactions on Pattern Analysis and Machine Intelligence **28**(11), 1768–1783 (2006)

162. Grady, L.: Minimal surfaces extend shortest path segmentation methods to 3D. IEEE Transactions on Pattern Analysis and Machine Intelligence **32**(2), 321–334 (2010)
163. Grady, L., Alvino, C.: The piecewise smooth Mumford-Shah functional on an arbitrary graph. IEEE Transactions on Image Processing **18**(11), 2547–2561 (2009)
164. Grady, L., Jolly, M.P.: Weights and topology: A study of the effects of graph construction on 3D image segmentation. In: Metaxas, D.N. et al. (eds.) Proc. of MICCAI 2008. Lecture Notes in Computer Science, vol. 5241, pp. 153–161. Springer, Berlin (2008)
165. Grady, L., Schiwietz, T., Aharon, S., Westermann, R.: Random walks for interactive alpha-matting. In: Villanueva, J.J. (ed.) Proc. of Fifth IASTED International Conference on Visualization, Imaging and Image Processing, pp. 423–429. Acta Press, Benidorm (2005)
166. Grady, L., Schwartz, E.L.: The Graph Analysis Toolbox: Image processing on arbitrary graphs. Technical Report TR-03-021, Boston University, Boston, MA (2003)
167. Grady, L., Schwartz, E.L.: Faster graph-theoretic image processing via small-world and quadtree topologies. In: Proc. of 2004 IEEE Computer Society Conference on Computer Vision and Pattern Recognition, vol. 2, pp. 360–365. IEEE Comput. Soc., Washington (2004)
168. Grady, L., Schwartz, E.L.: Isoperimetric graph partitioning for image segmentation. IEEE Transactions on Pattern Analysis and Machine Intelligence **28**(3), 469–475 (2006)
169. Grady, L., Schwartz, E.L.: Isoperimetric partitioning: A new algorithm for graph partitioning. SIAM Journal on Scientific Computing **27**(6), 1844–1866 (2006)
170. Graovac, A., Gutman, I., Trinajstić, N.: Topological Approach to the Chemistry of Conjugated Molecules. Springer, Berlin (1977)
171. Grassy, G., Calas, B., Yasri, A., Lahana, R., Woo, J., Iyer, S., Kaczorek, M., Floch, R., Buelow, R.: Computer-assisted rational design of immunosuppressive compounds. Nature Biotechnology **16**, 748–752 (1998)
172. Gray, R.M.: Toeplitz and Circulant Matrices: A Review. Now Publishers, Hanover (2006)
173. Greig, D., Porteous, B., Seheult, A.: Exact maximum *a posteriori* estimation for binary images. Journal of the Royal Statistical Society. Series B **51**(2), 271–279 (1989)
174. Gremban, K.: Combinatorial preconditioners for sparse, symmetric diagonally dominant linear systems. Ph.D. thesis, Carnegie Mellon University, Pittsburgh, PA (1996)
175. Gremban, K., Miller, G., Zagha, M.: Performance evaluation of a new parallel preconditioner. In: Proceedings of 9th International Parallel Processing Symposium, pp. 65–69. IEEE Comput. Soc., Santa Barbara (1995)
176. Gross, J., Tucker, T.: Topological Graph Theory. Dover, New York (2001)
177. Gross, J.L., Yellen, J.: Graph Theory and Its Applications. CRC Press, Boca Raton (1998)
178. Gross, P.W., Kotiuga, P.R.: Electromagnetic Theory and Computation: A Topological Approach. Cambridge University Press, Cambridge (2004)
179. Guattery, S., Miller, G.: On the quality of spectral separators. SIAM Journal on Matrix Analysis and Applications **19**(3), 701–719 (1998)
180. Gunduz, C., Yener, B., Gultekin, S.: The cell graphs of cancer. In: Proc. of International Symposium on Molecular Biology ISMB/ECCB, vol. 20, pp. 145–151 (2004)
181. Gutman, I., Kortvelyesi, T.: Wiener indices and molecular surfaces. Zeitschrift für Naturforschung A **50**(7), 669–671 (1995)
182. Gutman, I., Mohar, B.: The quasi-Wiener and the Kirchhoff indices coincide. Journal of Chemical Information and Computer Sciences **36**(5), 982–985 (1996)
183. Hadlock, F.: Finding a maximum cut of a planar graph in polynomial time. SIAM Journal on Computing **4**(3), 221–225 (1975)
184. Hall, K.M.: An r-dimensional quadratic placement algorithm. Management Science **17**(3), 219–229 (1970)
185. Hamilton, W.R.: On quaternions, or on a new system of imaginaries in algebra. Philosophical Magazine **25**(3), 489–495 (1844)
186. Hammer, P., Hansen, P., Simeone, B.: Roof duality, complementation and persistency in quadratic 0–1 optimization. Mathematical Programming **28**(2), 121–155 (1984)
187. Hansen, P., Jurs, P.: Chemical applications of graph theory: Part I. Fundamentals and topological indices. Journal of Chemical Education **65**(7), 574–580 (1988)

188. Harary, F.: Graph Theory. Addison-Wesley, Reading (1994)
189. Harrison, J.: Geometric Hodge star operator with applications to the theorems of Gauss and Green. Mathematical Proceedings of the Cambridge Philosophical Society **140**(1), 135–155 (2006). doi:10.1017/S0305004105008716
190. Harrison, L.M., Penny, W., Flandin, G., Ruff, C.C., Weiskopf, N., Friston, K.J.: Graph-partitioned spatial priors for functional magnetic resonance images. NeuroImage **43**(4), 694–707 (2008)
191. Hassin, R.: Maximum flow in (s, t) planar networks. Information Processing Letters **13**(3), 107 (1981)
192. Haupt, S.: Practical genetic algorithms. Wiley-Interscience, New York (2004)
193. He, X., Niyogi, P.: Locality preserving projections. In: Advances in Neural Information Processing Systems, vol. 16, pp. 153–160. MIT Press, Cambridge (2003)
194. Heiler, M., Keuchel, J., Schnorr, C.: Semidefinite clustering for image segmentation with a-priori knowledge. In: Proc. of the 27th DAGM Symposium, pp. 309–317. Springer, Berlin (2005)
195. von Helmholtz, H.: Über integral der hydronamischen gleichungen, welche den wirbelbewegungen entsprechen. Crelle's Journal für Mathematik **55**, 25–55 (1858). English translation by P. G. Tait, Philosophical Magazine and Journal of Science 4th Series, supplement to vol. 33, 1867, 485–511
196. Henle, M.: A Combinatorial Introduction to Topology. Dover, New York (1994)
197. Hestenes, D.: Space-Time Algebra. Gordon and Breach, New York (1966)
198. Higham, D., Kalna, G., Kibble, M.: Spectral clustering and its use in bioinformatics. Journal of Computational and Applied Mathematics **204**(1), 25–37 (2007)
199. Hiptmair, R.: Discrete Hodge-operators: an algebraic perspective. In: Teixeira, F.L. (ed.) Geometric Methods for Computational Electromagnetics. Progress in Electromagnetics Research, vol. 32, pp. 247–269. EMW Publishing, Cambridge (2001). Chap. 10
200. Hirani, A.N.: Discrete exterior calculus. Ph.D. thesis, California Institute of Technology (2003)
201. Hodge, W.V.D.: The Theory and Applications of Harmonic Integrals. Cambridge University Press, Cambridge (1941)
202. Hogan, N., Breedveld, P.: The physical basis of analogies in network models of physical system dynamics. In: Proc. of Conf. on Bond Graph Modeling and Simulation, vol. 31, pp. 96–104 (1999)
203. Hoppe, H., DeRose, T., Duchamp, T., McDonald, J., Stuetzle, W.: Surface reconstruction from unorganized points. In: Proc. of SIGGRAPH, vol. 26, pp. 19–26 (1992)
204. Horst, R., Hoang, T.: Global Optimization: Deterministic Approaches. Springer, Berlin (1996)
205. Horst, R., Pardalos, P., Thoai, N.: Introduction to Global Optimization, 2nd edn. Kluwer Academic, Dordrecht (2000)
206. Horton, J.: A polynomial-time algorithm to find the shortest cycle basis of a graph. SIAM Journal on Computing **16**, 358 (1987)
207. Hosoya, H.: A newly proposed quantity characterizing the topological nature of structural isomers of saturated hydrocarbons. Bulletin of the Chemical Society of Japan **44**, 2332–2339 (1971)
208. Howlett, P., Hewitt, W.: Mass-spring simulation using adaptive non-active points. In: Computer Graphics Forum, vol. 17, pp. 345–354 (1998)
209. Hughes, A.: The topography of vision in mammals of contrasting life style: Comparative optics and retinal organization. In: Crescitelli, F. (ed.) The Visual System in Vertebrates. The Handbook of Sensory Physiology, vol. 7, pp. 613–756. Springer, Berlin (1977). Chap. 11
210. Hutchinson, D., Preston, M., Hewitt, T.: Adaptive refinement for mass/spring simulations. In: 7th Eurographics Workshop on Animation and Simulation, vol. 45. Springer, Berlin (1996)
211. Ishikawa, H.: Exact optimization for Markov random fields with convex priors. IEEE Transactions on Pattern Analysis and Machine Intelligence **25**(10), 1333–1336 (2003)
212. Ishikawa, H.: Higher-order clique reduction in binary graph cut. In: Proc. of CVPR (2009)

213. Ivanciuc, O., Taraviras, S.L., Cabrol-Bass, D.: Quasi-orthogonal basis set of molecular graph descriptors as a chemical diversity measure. Journal of Chemical Information and Computer Sciences **40**, 126–134 (2000)
214. Jackson, J.D.: Classical Electrodynamics, 3rd edn. Wiley, New York (1999)
215. Jaeger, F.: A survey of the cycle double cover conjecture. In: Alspach, B.R., Godsil, C.D. (eds.) Cycles in Graphs. Annals of Discrete Mathematics, vol. 27, pp. 1–12. Elsevier, Amsterdam (1985)
216. Jain, A., Murty, M., Flynn, P.: Data clustering: A review. ACM Computing Surveys **31**(3), 264–323 (1999)
217. James, I.: History of Topology. North-Holland, Amsterdam (1999)
218. Jancewicz, B.: The extended Grassmann algebra of $\mathbb{R}^3$. In: Baylis, W.E. (ed.) Clifford (Geometric) Algebras with Applications to Physics, Mathematics, and Engineering, pp. 389–421. Birkhäuser, Boston (1996). Chap. 28.
219. Jiang, X., Lim, L.H., Yao, Y., Ye, Y.: Statistical ranking and combinatorial Hodge theory. Mathematical Programming (to appear)
220. Kac, M.: Can one hear the shape of a drum? American Mathematical Monthly **73**(4), 1–23 (1966)
221. Kakutani, S.: Markov processes and the Dirichlet problem. Proceedings of the Japan Academy **21**, 227–233 (1945)
222. Kass, M., Witkin, A., Terzopoulos, D.: Snakes: Active contour models. International Journal of Computer Vision **1**(4), 321–331 (1988)
223. Katz, L.: A new status index derived from sociometric analysis. Psychometrika **18**(1), 39–43 (1953)
224. Kaufhold, J.P.: Energy formulations of medical image segmentations. Ph.D. thesis, Boston University (2000)
225. Kavitha, S., Roomi, S., Ramaraj, N.: Lossy compression through segmentation on low depth-of-field images. Digital Signal Processing **19**(1), 59–65 (2009)
226. Kavitha, T., Mehlhorn, K., Michail, D., Paluch, K.: An algorithm for minimum cycle basis of graphs. Algorithmica **52**(3), 333–349 (2008)
227. Kelly, F.: Reversibility and Stochastic Networks. Wiley, New York (1979)
228. Kemeny, J., Snell, J., Knapp, A.: Denumerable Markov Chains. Springer, Berlin (1976)
229. Khaira, M.S., Miller, G.L., Sheffler, T.J.: Nested dissection: A survey and comparison of various nested dissection algorithms. Technical Report CMU-CS-92-106R, Computer Science Department, Carnegie Mellon University (1992)
230. Kim, H., Kim, J.: Cyclic topology in complex networks. Physical Review E **72**(3), 036109 (2005)
231. Kindermann, R., Snell, J.L.: Markov Random Fields and Their Applications. Am. Math. Soc., Providence (1980)
232. Kirchhoff, G.: Über die Auflösung der Gleichungen, auf welche man bei der Untersuchung der linearen Verteilung galvanischer Ströme geführt wird. Annalen der Physik und Chemie **72**, 497–508 (1847)
233. Kirkpatrick, S., Gelatt, C., Vecchi, M.: Optimization by simulated annealing. Science **220**(4598), 671–680 (1983)
234. Klein, D.: Resistance-distance sum rules. Croatica Chemica Acta **75**(2), 633–649 (2002)
235. Klein, D.J., Randić, M.: Resistance distance. Journal of Mathematical Chemistry **12**(1), 81–95 (1993)
236. Kleinberg, J.: Authoritative sources in a hyperlinked environment. Journal of the ACM **46**(5), 604–632 (1999)
237. Kodres, U.R.: Geometrical positioning of circuit elements in a computer. In: Proceedings of the 1959 AIEE Fall General Meeting. AIEE, New York (1959) No. CP59-1172
238. Kohlberger, T., Uzunbas, G., Alvino, C., Kadir, T., Slosman, D., Funka-Lea, G.: Organ segmentation with level sets using local shape and appearance priors. In: Yang, G.-Z. et al. (eds.) Int. Conf. on Medical Image Comp. and Comp.-Assisted Intervention (MICCAI '09). Lecture Notes in Computer Science, vol. 5762, pp. 34–42. Springer, Berlin (2009)

239. Kohli, P., Kumar, M., Torr, P.: P3 & beyond: Solving energies with higher order cliques. In: Proc. of IEEE Conference on Computer Vision and Pattern Recognition (2007)
240. Kohn, R.V., Vogelius, M.: Relaxation of a variational method for impedance computed tomography. Communications on Pure and Applied Mathematics **40**(6), 745–777 (1987)
241. Kolmogorov, V., Blossom, V.: A new implementation of a minimum cost perfect matching algorithm. Mathematical Programming Computation **1**(1), 43–67 (2009)
242. Kolmogorov, V., Boykov, Y., Rother, C.: Applications of parametric maxflow in computer vision. In: Proc. of ICCV (2007)
243. Kolmogorov, V., Rother, C.: Minimizing nonsubmodular functions with graph cuts—a review. IEEE Transactions on Pattern Analysis and Machine Intelligence **29**(7), 1274–1279 (2007)
244. Kolmogorov, V., Zabih, R.: What energy functions can be minimized via graph cuts? IEEE Transactions on Pattern Analysis and Machine Intelligence **26**(2), 147–159 (2004)
245. Komodakis, N., Paragios, N.: Higher-order clique reduction in binary graph cut. In: Proc. of CVPR (2009)
246. Konstantinos, T.: Maximum flow techniques for network clustering. Ph.D. thesis, Princeton University (2002)
247. Koschützki, D., Lehmann, K.A., Peeters, L., Richter, S., Tenfelde-Podehl, D., Zlotowski, O.: Centrality indices. In: Brandes, U., Erlebach, T. (eds.) Network Analysis. Lecture Notes in Computer Science, vol. 3418. Springer, Berlin (2005)
248. Kotiuga, P.: Theoretical limitations of discrete exterior calculus in the context of computational electromagnetics. IEEE Transactions on Magnetics **44**(6), 1162–1165 (2008). doi:10.1109/TMAG.2007.915998
249. Kotiuga, P.R.: Hodge decompositions and computational electromagnetics. Ph.D. thesis, McGill University, Montréal, Canada (1984)
250. Kron, G.: Diakoptics: The Piecewise Solution of Large Scale Systems. MacDonald, London (1963)
251. Laarhoven, P., Aarts, E.: Simulated Annealing: Theory and Applications. Springer, Berlin (1987)
252. Lan, X., Roth, S., Huttenlocher, D., Black, M.: Efficient belief propagation with learned higher-order Markov random fields. In: Proc. of ECCV (2006)
253. Lawler, E., Wood, D.: Branch-and-bound methods: A survey. Operations Research **14**(4), 699–719 (1966)
254. Lefschetz, S.: Algebraic Topology. Am. Math. Soc. Col. Pub., vol. 27. Am. Math. Soc., Providence (1942)
255. Lein, E., Hawrylycz, M., Ao, N., Ayres, M., Bensinger, A., Bernard, A., Boe, A., Boguski, M., Brockway, K., Byrnes, E., et al.: Genome-wide atlas of gene expression in the adult mouse brain. Nature **445**(7124), 168–176 (2006)
256. Lestrel, P.E. (ed.): Fourier Descriptors and Their Applications in Biology. Cambridge University Press, Cambridge (1997)
257. Levin, A., Lischinski, D., Weiss, Y.: A closed-form solution to natural image matting. IEEE Transactions on Pattern Analysis and Machine Intelligence **30**(2), 228–242 (2008)
258. Lévy, B.: Laplace-Beltrami eigenfunctions towards an algorithm that "understands" geometry. In: IEEE International Conference on Shape Modeling and Applications SMI 2006, p. 13 (2006)
259. Liang, Z., Hart, H.: Bayesian image processing of data from constrained source distributions—I. Non-valued, uncorrelated and correlated constraints. Bulletin of Mathematical Biology **49**(1), 51–74 (1987)
260. Lichnerowicz, A.: Théorie globale des connexions et des groupes d'holonomie. Edizioni Cremonese, Roma (1962)
261. Liu, H., Wang, J.: A new way to enumerate cycles in graph. In: Proc. of AICT/ICIW. IEEE (2006)
262. Liu, T.: A heuristic algorithm for finding circuits to double cover a bridgeless graph. Technical Report 57-2001, Rutgers (2001)

263. Lloyd, S.P.: Least square quantization in PCM. Technical Report, Bell Telephone Laboratories Paper (1957)
264. Lloyd, S.: Least squares quantization in PCM. IEEE Transactions on Information Theory **28**(2), 129–137 (1982)
265. Lukovits, I.: Decomposition of the Wiener topological index. Application to drug-receptor interactions. Journal of the Chemical Society. Perkin Transactions 2 **9**, 1667–1671 (1988)
266. Lukovits, I.: Correlation between components of the Wiener index and partition coefficients of hydrocarbons. International Journal of Quantum Chemistry **44**(s 19), 217–223 (1992)
267. Lyons, R., Peres, Y.: Probability on Trees and Networks (2005). http://php.indiana.edu/~rdlyons/prbtree/prbtree.html
268. MacLane, S.: A combinatorial condition for planar graphs. Fundamenta Mathematicae **28**, 22–32 (1937)
269. Mahajan, M., Nimbhorkar, P., Varadarajan, K.: The planar k-means problem is NP-hard. In: Proceedings of the 3rd International Workshop on Algorithms and Computation, pp. 274–285. Springer, Berlin (2009)
270. Marchal, M., Promayon, E., Troccaz, J.: Simulating complex organ interactions: Evaluation of a soft tissue discrete model. In: Lecture Notes in Computer Science, vol. 3804, p. 175. Springer, Berlin (2005)
271. Markowitz, H.: The elimination form of the inverse and its application to linear programming. Management Science **3**(3), 255–269 (1957)
272. Massey, W.: A Basic Course in Algebraic Topology. Springer, Berlin (1993)
273. Mateti, P., Deo, N.: On algorithms for enumerating all circuits of a graph. SIAM Journal on Computing **5**, 90 (1976)
274. Mateus, D., Horaud, R., Knossow, D., Cuzzolin, F., Boyer, E.: Articulated shape matching using Laplacian eigenfunctions and unsupervised point registration. In: IEEE Conference on Computer Vision and Pattern Recognition, 2008. CVPR 2008, pp. 1–8 (2008)
275. Mattiusi, C.: The finite volume, finite difference, and finite elements methods as numerical methods for physical field problems. Advances in Imaging and Electron Physics **113**, 1–146 (2000)
276. Matveev, S.V.: Lectures on Algebraic Topology. EMS Series of Lectures in Mathematics. European Mathematical Society, Zürich (2006)
277. Mead, C.: Analog VLSI and Neural Systems. Addison-Wesley, Reading (1989)
278. Mehlhorn, K., Michail, D.: Implementing minimum cycle basis algorithms. Journal of Experimental Algorithmics **11** (2007)
279. Merris, R.: An edge version of the matrix-tree theorem and the Wiener index. Linear and Multilinear Algebra **25**(4), 291–296 (1989)
280. Meyer, M., Desbrun, M., Schröder, P., Barr, A.H.: Discrete differential-geometry operators for triangulated 2-manifolds. In: Hege, H.C., Polthier, K. (eds.) Visualization and Mathematics III, pp. 35–57. Springer, Heidelberg (2003)
281. Michel, J., Pellegrini, F., Roman, J.: Unstructured graph partitioning for sparse linear system solving. In: Proc. of the 4th International Symposium, IRREGULAR'97, pp. 273–286 (1997)
282. Mihalić, Z., Trinajstić, N.: A graph-theoretical approach to structure: Property relationships. Journal of Chemical Education **69**(9), 701–712 (1992)
283. Milgram, S.: The small world problem. Psychology Today **2**(1), 60–67 (1967)
284. Milnor, J.: On the relationship between differentiable manifolds and combinatorial manifolds. In: Collected Papers of John Milnor, vol. III: Differential Topology, pp. 19–28. Am. Math. Soc., Providence (1956)
285. Mohar, B.: Isoperimetric inequalities, growth and the spectrum of graphs. Linear Algebra and Its Applications **103**, 119–131 (1988)
286. Mohar, B.: Isoperimetric numbers of graphs. Journal of Combinatorial Theory. Series B **47**, 274–291 (1989)
287. Mortensen, E., Barrett, W.: Interactive segmentation with intelligent scissors. Graphical Models in Image Processing **60**(5), 349–384 (1998)

288. Muhammad, A., Egerstedt, M.: Control using higher order Laplacians in network topologies. In: Proc. of the 17th Int. Symp. on Math. Theory of Networks and Systems, pp. 1024–1038 (2006)
289. Mumford, D., Shah, J.: Optimal approximations by piecewise smooth functions and associated variational problems. Communications on Pure and Applied Mathematics **42**, 577–685 (1989)
290. Munkres, J.R.: Elements of Algebraic Topology. Perseus Books, Cambridge (1986)
291. Nemhauser, G.L., Wolsey, L.A.: Integer and Combinatorial Optimization. Wiley-Interscience, New York (1999)
292. Newman, M.: Modularity and community structure in networks. Proceedings of the National Academy of Sciences of the United States of America **103**(23), 8577–8582 (2006)
293. Nicholls, F., Torr, P.H.S.: Discrete minimum ratio curves and surfaces. In: Proc. of CVPR (2010)
294. Nilsson, J.W., Riedel, S.A.: Electric Circuits, 8th edn. Pearson/Prentice-Hall, Upper Saddle River (2008)
295. Nocedal, J., Wright, S.: Numerical Optimization. Springer, Berlin (1999)
296. Okada, S.: On mesh and node determinants. Proceedings of the IRE **43**, 1527 (1955)
297. Oppenheim, A.V., Schafer, R.W.: Discrete-Time Signal Processing. Prentice-Hall, New York (1989)
298. Opsahl, T., Panzarasa, P.: Clustering in weighted networks. Social Networks **31**(2), 155–163 (2009)
299. Osborne, M.R.: Finite Algorithms in Optimization and Data Analysis. Wiley, New York (1985)
300. Osher, S., Shen, J.: Digitized PDE method for data restoration. In: Anastassiou, G.A. (ed.) Handbook of Analytic Computational Methods in Applied Mathematics, pp. 751–771. CRC Press, Boca Raton (2000). Chap. 16
301. Oster, G., Desoer, C.: Tellegen's theorem and thermodynamic inequalities. Journal of Theoretical Biology **32**, 219–241 (1971)
302. Oster, G., Perelson, A., Katchalsky, A.: Network thermodynamics. Nature **234**(5329), 393–399 (1971)
303. Pal, N., Pal, S.: A review on image segmentation techniques. Pattern Recognition **26**(9), 1277–1294 (1993)
304. Papadimitriou, C.H., Steiglitz, K.: Combinatorial Optimization. Dover, New York (1998)
305. Penfield, P., Spence, R., Duinker, S.: Tellegen's Theorem and Electrical Networks. MIT Press, Cambridge (1970)
306. Perona, P., Malik, J.: Scale-space and edge detection using anisotropic diffusion. IEEE Transactions on Pattern Analysis and Machine Intelligence **12**(7), 629–639 (1990)
307. Poincaré, H.: Analysis situs. Journal de l'École Polytechnique **2**(1), 1–123 (1895)
308. Polya, G.: Über eine Aufgabe betreffend die Irrfahrt im Strassennetz. Mathematische Annalen **84**, 149–160 (1921)
309. Potetz, B.: Efficient belief propagation for vision using linear constraint nodes. In: Proc. of CVPR (2007)
310. Pothen, A., Simon, H., Liou, K.P.: Partitioning sparse matrices with eigenvectors of graphs. SIAM Journal on Matrix Analysis and Applications **11**(3), 430–452 (1990)
311. Potts, R.B.: Some generalized order-disorder transformations. Proceedings of the Cambridge Philosophical Society **48**, 106–109 (1952)
312. Press, W.H., Teukolsky, S.A., Vetterling, W.T., Flannery, B.P.: Numerical Recipes: The Art of Scientific Computing, 3rd edn. Cambridge University Press, Cambridge (2007)
313. Qiu, H., Hancock, E.: Image segmentation using commute times. In: Proceedings of the 16th British Machine Vision Conference (BMVC 2005), pp. 929–938 (2005)
314. Qiu, H., Hancock, E.R.: Commute times, discrete Green's functions and graph matching. In: Proc. of 13th Int. Conf. on Image Analysis and Processing—ICIAP 2005, Cagliari, Italy, September 6–8, 2005, pp. 454–462. Springer, Berlin (2005)
315. Rand, W.M.: Objective criteria for the evaluation of clustering methods. Journal of the American Statistical Association **66**, 846–850 (1971)

316. Rao, V., Murti, V.: Enumeration of all circuits of a graph. Proceedings of the IEEE **57**, 700–701 (1969)
317. Ripley, B.D.: Spatial Statistics. Wiley-Interscience, New York (2004)
318. Robinson, S.: The ongoing search for efficient web search algorithms. SIAM News **37**(9) (2004)
319. Rockafellar, R.: Lagrange multipliers and optimality. SIAM Review **35**(2), 183–238 (1993)
320. Roerdink, J., Meijster, A.: The watershed transform: definitions, algorithms, and parallelization strategies. Fundamenta Informaticae **41**, 187–228 (2000)
321. Rojer, A.S., Schwartz, E.L.: Design considerations for a space-variant visual sensor with complex-logarithmic geometry. In: Proc. ICPR, vol. 2, pp. 278–285. IEEE Comput. Soc., Los Alamitos (1990)
322. Rose, D.: A graph-theoretic study of the numerical solution of sparse positive definite systems of linear equations. In: Graph Theory and Computing, pp. 183–217. Academic Press, New York (1973)
323. Roth, J.P.: An application of algebraic topology to numerical analysis: On the existence of a solution to the network problem. Proceedings of the National Academy of Sciences of the United States of America **41**(7), 518–521 (1955)
324. Rother, C., Kohli, P., Feng, W., Jia, J.: Minimizing sparse higher order energy functions of discrete variables. In: Proc. of CVPR (2009)
325. Rouvray, D.H.: The prediction of biological activity using molecular connectivity indices. Acta Pharmaceutica Jugoslavica **36**, 239–251 (1986)
326. Rudin, L., Osher, S., Fatemi, E.: Nonlinear total variation based noise removal algorithms. Physica D **60**(1–4), 259–268 (1992)
327. Saerens, M., Fouss, F., Yen, L., Dupont, P.: The principal components analysis of a graph, and its relationships to spectral clustering. In: Proceedings of the 15th European Conference on Machine Learning (ECML 2004). Lecture Notes in Artificial Intelligence, pp. 371–383. Springer, Berlin (2004)
328. Sandini, G., Questa, P., Scheffer, D., Mannucci, A.: A retina-like CMOS sensor and its applications. In: IEEE Sensor Array and Multichannel Signal Processing Workshop, IEEE Comput. Soc., Cambridge (2000)
329. Scannell, J.W., Burns, G.A.P.C., Hilgetag, C.C., O'Neil, M.A., Young, M.P.: The connectional organization of the cortico-thalamic system of the cat. Cerebral Cortex **9**, 277–299 (1999)
330. Schaeffer, S.: Graph clustering. Computer Science Review **1**(1), 27–64 (2007)
331. Schlesinger, D., Flach, B.: Transforming an arbitrary MinSum problem into a binary one. Technical Report TUD-FI06-01, Dresden University of Technology (2006)
332. Schmalz, M.S., Ritter, G.X.: Region segmentation techniques for object-based image compression: A review. In: Schmalz, M.S. (ed.) Mathematics of Data/Image Coding, Compression, and Encryption VII, with Applications, vol. 5561, pp. 62–75. SPIE, Bellingham (2004)
333. Schoenberg, I.: Remarks to Maurice Frechet's article "Sur la définition axiomatique d'une classe d'espace distanciés vectoriellement applicable sur l'espace de Hilbert". Annals of Mathematics **36**, 724–732 (1935)
334. Schoenemann, T., Kahl, F., Cremers, D.: Curvature regularity for region-based image segmentation and inpainting: A linear programming relaxation. In: IEEE International Conference on Computer Vision (ICCV), Kyoto, Japan (2009)
335. Schutz, B.F.: Geometrical Methods of Mathematical Physics. Cambridge University Press, Cambridge (1980)
336. Schwartz, E.L.: Computational anatomy and functional architecture of striate cortex: a spatial mapping approach to perceptual coding. Vision Research **20**(8), 645–669 (1980)
337. Serra, J.: Image Analysis and Mathematical Morphology. Academic Press, London (1982)
338. Seshu, S.: The mesh counterpart of Shekel's theorem. Proceedings of the IRE **43**, 342 (1955)
339. Sethian, J.: Level Set Methods and Fast Marching Methods. Cambridge University Press, Cambridge (1999)
340. Seymour, P.: Sums of circuits. In: Bondy, J.A., Murty, U.R.S. (eds.) Graph Theory and Related Topics, pp. 341–355. Academic Press, New York (1979)

341. Shearer, J., Murphy, A., Richardson, H.: Introduction to System Dynamics. Addison-Wesley, Reading (1967)
342. Shen, J.: The Mumford–Shah digital filter pair (MS-DFP) and applications. In: Proc. of ICIP, vol. 2, pp. 849–852 (2002)
343. Shewchuk, J.R.: Triangle: Engineering a 2D quality mesh generator and Delaunay triangulator. In: Lin, M.C., Manocha, D. (eds.) Applied Computational Geometry: Towards Geometric Engineering. Lecture Notes in Computer Science, vol. 1148, pp. 203–222. Springer, Berlin (1996)
344. Shewchuk, J.: Delaunay refinement algorithms for triangular mesh generation. Computational Geometry: Theory and Applications 22(1–3), 21–74 (2002)
345. Shi, J., Malik, J.: Normalized cuts and image segmentation. IEEE Transactions on Pattern Analysis and Machine Intelligence 22(8), 888–905 (2000)
346. Shih, W.K., Wu, S., Kuo, Y.S.: Unifying maximum cut and minimum cut of a planar graph. IEEE Transactions on Computers 39(5), 694–697 (1990)
347. Simon, H.D., Teng, S.H.: How good is recursive bisection? SIAM Journal of Scientific Computing 18(5), 1436–1445 (1997)
348. Singaraju, D., Grady, L., Sinop, A.K., Vidal, R.: P-brush: A continuous valued MRF for image segmentation. In: Blake, A., Kohli, P., Rother, C. (eds.) Advances in Markov Random Fields for Vision and Image Processing. MIT Press, Cambridge (2010)
349. Singaraju, D., Grady, L., Vidal, R.: Interactive image segmentation of quadratic energies on directed graphs. In: Proc. of CVPR 2008. IEEE Comput. Soc., Los Alamitos (2008)
350. Sinop, A.K., Grady, L.: A seeded image segmentation framework unifying graph cuts and random walker which yields a new algorithm. In: Proc. of ICCV 2007. IEEE Comput. Soc., Los Alamitos (2007)
351. Smith, B.F., Bjørstad, P.E., Gropp, W.: Domain Decomposition: Parallel Multilevel Methods for Elliptic Partial Differential Equations. Cambridge University Press, Cambridge (1996)
352. Spielman, D.A., Teng, S.H.: Spectral partitioning works: Planar graphs and finite element meshes. Technical Report UCB CSD-96-898, University of California, Berkeley (1996)
353. Spivak, M.: A Comprehensive Introduction to Differential Geometry, vol. 1, 3rd edn. Publish or Perish, Houston (2005)
354. Sporns, O., Kotter, R.: Motifs in brain networks. PLoS Biology 2, 1910–1918 (2004)
355. Stahl, S.: Generalized embedding schemes. Journal of Graph Theory 2, 41–52 (1978)
356. Stakgold, I.: Boundary Value Problem of Mathematical Physics, vol. II. Macmillan, New York (1968)
357. Stocco, L.J., Yedlin, M.J.: Modelling robot dynamics with masses and pulleys. In: Informatics in Control, Automation and Robotics. Lecture Notes in Electrical Engineering, pp. 225–238. Springer, Berlin (2008)
358. Strang, G.: Maximum flows through a domain. Mathematical Programming 26, 123–143 (1983)
359. Strang, G.: Introduction to Applied Mathematics. Wellesley-Cambridge Press, Wellesley (1986)
360. Strang, G.: Computational Science and Engineering. Wellesley-Cambridge Press, Wellesley (2007)
361. Strang, G., Fix, G.J.: An Analysis of the Finite Element Method. Prentice-Hall, Englewood Cliffs (1973)
362. Strogatz, S.: Exploring complex networks. Nature 410(6825), 268–276 (2001)
363. Stuwe, M.: Plateau's Problem and the Calculus of Variations. Princeton University Press, Princeton (1989)
364. Sullivan, J.M.: A crystalline approximation theorem for hypersurfaces. Ph.D. thesis, Princeton University, Princeton, NJ (1990)
365. Sumner, R., Popovic, J.: Deformation transfer for triangle meshes. Proceedings of SIGGRAPH 23(3), 399–405 (2004)
366. Surazhsky, T., Magid, E., Soldea, O., Elber, G., Rivlin, E.: A comparison of Gaussian and mean curvatures estimation methods on triangular meshes. In: Proc. of IEEE International Conference on Robotics and Automation ICRA'03, vol. 1 (2003)

367. Sylvester, J.J.: Chemistry and algebra. Nature **17**(432), 284 (1878)
368. Szallasi, Z., Somogyi, R.: Genetic network analysis—The millennium opening version. In: Proc. Pacific Symposium of Biocomputing Tutorial (2001)
369. Szeliski, R., Zabih, R., Scharstein, D., Veksler, O., Kolmogorov, V., Agarwala, A., Tappen, M., Rother, C.: A comparative study of energy minimization methods for Markov random fields with smoothness-based priors. IEEE Transactions on Pattern Analysis and Machine Intelligence **30**(6), 1068–1080 (2008)
370. Taraviras, S.L., Ivanciuc, O., Cabrol-Bass, D.: Identification of groupings of graph theoretical molecular descriptors using a hybrid cluster analysis approach. Journal of Chemical Information and Computer Sciences **40**, 1128–1146 (2000)
371. Taubin, G.: A signal processing approach to fair surface design. In: Cook, R. (ed.) Computer Graphics Proceedings. Special Interest Group in Computer Graphics (SIGGRAPH) 95, pp. 351–358. ACM, Los Angeles (1995)
372. Taubin, G., Zhang, T., Golub, G.: Optimal surface smoothing as filter design. In: Proc. of ECCV 1996, pp. 283–292 (1996)
373. Tenenbaum, J., de Silva, V., Langford, J.: A global geometric framework for nonlinear dimensionality reduction. Science **290**(5500), 2319–2323 (2000)
374. Tewari, G., Gotsman, C., Gortler, S.: Meshing genus-1 point clouds using discrete one-forms. Computers & Graphics **30**(6), 917–926 (2006)
375. Tiernan, J.: An efficient search algorithm to find the elementary circuits of a graph. Comm. of the ACM **13**(12), 726 (1970)
376. Tinney, W., Walker, J.: Direct solutions of sparse network equations by optimally ordered triangular factorization. Proceedings of the IEEE **55**(11), 1801–1809 (1967)
377. Tobler, W.R.: A computer movie simulating urban growth in the Detroit region. Economic Geography **46**, 234–240 (1970)
378. Tonti, E.: The reason for analogies between physical theories. Applied Mathematical Modelling **I**, 37–50 (1976)
379. Tonti, E.: On the geometrical structure of the electromagnetism. In: Gravitation, Electromagnetism and Geometrical Structures, for the 80th Birthday of A. Lichnerowicz, pp. 281–308. Pitagora Editrice, Bologna (1995)
380. Tonti, E.: A direct discrete formulation of field laws: The cell method. Computer Modeling in Engineering and Sciences **2**(2), 237–258 (2001)
381. Toselli, A., Widlund, O.: Domain Decomposition Methods—Algorithms and Theory. Springer Series in Computational Mathematics, vol. 34. Springer, Berlin (2004)
382. Trichili, H., Bouhlel, M.S., Kammoun, F.: Review and evaluation of medical image segmentation using methods of optimal filtering. Journal of Testing and Evaluation **31**(5), 398–404 (2003)
383. Truemper, K.: Algebraic characterizations of unimodular matrices. SIAM Journal on Applied Mathematics **35**(2), 328–332 (1978)
384. Tsai, A., Yezzi, A., Willsky, A.: Curve evolution implementation of the Mumford–Shah functional for image segmentation, denoising, interpolation, and magnification. IEEE Transactions on Image Processing **10**(8), 1169–1186 (2001)
385. Unger, M., Pock, T., Bischof, H.: Interactive globally optimal image segmentation. Technical Report 08/02, Inst. for Computer Graphics and Vision, Graz University of Technology (2008)
386. Unger, M., Pock, T., Trobin, W., Cremers, D., Bischof, H.: TVSeg—Interactive total variation based image segmentation. In: Proc. of British Machine Vision Conference (2008)
387. Van Dam, E., Haemers, W.: Spectral characterizations of some distance-regular graphs. Journal of Algebraic Combinatorics **15**(2), 189–202 (2002)
388. Veblen, O., Whitehead, J.: The Foundations of Differential Geometry. Cambridge University Press, Cambridge (1932)
389. Vese, L., Chan, T.: A multiphase level set framework for image segmentation using the Mumford and Shah model. International Journal of Computer Vision **50**(3), 271–293 (2002)

390. Wainwright, M., Jaakkola, T., Willsky, A.: MAP estimation via agreement on trees: Message-passing and linear programming. IEEE Transactions on Information Theory **51**(11), 3697–3717 (2005)
391. Wallace, R., Ong, P.W., Bederson, B., Schwartz, E.: Space variant image processing. International Journal of Computer Vision **13**(1), 71–90 (1994)
392. Walshaw, C., Cross, M., Everett, M.: Mesh partitioning and load-balancing for distributed memory parallel systems. In: Topping, B. (ed.) Proc. Parallel & Distributed Computing for Computational Mechanics (1997)
393. Wang, H., Chen, Y., Fang, T., Tyan, J., Ahuja, N.: Gradient adaptive image restoration and enhancement. In: Proc. of Int. Conf. on Image Procession, pp. 2893–2896. IEEE Press, New York (2006)
394. Warner, F.W.: Foundations of Differentiable Manifolds and Lie Groups, Reprint edn. Springer, New York (1983)
395. Warnick, K.F., Selfridge, R.H., Arnold, D.V.: Teaching electromagnetic field theory using differential forms. IEEE Transactions on Education **40**(1), 53–68 (1997)
396. Watts, D., Strogatz, S.: Collective dynamics of 'small-world' networks. Nature **393**(6684), 440–442 (1998)
397. Watts, D.J.: Small Worlds: The Dynamics of Networks Between Order and Randomness. Princeton Studies in Complexity. Princeton University Press, Princeton (1999)
398. Watts, D.J.: Six Degrees: The Science of a Connected Age. Norton, New York (2004)
399. Weinreich, G.: Geometrical Vectors. University of Chicago Press, Chicago (1998)
400. Weyl, H.: Repartición de corriente en una red conductora. Revista Matemática Hispano-Americana **5**(6), 153–164 (1923)
401. Weyl, H.: On Hodge's theory of harmonic integrals. Annals of Mathematics. Second Series **44**(1), 1–6 (1943)
402. Whitney, H.: Non-separable and planar graphs. Transactions of the American Mathematical Society **34**, 339–362 (1932)
403. Whitney, H.: Geometric Integration Theory. Princeton University Press, Princeton (1957)
404. Wiener, H.: Correlation of heats of isomerization and differences in heats of vaporization of isomer among the paraffin hydrocarbons. Journal of the American Chemical Society **69**, 2636–2638 (1947)
405. Wiener, H.: Structural determination of paraffin boiling points. Journal of the American Chemical Society **69**, 17–20 (1947)
406. Worsley, K., Friston, K.: Analysis of fMRI time-series revisited—again. NeuroImage **2**(3), 173–181 (1995)
407. Wu, Z., Leahy, R.: An optimal graph theoretic approach to data clustering: Theory and its application to image segmentation. IEEE Transactions on Pattern Analysis and Machine Intelligence **15**(11), 1101–1113 (1993)
408. Xing, E., Karp, R.: CLIFF: Clustering of high-dimensional microarray data via iterative feature filtering using normalized cuts. Bioinformatics **17**, 306–315 (2001)
409. Xu, C., Prince, J.: Snakes, shapes, and gradient vector flow. IEEE Transactions on Image Processing **7**(3), 359–369 (1998)
410. Xu, W., Zhou, K., Yu, Y., Tan, Q., Peng, Q., Guo, B.: Gradient domain editing of deforming mesh sequences. In: Proc. of SIGGRAPH, vol. 26 (2007)
411. Yedidia, J., Freeman, W., Weiss, Y.: Generalized belief propagation. In: Proc. of NIPS, pp. 689–695 (2001)
412. Yen, L., Vanvyve, D., Wouters, F., Fouss, F., Verleysen, M., Saerens, M.: Clustering using a random walk based distance measure. In: Proceedings of the 13th Symposium on Artificial Neural Networks (ESANN 2005), pp. 317–324 (2005)
413. Yu, S.X., Shi, J.: Segmentation with pairwise attraction and repulsion. In: Proc. of ICCV, vol. 1. IEEE Comput. Soc., Los Alamitos (2001)
414. Yu, S.X., Shi, J.: Understanding popout through repulsion. In: Proc. of CVPR, vol. 2. IEEE Comput. Soc., Los Alamitos (2001)
415. Zachary, W.W.: An information flow model for conflict and fission in small groups. Journal of Anthropological Research **33**, 452–473 (1977)

416. Zahn, C.: Graph theoretical methods for detecting and describing Gestalt clusters. IEEE Transactions on Computers **20**, 68–86 (1971)
417. Zahn, C.T., Roskies, R.Z.: Fourier descriptors for plane closed curves. IEEE Transactions on Computers **C-21**(3), 269–281 (1972)
418. Zeng, X., Chen, W., Peng, Q.: Efficiently solving the piecewise constant Mumford–Shah model using graph cuts. Technical Report, Zhejiang University (2006)
419. Zhang, F., Hancock, E.R.: Graph spectral image smoothing using the heat kernel. Pattern Recognition **41**(11), 3328–3342 (2008)
420. Zhang, F., Li, H., Jiang, A., Chen, J., Luo, P.: Face tracing based geographic routing in nonplanar wireless networks. In: Proc. of the 26th IEEE INFOCOM (2007)
421. Zhang, H., van Kaick, O., Dyer, R.: Spectral mesh processing. In: Computer Graphics Forum, pp. 1–29 (2008)
422. Zhang, Z.: Parameter estimation techniques: A tutorial with application to conic fitting. Image and Vision Computing **15**(1), 59–76 (1997)
423. Zhou, D., Schölkopf, B.: Regularization on discrete spaces. In: Proc. of the 27th DAGM Symp. Lecture Notes in Computer Science, vol. 3663, pp. 361–368. Springer, Berlin (2005)
424. Zhu, X., Ghahramani, Z., Lafferty, J.: Semi-supervised learning using Gaussian fields and harmonic functions. In: Machine Learning: Proceedings of the Twentieth International Conference on Machine Learning, pp. 912–919 (2003)
425. Zomorodian, A.: Topology for Computing. Cambridge University Press, Cambridge (2005)

Index

A

Across and through variables, 32
Activation maps, 188
Active contours, 7
Adjacency matrix, **61**
Adjointness, 30
Advection, **86**, 87, 88, 255
Advection equation, **87**
Affinity weights, **58** 62, 80, 93, 135, 149, 235, 248, 249, 271
Algebraic connectivity, *see* Fiedler value
Allen Brain Atlas, 240
Antiderivation, **27**
Antiderivative, **14**
Argand plane, **2**
Authority score, 256
Average path length, **270**

B

Balloon force, 217
Barack Obama, 192
Basic Energy Model, 101, **164**, 168, 178, 202, 251
　optimization, 310
Basic p-chain, **45**
Betweenness, **269**
Biharmonic, **98**, 99
Biological vision, 182
Blossom algorithm, 315
Boundary, 47
Boundary conditions, **298**
　Dirichlet boundary conditions, **298**
　Neumann boundary conditions, **298**
Boundary length, 136, 206, 223

Boundary operator, 16
　continuous, 30
　discrete, **46**
Brain data, 187
Brain matching, 260
Branch cuts, 31
Branch-and-bound, 318
Buser's inequality, 81, **275**

C

Canonical volume element, **33**, 57
Cauchy functions, 140
Cell complex, **39**, **40**
Chain, 45, 46
Chan–Vese model, **223**
　multi-phase, 223
Cheeger constant, *see* isoperimetric constant
Cheeger's inequality, 81, **275**
Chemical graph theory, 286–288
　quantitative structure–activity relationships, **286**
　quantitative structure–property relationships, **286**
Circuit
　superposition, method of, 115
Circuit theory, 91–122
　admittance, **103**
　alternating voltage source, **102**
　apparent power, **103**
　capacitance, **102**
　capacitor, **102**
　complex power, **103**
　conductance, **93**
　current, **92**

Circuit theory (*cont.*)
  current source, **94**
  delta–wye, **117**
  dependent sources, 98
  electric potential, **92**
  equivalent circuit, **117**
  ground node, **92**, **219**
  impedance, **103**
  inductor, **102**
  mesh analysis, **100**
  node analysis, **100**
  nonlinear resistors, 101
  Ohm's Law, **93**
  reactive power, **103**
  real power, **103**
  resistance, **92**
  star–mesh, **117**
  voltage, **92**
  voltage source, **94**
  voltage-controlled current sources, 98
Circulant matrix, **156**
Citation networks, 264–266
Clique, **40**
Closed form, **29**
Clustering coefficient, **276**
Coboundary operator, **30**, **49**
Cochain, 48–50
Codifferential operator, **36**
Coherent orientation
  complexes, 42
Cohomologous, **29**
Cohomology, 319, 330
Cohomology theory, **31**
Combinatorial manifold, 67
Commute time, **108**, 247
Complex plane, **2**
Conjugate gradients, 129
Constitutive, 105
Constitutive Determination Problem, **144**
Constitutive laws, 35, **57**
  homogeneous, 83
  inhomogeneous, 83
Constitutive relation, 144
Content extraction, **7**
Continuous max-flow, 209, **209**
Contravariant version, **25**
Convexity, **296**
  strictly convex, **297**
Convolution, 157

Coordinate invariance, 20
Cortical surface, 189, 355
Cotangent bundle, 22
Cotangent space, **22**
Cotree, **132**
Covariance matrix, 142
Covariant version, **24**
Current source, 94
Curvature, 136, 213
  Gaussian, **282**
  mean, **282**
  total, **283**
Cycle, 47
Cycle basis, 126–134
Cycle double cover, **134**
Cycle double cover conjecture, 52, 134
Cycle set, 51, 129

D

Dampening factor, **254**
Data attachment term, 167
Data discovery, **243**
Data tuple, **222**
DC, *see* direct current
Degree, **18**, **61**
Dependent source, **97**, 98, 99
Differentiable manifold, *see* manifold
Differential, **22**
Differential form, **21**
Differential forms, 16
Diffusion, 156
Diffusion equation, 82, **83**, 84–86
  diffusion constant, 83
Dijkstra's shortest path algorithm, 153, 212
Direct current, 94
Directed edges, 42
Directed graph, 85, 86
Directional derivative, **21**
Dirichlet boundary conditions, 167
Dirichlet's Principle, 82, 100
Discrete Fourier Transform, 156
Distance operator, **150**, 151–153, 268
Distance operator, on chains, 151, 152
Distance weights, **58**, 135, 271
Divergence Theorem, *see* Gauss's Theorem
Dongle nodes, **167**
Dual basis, **19**
Dual coboundary operator, **59**
Dual metric tensor, 54, 149, 151

Dual space, **19**
Duality, 50–54

E

Edge expansion, **275**
Edge Laplacian, **62**, 147, 148
Edge set augmentation
    via the Watts–Strogatz model, 128
Edge set generation, 126–129
    by k-nearest neighbors, 128
    from a Delaunay triangulation, 128
    from an ambient metric, 127
Effective resistance, **108**, 109–118, 152, 206, 271, 272
Elliptic equations, 79–81
Elongation, **105**
Energy, 163
Euclidean distance matrix, **246**
Euler characteristic, 131, **280**
Exact form, **29**
Expander graphs, **275**
Extended Basic Energy Model, 168, 170, 171, 180, 204
Exterior algebra, **19**
Exterior derivative, **26**
Exterior face, *see* outside cell
Exterior product, **17**

F

Fick's Law, **83**
Fiedler value, **217**, **275**
Fiedler vector, **218**
Finite differences, **3**, 69, 70
Finite element method, **3**
Fisher iris data, 238
Flow, 71
fMRI, 187
Forms, **19**
Fourier descriptors, 159
Fourier transform, 156–161
Foveal images, 182
Fredholm Alternative, **327**
Fredholm operator, **327**
FreeSurfer, 189, 262, 355, 359

Fundamental Theorem of Calculus, **14**, 14–16, 29–31
    First Fundamental Theorem of Calculus, 14
    Second Fundamental Theorem of Calculus, 14

G

Gauss–Green Theorem, **77**
Gauss–Jordan elimination, 117
Gauss's Theorem, 15, 30, 72
Gene expression, 240–242
Generalized Stokes' Theorem, 16, **29**, 38
Genetic algorithms, 312
Genus, **281**
Geodesic segmentation, **207**
Geographical information system, 192
Geometric cycles, 130
Geometric embedding, **127**
Geospatial data, 192, 193
Gibbs distribution, 111
Global metric, **152**
GPU, 127
Gradient descent, **308**
Gradient manipulation, 174
Gradient Theorem, 15, 30
Graph, **39**
    arcs, **92**
    branches, **92**
    bridge, **130**
    bridgeless graph, 52, 134
    center, **272**
    conductance, 275
    degree distribution, **277**
    edges, **39**
    faces, **39**
    fully connected graph, **127**
    genus, **131**
    nodes, **39**
    volume, **274**
    weighted graph, **57**
Graph cuts, 206
Graph diameter, **272**
Graph partitioning problem, **275**
Graph radius, **272**
Grassmann algebra, *see* exterior algebra
Green's Theorem, 15, 30
Gridding artifacts, 168, 204

H

Harmonic, **79**
Harmonic form, **37**
Harnack Inequality, **81**
Heat equation, *see* diffusion equation
Heffter–Edmonds Principle, **131**
Helmholtz decomposition, 37, 76, 77, 319–326
Helmholtz equation, 218
Hessian matrix, **296**
HITS, *see* Hyperlink-Induced Topic Search
Hitting time, **107**
Hodge decomposition, 37, 326–328, **329**, 330
Hodge star operator, **32**, 33–36
  Hodge dual, **33**
Hodge's Theorem, **37**
Homology theory, **31**
Hooke's Law, 105
Hubs, 256
Human eye, 140
Hyperlink-Induced Topic Search, 253, 256, 257

I

Image filtering, 180
Image segmentation, 229–234
Incidence matrix, **44**, 45
  edge–node, 93
  face–edge, 100
  reduced, **95**, 113
Incompressible fluids, 148
Independent source, **97**
Indicator vector, **46**
Influence function, 139
Information matrix, **110**
Inner product, **24**
Integration by parts, **15**, 30, 77, 78
Intelligent scissors, 212
Interior point methods, 311
Isomap algorithm, 246–249, 258
Isoperimetric algorithm, **220**
Isoperimetric problem, **274**
  isoperimetric constant, 218, **275**
  isoperimetric number, **275**
  isoperimetric ratio, **274**
  isoperimetric sets, **274**
Iterated Conditional Modes, **317**
Iteratively reweighted least squares, **310**

J

Joint statistics, 144
Judicial citation networks, 264–266

K

k-means model, 221
k-nearest neighbors, **128**
Kirchhoff index, **271**
Kirchhoff's Current Law, **93**
Kirchhoff's Voltage Law, **94**
Köningsberg bridge problem, 126

L

"$\lambda-\mu$" filter, 161
Lanczos algorithm, **307**
Laplace equation, **81**
Laplace–Beltrami operator, *see* Laplace–de Rham operator
Laplace–de Rham operator, **36**, 61, 149
Laplacian Eigenmaps, 142, 247, 258
  algorithm, 249
Laplacian matrix, **61**, 94
  normalized, 62, 161, 276
  pseudoinverse, 152
  reduced, 113
Laplacian smoothing, *see* spectral filtering
LE, *see* Laplacian Eigenmaps
Level sets, 7
Linear functional, **19**
Linear programming, **303**
Live wire, 212
Liver, 258
Lloyd's algorithm, 216, 222, 223
Local metric, **152**
Locality Preserving Projections, 142, 249
  algorithm, 249–251
LPP, *see* Locality Preserving Projections
LTI system theory, **159**

M

M-estimator, 140, 166
Machine learning, 236–240
MacLane's Planarity Criterion, 132
Mahalanobis distance, **142**
Manifold, **20**, 243
  orientable, 281
Manifold learning, 243, **244**, 245–266
Markov random fields, 110–112, 138
Mathematical morphology, 208

Matrix ordering, 121
Matrix-Tree Theorem, **113**
Max-flow/min-cut, 101, 206, 314–318
Maximum spanning forest, 208
MDS, *see* Multidimensional Scaling
Mean filter, 164
Mean-value theorem, **80**
    local maximum principle, **80**
    local minimum principle, **80**
    maximum principle, **80**
    minimum principle, **80**
    strong local maximum principle, **80**
    strong maximum principle, **81**
Median filter, 164
Mesh analysis, 100
    mesh variables, 100
Mesh fairing, **185**
Metric, **26**, **150**, **268**
Metric tensor, **25**, 54
    dual metric tensor, **25**
    primal metric tensor, **25**
Metrication artifacts, 168
Minimal Lipschitz extension, 207
Minimal surface, 213–215
Minimax filter, 164
Minimum cycle basis, **133**
Minimum-degree orderings, **119**
Mode filter, 164
Multidimensional Scaling, **245**
Multivariate data, 141

N
NCuts, *see* Normalized Cuts
Negative weights, 143, 209
Neural connectivity, 193
Neuroscience, 187
Newtonian fluid, 148
Newton's method, **308**
Node analysis, 100
Node centrality, **269**
    closeness, **269**
    eccentricity, **269**
    total distance, **269**
Non-local means, 143
Nonlinear anisotropic diffusion, 166, 167, 197
Nonlocal filtering, 174, 175
Norm, **26**
Normalized Cuts, **218**, 251

Nullspace approach, **302**
Nullspace transformation, 100

O
Object File Format, 291
Optimal matching, 315
Optimization techniques
    genetic algorithms, 318
    simulated annealing, 318
    tabu search, 318
Orientation, 42–44, 92
Outside cell, **51**, 52, 134
Outside face, 132, 213
Over-representation, 153

P
P-Brush, **207**
p-cell, **39**
p-chain, **45**
p-cochain, **48**
p-domain, **29**
p-form, **19**
p-simplex, **40**
p-vector, **18**
PageRank, **253**, 254–257
Penalty method, **302**
Peripheral node, **272**
Phantom seed, 206
Piecewise constant Mumford–Shah, 223
Pink noise, 163
Plateau's problem, 213, 314
Poincaré duality, **51**, 55, 60
Poincaré lemma, **31**
Point correspondence, 260–262
Poisson equation, **81**, 82, 174, 220
Polar and axial vectors, 32
Polling, 192
Polya's theorem, 110
Potential matrix, *see* information matrix
Potts Model, 317
Power method, 254, 257, **307**
Power watershed, 208, **209**
Precision matrix, *see* information matrix
Preferential attachment, **279**
Primal metric tensor, 151
Primal–dual algorithm, **305**
Pseudocentroid, 224
Pseudoform, 32

Q
Quadratic pseudo-boolean functions, 315
Quaternions, **2**
Quincunx, 159

R
Rand index, 239
Random graph, 277
Random surfer, *see* random walker
Random walker, 206, 209
Random walks, 106–110
Rayleigh quotient, **306**
    generalized Rayleigh quotient, **306**
Reduced Laplacian matrix, *see* Laplacian matrix
Regular graphs, **6**
Representations
    cells list representation, **291**
    Compressed Column (or Row) format, **293**
    Compressed Diagonal Storage format, **293**
    neighbor list representation, **293**
    operator representation, **291**
Resistance distance, *see* effective resistance
Riemannian metric, 55
Right-hand rule, 43
Robust error functions, 138
Rotation system, **130**, 131
    rotate operation, 130
    rotation, **130**
    rotation table, **130**
    trace operation, 130
Roth diagram, **63**
Rudin, Osher, Fatemi, *see* total variation

S
Scale-free network, **277**
Schur factorization, **302**
Screw-sense, 42
Seeds, **205**
Semi-supervised learning, 201
Separable Hilbert space, **326**
Shape characterization, 257–260
Shift invariant, 143, **158**
Shortest path, 153, 207, **268**
Simplicial complex
    structured complex, 292
Simplicial decomposition, **44**

Simulated annealing, 312
Six degrees of separation, 277
Small-world network, 128, **277**
Smith Normal Form, 281
Smoothness term, 167
Social networks, 235, 236, 285, 286
Space-variant images, 182
Spatial statistics, **192**
Spectral clustering, **218**
Spectral coordinates, **247**
Spectral filtering, **160**, 165
Spring constant, **105**
Spring networks, 104–106
Stereographic projection, 280
Stiffness matrix, **106**
Stokes' Theorem, 15, 30
Straight and twisted forms, 32
Submanifold, *see* manifold
Submodular functional, **314**
Supervised learning, 201

T
t-links, *see* dongle nodes
Tabu search, 312
Tangent bundle, 21
Tangent space, **21**
Targeted clustering, 199
Teleportation parameter, **254**
Tellegen's Theorem, **94**
Tensors
    antisymmetric tensors, **23**
    contravariant tensors, **23**
    covariant tensors, **23**
    index lowering, 26
    index raising, 26
    musical isomorphisms, 26
The First Law of Geography, **193**
Tikhonov regularization, **298**
Tobler's Law, **193**
Toeplitz matrix, **157**
Tonti diagram, **63**
Torsion coefficients, **281**
Total unimodularity, **48**
Total Variation, **169**, 204
    segmentation, 209
Totally unimodular matrix, 221, 313
    pre-unimodular matrix, 214, 313
Touch sensor, 187
Transductive learning, 201, 236

Transport equation, *see* advection equation
Tree, 113, 132
    2-tree, **114**
    spanning tree, **47**, 113
Tree counting, 112–117
Triadic closure, **276**
Tukey biweight function, 140
Tuple, 64, 141
TVSeg, *see* total variation segmentation

U
Untargeted clustering, 199

V
Vector data, 141
Vector Laplacian, 62, 176
Viscosity, 63, 148
    bulk viscosity, 148
    shear viscosity, 148
Voltage source, 94

Volume cochain, **57**, 150
Vorticity, 148

W
Wall filter, 161
Watershed algorithm, 208
Web search, 262, 263
Weber's Law, 140
Wedge product, *see* exterior product
Weights, **57**, 58, 135–153
    edge weights, 135–146
    higher-order weights, 147–150
Welsch function, 140, 208, 230
Wheatstone bridge circuit, 115
Wiener index, **270**
    quasi-Wiener index, **272**
Wiener number, *see* Wiener index

Z
Zachary's karate club, 235, 236, 285, 286

Color Plates

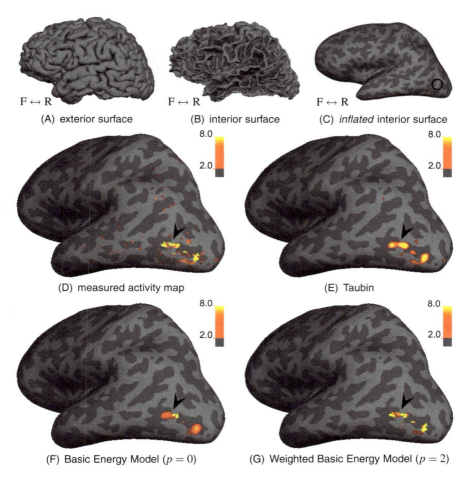

Fig. 5.10 Filtering fMRI data along a cortical surface model. Surface models of the (**A**) exterior surface, (**B**) interior surface, and (**C**) the "inflated" interior surface of the cortical gray matter of the left cerebral hemisphere, with approximate location of area MT indicated by *circle*. (Surfaces generated with FREESURFER [96, 136].) *Legend* indicates Front–Rear axis of brain. Locations of negative mean curvature (within sulci) are rendered in *dark gray* and locations of positive mean curvature (within gyri) are rendered in *light gray*. Measured activity map plotted as *z*-statistics, with color scale provided at *upper right*. The threshold is set to exclude nodes where the activity is not statistically significant, which leads to many isolated points or small clusters of activation appearing in the map—likely false positives due to noise. The results of filtering the data using (**E**) spectral filtering, (**F**) the Basic Energy Model with $p = 0$ and (**G**) $p = 2$, (**H**) Total Variation, and (**I**) the Mumford–Shah algorithm are provided with the same color scale representing the statistical significance. Note that many of the false positives are removed with the filtering. *Arrows* indicate site of MT activation

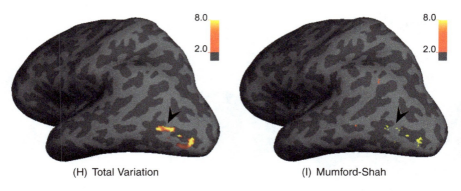

(H) Total Variation (I) Mumford-Shah

Fig. 5.10 (Continued)

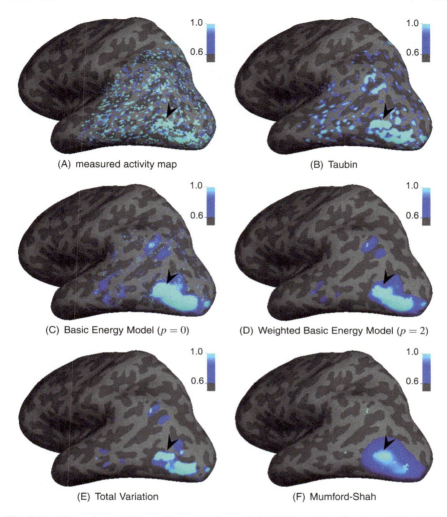

Fig. 5.11 Effect of smoothing methods on sub-threshold fMRI activity. The data of Fig. 5.10 is re-plotted with a color scale that highlights the relative performance and behavior of the filtering methods on activity below the significance threshold. *Reference arrows* are positioned as in Fig. 5.10

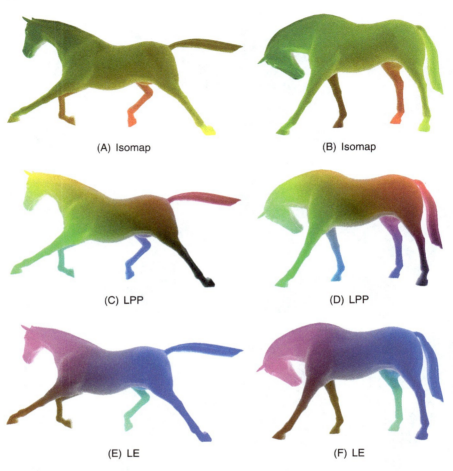

Fig. 7.5 Two three-dimensional meshes of a horse in different poses. Each mesh was mapped to three dimensions using the various manifold learning techniques. The mapped three-dimensional coordinates of each point were mapped to RGB space for display. Two vertices having similar colors should therefore be identified as the same coordinate location in the two meshes

Color Plates 365

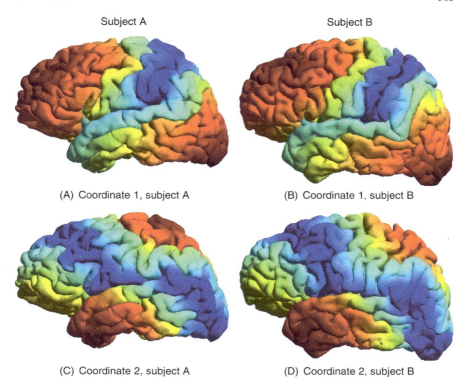

(A) Coordinate 1, subject A (B) Coordinate 1, subject B

(C) Coordinate 2, subject A (D) Coordinate 2, subject B

Fig. 7.6 Brain surface matching via the Laplacian Eigenmaps method. The first two eigenvectors of the graph Laplacian are used to establish corresponding two-dimensional coordinate parameterizations of the surfaces which can be utilized as a correspondence map providing the matching. The surfaces represent the outer surface of the cortical gray matter (i.e., the "pial surface") and each surface was generated automatically with the FreeSurfer software environment [96, 136]. The surface meshes for subjects A and B contained 248,868 and 259,792 triangles, respectively

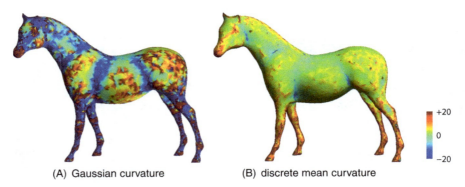

(A) Gaussian curvature (B) discrete mean curvature

Fig. 8.5 Example of discrete curvature measures applied to horse triangular mesh. (**A**) Gaussian curvature calculated from the Gauss–Bonnet theorem. (**B**) Discrete mean curvature computed from the method of Meyer et al. [280]. Both measures calculate curvature as a node quantity measured for each vertex in the polygonal mesh based on the embedding of the neighboring vertices and the incident faces. Both curvatures are visualized with a common color scale provided on the *lower right*

CPSIA information can be obtained
at www.ICGtesting.com
Printed in the USA
LVHW081819130621
690119LV00003B/119